TRANSLATING MicroRNAs TO THE CLINIC

T0326379

TRANSLATING MicroRNAs TO THE CLINIC

Edited by

JEFFREY LAURENCE

Division of Hematology-Medical Oncology, Weill Cornell
Medical College and New York Presbyterian Hospital,
New York, NY, United States

Guest Editor

MARY VAN BEUSEKOM

Health Partners Institute for Education and Research and
Synapse Writing and Editing, Exelsior, MN, United States

Amsterdam • Boston • Heidelberg • London
New York • Oxford • Paris • San Diego
San Francisco • Singapore • Sydney • Tokyo

Academic Press is an imprint of Elsevier

Academic Press is an imprint of Elsevier
125 London Wall, London EC2Y 5AS, United Kingdom
525 B Street, Suite 1800, San Diego, CA 92101-4495, United States
50 Hampshire Street, 5th Floor, Cambridge, MA 02139, United States
The Boulevard, Langford Lane, Kidlington, Oxford OX5 1GB, United Kingdom

Notices
Knowledge and best practice in this field are constantly changing. As new research and
experience broaden our understanding, changes in research methods, professional practices,
or medical treatment may become necessary.

Practitioners and researchers may always rely on their own experience and knowledge in
evaluating and using any information, methods, compounds, or experiments described
herein. In using such information or methods they should be mindful of their own safety
and the safety of others, including parties for whom they have a professional responsibility.

To the fullest extent of the law, neither the Publisher nor the authors, contributors, or
editors, assume any liability for any injury and/or damage to persons or property as a
matter of products liability, negligence or otherwise, or from any use or operation of any
methods, products, instructions, or ideas contained in the material herein.

Library of Congress Cataloging-in-Publication Data
A catalog record for this book is available from the Library of Congress

British Library Cataloguing-in-Publication Data
A catalogue record for this book is available from the British Library

ISBN: 978-0-12-800553-8

For information on all Academic Press publications
visit our website at https://www.elsevier.com

Working together
to grow libraries in
developing countries

www.elsevier.com • www.bookaid.org

Publisher: Mica Haley
Acquisitions Editor: Mica Haley
Editorial Project Manager: Sam W. Young
Production Project Manager: Chris Wortley
Designer: Mark Rogers

Typeset by TNQ Books and Journals

CONTENTS

LIST OF CONTRIBUTORS

S. Akkina
University of Illinois at Chicago, Chicago, IL, United States

S. Ansar Ahmed
Virginia Tech, Blacksburg, VA, United States

B.N. Becker
University of Chicago, Chicago, IL, United States

C. Bime
The University of Arizona, Tucson, AZ, United States

D.M. Brown
Ohio State University Medical Center, Columbus, OH, United States

E.E. Creemers
University of Amsterdam, Amsterdam, The Netherlands

C. Croce
Ohio State University Medical Center, Columbus, OH, United States

R. Dai
Virginia Tech, Blacksburg, VA, United States

A.A. Desai
The University of Arizona, Tucson, AZ, United States

A. Esquela-Kerscher
Eastern Virginia Medical School, Norfolk, VA, United states

J.G.N. Garcia
The University of Arizona, Tucson, AZ, United States

C.I. Gurguis
The University of Arizona, Tucson, AZ, United States

T. Hasegawa
Eastern Virginia Medical School, Norfolk, VA, United states

L. Hecker
The University of Arizona, Tucson, AZ, United States

C.D. Hoang
National Institutes of Health, Bethesda, MD, United States

N. Kaminski
Yale University, New Haven, CT, United States

T.A. Kerr
University of Texas Southwestern Medical Center, Dallas, TX, United States

R.A. Kratzke
University of Minnesota, Minneapolis, MN, United States

H. Lewis
Eastern Virginia Medical School, Norfolk, VA, United states

B.B. Madison
Washington University, Saint Louis, MO, United States

M.A. Montano
Harvard Medical School, Boston, MA, United States

S.P. Nana-Sinkam
Ohio State University Medical Center, Columbus, OH, United States

K. Ono
Kyoto University, Kyoto, Japan

K.V. Pandit
University of Pittsburgh, Pittsburgh, PA, United States

Y.M. Pinto
University of Amsterdam, Amsterdam, The Netherlands

G. Song
University of Minnesota, Minneapolis, MN, United States

A.J. Tijsen
Technion-Israel Institute of Technology, Haifa, Israel; University of Amsterdam, Amsterdam, The Netherlands

T.H. Tran
Harvard Medical School, Boston, MA, United States

H. Vollbrecht
University of Minnesota, Minneapolis, MN, United States

T. Wang
The University of Arizona, Tucson, AZ, United States

CHAPTER 1

MicroRNAs: Mirrors of Health and Disease

T.H. Tran, M.A. Montano

Key Concepts

This chapter gives examples of tissue and cell type-specific knockdown of components of the microRNA machinery and their deleterious effects. It discusses the genetic organization of microRNAs and how that architecture is designed to regulate multiple targets within regulatory pathways, and their role in adaptive regulation in response to changes in the microenvironment. It looks at the role of MicroRNA networks in specific tissue and organ diseases, including lung, heart, and skeletal muscle, as well as in modulating cancers in those tissues. Newer research is introduced linking shifts in microRNA expression with biological aging and diseases associated with aging. The potential role for MicroRNAs as therapeutic tools to modulate target RNA sequences implicated in disease states is explored. Finally, the utility of MicroRNAs as circulating biomarkers for disease risk and severity, as well as in the therapeutic response to candidate drug interventions is discussed.

1. INTRODUCTION

Since their discovery as a novel class of small noncoding RNAs capable of regulating protein translation and/or messenger RNA (mRNA) stability, microRNAs (miRNAs) have been implicated as key regulators in the homeostasis of multiple biological systems that include organs such as heart, lung, and skeletal muscle, as well as modulation of the pathobiology processes such as cancer and aging. Experimental loss of key regulators of miRNA biogenesis has revealed the crucial role for the control of posttranscriptional gene expression. The genetic organization, the variability of their loci, and the tissue specificity of their expression further illustrate an adaptive capacity of the miRNA machinery to modify and fine-tune the transcriptome and translated protein targets in response to physiologic environmental cues, as well as their vulnerability to becoming co-opted in diseases such as cancer, fibrosis, sepsis, aging, and autoimmune disease. Current efforts to leverage knowledge of the miRNA regulatory system to

1

diagnose, track, and attenuate disease progression, represent a major new research opportunity, and challenge in this rapidly growing area of translational medicine.

2. KNOCKING OUT MICRORNA: SMALL RNAs WITH BIG EFFECTS

As our understanding of miRNA biogenesis and downstream regulatory activities unfold, so too does our appreciation for the extent and scope of influence these small RNAs have on multiple biological processes across human health and disease. The endonuclease Dicer is an obligatory first step in the RNA interference pathway and formation of the RNA-inducing silencing complex (RISC). Without RISC formation, argonaute-mediated degradation of target mRNA is lost [1,2]. Therefore, Dicer knockdown, in effect, disables the processing of miRNA from precursors of miRNA, ie, premiRNA. This can have deleterious effects. In murine models of cancer, global gene-disruption of Dicer can enhance tumor susceptibility [3]. Tissue- and cell type–specific knockdown of Dicer can also lead to deleterious effects. For example, cardiomyocyte-specific ablation of Dicer can result in cardiomyopathy [4]. Kidney podocyte-specific knockdown of Dicer can lead to systemic proteinuria and glomeruli abnormalities in mice [5]. Moreover, conditional knockdown of Dicer in the kidney impairs its normal development, further emphasizing the normal and dysregulatory potential in miRNA processing. Dicer deletion in skeletal muscle can result in increased myocyte apoptosis, skeletal muscle hypoplasia, and eventually perinatal death [6]. Hepatocytes obtained from Dicer-null mice exhibit abnormal lipid buildup and eventual renal steatosis [7]. Liver-specific Dicer deletion at 3 weeks leads to progressive hepatic steatosis [8]. This phenotype, however, has not been universally observed. While unlikely to change the overall conclusion that miRNAs clearly play a significant role in regulating multiple pathways, the subtleties or disparities in observation of distinct phenotypes in these conditional knockdowns may hinge upon the buildup of precursor miRNAs that activate other RNA surveillance pathways, such as RNA editing, which would then introduce an additional variable [9,10]. Comparative transcriptome profiling with high throughput RNA sequencing will likely provide more insight into the global regulatory profile of these knockdowns.

3. GENETIC ORGANIZATION, VARIATION IN MICRORNAs, AND TISSUE-SPECIFIC EXPRESSION OF MICRORNAs

MicroRNAs (miRNAs) undergo multiple processing events to reach their functional 21–23 ribonucleotide RNA sequence. Canonical miRNAs are generated from protein-coding transcriptional units; whereas, other miRNAs (ie, noncanonical miRNAs) are generated from nonprotein-coding transcriptional units. In both cases, the miRNAs can be located either within intronic or exonic regions. A noteworthy mechanistic distinction in canonical versus noncanonical miRNAs is that canonical intronic miRNAs are Drosha dependent and are thus processed cotranscriptionally with protein-coding transcripts in the nucleus. The premiRNA then enters the miRNA pathway, whereas the rest of the transcript undergoes premRNA splicing to produce mature mRNA which will then direct protein synthesis. Noncanonical intronic small RNAs (also called mirtrons) can derive from small introns that resemble premiRNAs, and bypass the Drosha-processing step [11]. miRNAs tend to be organized in a related cluster and also tend to target multiple mRNA transcripts within common cellular response pathways (eg, proliferation, apoptosis). This organizational thematic provides miR clusters with a capacity for coordinate regulation of multiple steps within a pathway, providing an opportunity for complex and adaptive regulatory control of entire pathways. An interesting class of miRNAs is myomiRs—so-called because they are coded within myosin heavy chain (MYH) genes. myomiRs are transcribed in the same precursor mRNA as the parental MYH gene [4]. Of special note is the myomiR-499, which despite the absence of a parent mRNA, is one of the most highly expressed miRNAs in heart tissue. In an apparently novel evolutionary phenomenon, alternative splicing in the heart uncouples production of mature miR-499 from expression of parent MYH7b mRNA, meaning that the mRNA has perhaps evolved into a nonfunctional host mRNA for its intronic miR (ie, miR-499).

Comparative studies evaluating the organizational structure of the mammalian genome have identified a wealth of chromosomal insertion–deletions, copy number variants, and single nucleotide polymorphisms (SNPs) that, depending on the environmental context, contribute to the genetic variation that can underlay phenotypic diversity. This diversity is evident in nearly every aspect of human health and disease that has been investigated. Perhaps not surprisingly, there is now a growing recognition that variation in miRNAs and their target genes also contribute to this phenotypic variability. Several solid and hematologic malignancies can be linked to miRNAs located at amplified, deleted, or translocated chromosomal regions in the mammalian genome [12]. Variation in gene expression and regulation is likely influenced by genetic variants in *cis-* and *trans*-acting SNPs (also known as expression quantitative trait loci) [13]. An interesting observation of miRNA binding is their ability to recognize binding site polymorphism (miRSNPs) in transcribed functional genes. For example, miR-24 appears to be deregulated in human colorectal tumor through a target site polymorphism in the dihydrofolate reductase gene. In another example, a polymorphism within the myostatin gene creates a target site for miR-1 and miR-206, which are highly expressed in skeletal muscle. The binding of these miRs to the polymorphism in myostatin causes translational inhibition of myostatin transcripts and can phenocopy the observed muscle hypertrophy that is observed with genetic knockouts of the myostatin gene [14]. Given the significant differences in gene expression and genetic variation across human populations, analysis of the role of miRNAs in contributing to population differences in gene expression is likely to provide substantial insights in population based health disparities and physical functionalities [13]. Indeed, comparative genomic studies indicate that the target mRNA sequences for miRNAs: untranslated regions (UTRs) on mRNAs often display sequence diversity. This may suggest adaptive evolution of coexpressed miRNAs and cognate mRNAs with these UTR variants. Depending on whether the dampening of protein output is beneficial, inconsequential, or harmful, the UTR sites may be selectively conserved, neutral, or avoided during miRNA:mRNA coevolution [1].

Studies evaluating the tissue-specific expression of miRNAs illustrate a "cross-regulation" feature of miRNAs that contributes to cell fate specification by repressing alternate cell fates to facilitate commitment to one cell fate and to maintain stability of a differentiated phenotype [15]. For example the myomiRs miR-1, miR-133, miR-206, miR-208, miR-486, and miR-499 are enriched in skeletal muscle and play crucial roles in the development, growth, and maintenance of skeletal muscle [16]. Notably, miR-133 prevents osteogenic cell lineage differentiation by repressing Runx2, a factor essential for bone formation [15]. MiR-7, miR-24, miR-128, miR-134, miR-219, and others are highly expressed in the mammalian brain and regulate neurite outgrowth, neuronal differentiation, and dendritic spine size [17]. Regionally enriched miRNAs in specific tissues can exert specialized function. For example, miR-7 and miR-24 are highly expressed in the hypothalamus and fine-tuning expression of Fos and oxytocin which play vital roles in the control of water, lactation, and parturition [18].

4. MICRORNA REGULATORY COMPLEXITY: FEEDBACK LOOPS AND ENVIRONMENTAL SENSING

Perhaps one of the most salient and unexpected features of miRNA is the role for miRNA in amplifying or tempering cell signaling. The modulatory capacity provides an adaptive tool that allows the initiation of cellular signaling to be calibrated to accommodate cues from the microenvironment, as well as to potentially buffer signaling. This buffering function has been proposed to function as a network stabilizing effect in the context of dynamic and interlocking feedback loops [19]. In the lung, miR-21 promotes transforming growth factor (TGF-β) amplification and fibrosis in a feed-forward loop by relieving the repression of the inhibitor Smad 7 through targeting the mRNA [20]. Interestingly, TGF-β both inhibits miR let-7 and upregulates miR-21. These two miRNAs appear to be functionally opposed in lung tissue from human subjects with idiopathic pulmonary lungs (let-7 acts as a negative regulator and miR-21 as a positive regulator) that in effect can balance the fibrotic phenotype. In the context of sepsis, monocyte expression of miR-146 and miR-155 are rapidly upregulated during endotoxin/LPS exposure, representing an innate response regulatory role for these miRNAs [11,13]. MiR-146a appears to function as a negative feedback regulator by targeting IL-1 receptor-associated kinase 1 (IRAK-1) and TNF receptor-associated factor 6 (TRAF6). A reciprocal negative feedback is achieved through MAP kinase phosphatase-1 (MKP-1)-mediated suppression of miR-155, which in turn targets suppressor of cytokine signaling-1. Thus, MKP-1 appears to function as a derepressor of miR-155-mediated suppression to modulate LPS response in monocyte/macrophages. In the context of skeletal muscle, C2C12, a skeletal muscle precursor cell line, miR-1 targets mRNAs for the insulin growth factor-1 (IGF-1) and the IGF-1 receptor. In a familiar theme of regulatory loops, IGF-1 reciprocally regulates miR-1 via the transcription factor Foxo3a, apparently through an enhancer-binding element within the miR-1 promoter. Interestingly, skeletal muscle response to endurance or resistance exercise stimuli has been linked to changes in miRNA levels [21,22]. However, such changes are depending on the age of the subjects, the mode of exercise, and the duration of exercise. For example, when anabolic response is dichotomized into so-called "low responders" versus "high responders" to resistance exercise-induced muscle hypertrophy, the former, but not the latter, exhibit a decrease in miR-26a, miR-29, and miR-378. Since miR-29 has been implicated as positive regulator of myogenesis, its reduction in "low responders" may contribute to the attenuated response in muscle hypertrophy. Interestingly, physical inactivity such as prolonged bed rest (10 days) leads to muscle atrophy and a reduction in the levels miR-206, miR23a, and several members of let-7 family of miRNAs [23]. These miRNAs contribute to a wide array of functions in that influence muscle function and maintenance, insulin response, growth and atrophy, cell cycle, differentiation, and glucose homeostasis. Other anabolic modulators such as estrogen and androgen also alter miRNA levels [11].

5. ORGAN DISEASES AND MICRORNAs

A potentially important and therapeutically exploitable distinction between disease and healthy states in tissue homeostasis may reside in the regulatory miRNA networks that reflect healthy and disease states. There is a growing recognition that miRNA networks are often associated with tissue dysfunction and are likely to be a key source of altered gene expression that underlays and distinguishes healthy versus disease states. A better understanding of the key perturbations that lead to these alternative regulatory network states may help to inform therapeutics. A reduction in Dicer expression has been implicated in cardiac disease. In skeletal muscle, a reduction of miR-29 expression has been associated with decreased muscle regeneration and Duchenne's muscular dystrophy, but also chronic kidney disease, pointing to the complexity of miRNA control networks [21]. To add to this complexity, miR-29 expression increases with age and this induction may inhibit muscle regeneration by affecting overall protein translation via modulating IGF-1 and p85α. In the context of lung biology, profiling studies of bronchial airway epithelia reveal an array of changes, wherein multiple miRNAs were downregulated in smokers that could be correlated with mRNA target expression in vivo [24]. In the kidney, miR-155 is differentially expressed in different tissue compartments (ie, kidney cortex versus kidney medulla regions) [5]. In individuals with trisomy 21, the chromosomal disorder also known as Down's syndrome, fibroblasts display higher levels of miR-155. In kidney, the expression of miR-192 is dysregulated in diabetic nephropathy, with a suggestive role in mediating TGF-β-induced collagen expression. In that study, deletion of the inhibitory Smad 7 promoted miR-192 expression, in a model for obstructive kidney disease. In diabetes, mRNA profiles are altered based on the observation that there is an apparent discordance between proteomic profiles and their respective mRNA levels, suggesting a potential role for miRNAs [7]. Interestingly, insulin-secreting cells exposed to proinflammatory cytokines display elevated miR-21 levels, and miR-146 levels are increased in a murine model for type 2 diabetes (db/db mice). Studies on insulin effects in skeletal muscle indicate that insulin expression is associated with a reduction in multiple miRNAs including miR-1, miR-206, and miR-133a (all canonical muscle regulatory miRNAs), in a process potentially mediated in part by the transcription factor sterol regulatory element-binding protein-1c (SREBp-1c) and myocyte enhancer factor 2C. In the liver, miR-122 is one of most abundant miRNAs and therapeutic knockdown of miR-122 can decrease hepatic lipogenesis, resulting in a protection from a high-fat diet–induced hepatic steatosis [8]. In the context of immune system regulatory networks, miR-155 was upregulated in T and B cells upon pathogen activation and required for the maintenance of lymphocyte homeostasis, notably in promoting Th17 cytokine expression [11]. Mice with an ablation of miR-146a specifically in regulatory T cells display elevated levels of the T helper-1 (Th1) cytokine interferon-gamma (IFN-γ) and a breakdown of immune tolerance with an increased risk for autoimmune disease. Several additional studies describe a dysregulated miRNA profile in blood cells from individuals with lupus, an autoimmune disease characterized by production of autoantibodies. The upregulation of miR-155 in lupus B and T cells may lead to the characteristic abnormal B-cell responses in germinal centers. Dysregulated expression of the miRNAs miR-146a and miR-155 in synovial fibroblasts is also implicated in rheumatoid arthritis (RA). Notably, although miR-146a was upregulated in active RA disease, the target genes, IRAK-1 and TRAF6, had no apparent change in their levels when compared to healthy controls [11]. This result may indicate that miRNAs may lose their capacity to function as negative regulators of inflammation in the context of RA, underscoring the contextually specific effects of miRNA networks and the need for global profiling approaches to define disease states and their contradistinction from healthy states.

6. CANCER AND MICRORNAs

Perhaps one of the most critical developments in the pathobiology of miRNAs has been the recognition that miRNAs modulate the transcriptome and can reflect a cancerous state, and indeed promote or attenuate cancer risk. A reduction in miRNA expression has been associated with dedifferentiated tumors in lung epithelial cells [25] and downregulation of epithelial markers has been linked to an epithelial–mesenchymal transition [20], a key developmental program that is often activated during cancer invasion and metastasis, and in this case diagnostic of lung cancer. Global repression of miRNA expression may in fact be a common feature in cancer [12]. In lung tumors, as well as in advanced adenocarcinoma, Dicer expression is often reduced. In rhabdomyosarcoma, a muscle stem cell tumor affecting the pediatric population, myomiRs such as miR-1 and -133 tend to be significantly downregulated [26]. miRNAs may promote or attenuate the oncogenic phenotype by either decreasing expression of tumor-suppressor genes (oncomiRs) or alternatively by targeting oncogenic mRNA for silencing (tumor-suppressor miRNAs). Perhaps one of the most studies miRNAs is let-7, a tumor suppressor [12]. Let-7 overexpression has been shown to inhibit tumor development and growth. miRNAs (eg, miR-17-92, miR-21) can function as oncomiRs by modulating cell cycle regulators that in effect promote proliferation, eg, CDKN1A (p21), E2F family, and PTEN [12,20]. Diagnostically, low let-7a-2 levels and high miR-155 levels have been correlated with poor survival in adenocarcinoma of the lung [12]. Comparative miRNA expression profiling in adenocarcinoma and squamous cell carcinoma revealed a signature of miRNAs in male smokers and a small set of miRNAs that predict survival in a cohort of smokers with early stage squamous cell carcinoma. An additional complexity described in Liu et al. is that the cell and tissue context of miRNA expression can be critical to understanding the overall pathobiology of cancer. For example, while miR-31 appears to be prooncogenic in lung cancer, it can alternatively be associated with protective affects in breast cancer.

7. AGING AND MICRORNAs

Recent studies have uncovered alterations in miRNA expression during the normal aging process, and also in extreme longevity [2], perhaps collectively providing more insights into mechanisms governing age-related regenerative disorders in lung, heart, skeletal muscle, liver, and in neurodegenerative disease. Deep sequencing analysis of miRNAs in mouse serum reveals an increase in multiple miRNAs and a decrease in some miRNAs with age [27]. Some of these miRNAs behave as functional clusters to putatively target regulatory networks, eg, the regulation of macromolecule biosynthesis (miR-134-5p, -148a-5p, -192-5p, -217-5p, -298-5p, -365-3p, and -434-3p), apoptosis (miR-3096b-5p, -376b-3p, -431-5p, and -138-5p), and the Wnt-signaling pathway (miR-592-5p, -667-3p, and -668-3p). In separate studies, increased expression miRNAs (miR-146a-5p, -148-5p, and -182-5p) and decreased miRNA (miR-144-3p) have been implicated in regulating the inflammatory pathway [28–31], known to be progressively deregulated during aging. Interestingly, three miRNAs (miR-151a-5p, -181a-5p, and -1248) associated with modulating the inflammatory pathway are significantly decreased in the serum of older individuals as compared to those of young individuals [32]. However, whether the change in the circulating miRNA abundance is the cause or the consequence of aging requires more comprehensive studies to establish the causal pathways that miRNAs engage during the aging process. There is also evidence of age-related miRNA alterations that are tissue specific. For example, in the liver, levels for miRNAs targeting the oxidative stress defense such as miR-93, -214, -669c, and -709 increase with age [33]. Similarly, in the brain, miRNAs (miR-101a, -720, and -721) regulating oxidative phosphorylation are upregulated during aging. In mouse skeletal muscle, among the 57 differentially expressed miRNAs during aging, miR-7, -486, -542, and -698 were significantly increased, whereas miR-124a, -181a, -221, -382, -434, and -455 were substantially decreased [33]. In human skeletal muscle, let-7, known to regulate cell cycle and perhaps satellite cell turnover, is among the 18 miRNAs that are upregulated during aging [34]. Cellular senescence is one of the hallmarks of cellular aging that hinders stem cell regenerative capacity. Signaling pathways regulating cellular senescence include the insulin/IGF-1 (IIS), reactive oxygen species, mitogen-activated protein kinase, p53, retinoblastoma, and inflammatory pathways [33]. As such, changes in the abundance of miRNAs targeting these pathways profoundly affect life span based on studies in the nematode *Caenorhabditis elegans*. For example, lin-4, miR-71, miR-238, and miR-246 promote longevity, whereas, miR-239 reduces life span [32,35]. In humans, the deregulation of miR-126 which targets the IGF/IIS pathway is proposed as one of the possible contributors to the observed blunted response of aging muscle following anabolic stimulation, in a phenomenon described as anabolic resistance [36]. Furthermore, miRNAs known to regulate IGF/IIS pathway have been associated with heart disease (miR-1, miR-122, and miR-375), skeletal muscle growth (miR-1 and -206), and neurodegenerative disease (let-7 and miR-320) [35]. As high-throughput technologies for identifying functional miRNA targets such as HITS-CLIP (high-throughput sequencing of RNA isolated by cross-linking immunoprecipitation) and PAR-CLIP (photoactivatable-ribonucleoside-enhanced cross-linking and immunoprecipitation) become more advanced and affordable, so will our understanding of the regulatory network of miRNAs in modulating aging and age-related disorders [35].

8. MICRORNA THERAPEUTICS AND TARGET PREDICTION: SUCCESSES AND CHALLENGES

The prospect of leveraging the growing knowledge of miRNAs and their respective targets to achieve therapeutic outcomes has prompted an explosion of research to identify miRNA modulators with applications to a broad range of disease conditions. miRNAs can be chemically modified to act either as miRNA antagonists or agonists to counteract or reduce miR expression associated with specific disease states. miRNA antagonist approaches include locked nucleic acid (LNA), antimiRs, antagomirs, morpholinos, and miRNA sponges—with each there is the promise of efficient inhibition of miRNA function and evidence for effective activity in vivo [7]. A particularly intriguing example of their use to date involves the use of miRNA sponges as decoys against specific miRNAs linked to type 1 and type 2 diabetes. In other studies the use of an in vivo antagomir antisense oligonucleotide for miR-122 (one of the most abundant miRs in liver that has been associated with cholesterol regulation) resulted in almost compete depletion of miR-122 in hepatocytes and a decrease in hepatic fatty acids, cholesterol synthesis, and plasma cholesterol levels. In their review, Nana-Sinkham and colleagues [12] have described additional successes in the use of miR-based therapeutics; wherein the achieved systemic delivery of an oligonucleotide-LNA of miR-122 in a non-human primate model for chronic HCV. In those experiments there was evidence for early success, with a potent reduction in the HCV burden. Nevertheless, an important caveat and challenge in the design of miRNA therapeutics will be estimating the likelihood for "off-target" binding or quenching (in the case of miRNA sponges) and designing safeguards against those undesirable off-target effects [4]. Targeted delivery approaches for miRNA therapeutics include both passive and active delivery systems that have been developed to minimize off-target effects [37]. Passive targeted delivery includes utilizing the cell-penetrating peptide (penetratin) in the biodegradable poly(lactide-co-glycolide) nanoparticles, the enhanced permeability and retention of solid tumors, and solid lipid nanoparticles. Active targeted delivery includes utilizing peptide or protein ligands, antibodies, aptamers, other ligands (ie, hyaluronic acid, folate acid), and magnetic nanoparticles. Currently, the antimiR-122 for HCV and the miR-34 mimic to treat liver cancer are in clinical trials [38].

With the increasing sophistication of bioinformatics tools for probing genome-wide sequence features, and the wide accessibility of reference genomes for several organisms, most notably human, mouse, and nematode, there has been an explosion of Web-accessible algorithms designed to predict genomic interactions between miRNA and putative mRNA. This has been coupled with substantial progress in the in vitro validation of sequence recognition criteria for miRNA:mRNA target interaction. Balancing the many algorithmic successes, miRNA-binding predictions based on the use of the most popular computational tools (ie, miRanda, TargetScan, PicTar) have not generally been correlated to each other and their predictions for interaction are often not supported by experimental evidence [13]. Clearly refinements, based on more empirical evidence and theoretical considerations, will be needed to expand the utility of these tools to interrogate complex rules for genomic interaction. One level of improved sophistication will be the implementation of additional limitations in these bioinformatics tools to incorporate critical information obtained from empirical data for miRNA and mRNA expression differences [4]. In the context of different states, eg, healthy and disease tissue, weighted differences in miRNA recognition will likely influence global target recognition and overall predicted expression profiles. One approach to this concern of contextual expression of miRNA is discussed by Dorn and colleagues [4] who have championed the use of high-throughput sequencing procedures coupled to immunoprecipitation of miRNA:mRNA complexes bound within the RISC complex. In a technique termed "RISC-Seq," RISC complexes are immunoprecipitated (eg, using anti-Argonaute 2 antibodies), followed by RNA sequencing. This approach allows for precise mapping of target sequences in the context of overexpressed miRNAs and will likely be instrumental in profiling tissue and disease specific miRNA expression.

9. CIRCULATING MICRORNA AS BIOMARKERS

There is substantial interest in identifying circulating biomarkers that are useful in classifying disease severity and in gauging therapeutic response to candidate drug interventions. With the surprising observation that miRNAs are detectable in serum and are relatively stable—due to protection in extracellular vesicular structures called exosomes, there has been a substantial new growth and enthusiasm for this line of research. In skeletal muscle, serum level of myomiRs (miR-1, miR-133, and miR-206) are increased in the muscular dystrophic mouse and dog (canine X-linked muscular dystrophy in Japan) compared to expression of these miRs in control animals [39]. Notably, the myomiR miR-206 is significantly increased not only in rhabdomyosarcoma (RMS) cell lines, but also in serum of patients with RMS tumors [40]. Interestingly, miR-206 is observed in serum of patients with RMS, but not in other types of cancer [41]. However, miR-206 is increased in serum of patients with chronic obstructive pulmonary disease [42]. Furthermore, physical activity also alters circulating myomiR levels. In nonsmall cell–lung cancer, a four-miRNA serum signature has been identified [3]. In the context of heart disease, serum levels of miR-208b and miR-499 were strikingly increased (~1000-fold) after myocardial infarction [4]. Despite these early successes, there are many validation studies required [25]. Foremost among these validation studies is the need for RT-PCR as a validation for miR profiling as well as novel solutions to the difficulty of normalizing biomarker levels in serum. In the immune system, miRNAs in plasma and serum including miR-150, miR-146a, and miR-223 have been proposed as potential biomarkers for sepsis [13,20]. MiR-150 is correlated with aggressiveness of sepsis, and as such, may be a prognostic marker in patients with sepsis. In diabetes, serum miRNA profiles from diabetic and healthy subjects identified 65 common and 42 differently expressed miRNAs [7]. Studies using pooled sera have identified 13 miRNAs in diabetic patients compared with healthy subjects, further highlighting the promise of miRNAs as serum and plasma biomarkers. In the mouse, circulating miRNAs targeting macromolecule biosynthetic process, apoptosis regulators, and the Wnt signaling pathway are increased with age and intriguingly are decreased by calorie restriction [27]. Furthermore, miRNAs regulating the inflammatory pathways in detectable in serum of young and old individuals are altered with age [32]. However, whether changes in circulating miRNAs could be used as biomarkers for disease diagnosis and/or prognosis is a work still in progress. One of the technical challenges involves standardize approaches for sample processing to minimize variations, analytically and biologically. Analytical variation includes blood draw technique (ie, epithelial cell contamination due to skin plug), specimen processing (ie, hemolysis, lipemia, drugs, and antibodies), and normalization techniques [43,44]. For example, Wang et al. [45] have shown discrepancies in miRNA levels in serum and plasma of corresponding subjects in which miRNA levels in serum are consistently higher than those in plasma, possibly due to contamination of miRNAs from platelets. Biological variation can consists of extrinsic variation among individuals at the time of sampling (ie, diurnal, fasting/fed, health state) and intrinsic differences between individual (ie, age/gender, genetic/ethnicity, chronic disease) [44]. Another challenge involves establishing the source(s) of circulating miRNAs and their role in disease. Blood cells, endothelium lining the blood, and cells from other tissues all are likely contribute to circulating miRNAs [44]; therefore, determining tissue-specific miRNAs in the plasma and serum would further improve biomarker specificity and diagnostic utility. Collectively, in the coming years, it will be interesting to see how sensitive and stable these miRNA circulating biomarkers are and to what extent these peripheral markers can reflect organ health and disease interventions.

10. CONCLUDING REMARKS

In summary, miRNA is an essential feature of lineage commitment and regulatory guidance during tissue development such that when absent or hampered, often leads to disease states. In the coming years, there is much to be learned about adaptive (and maladaptive) states by examining how expression of miRNAs is influenced by the genetic architecture of miRNA genes, clusters, and mirtrons, as well as miRNA polymorphism and polymorphism in their mRNA targets. There are manifold modes of miRNA regulation (negative feedback, positive feedback, cross-regulatory) that monitor, modulate, or resolve signaling pathways in a variety of biological processes that include sepsis response, fibrosis, acute exercise, aging, and steroid biology. Perhaps the homeostasis and micromanagement of these miRNA regulatory systems, when perturbed, can arrive at new quasistable state of networked interactions that have an undesired effect of promoting or antagonizing disease severity and cancer progression. Clearly, a better understanding of these miRNA regulatory networks, as well as improved therapeutic tools for guiding miRNA expression and their targets toward desired outcomes, and healthy regulatory states, as a form of pathway therapeutics, will be the subject of many advances in miRNA biology over the coming years. Undoubtedly, recent advance of miRNA-based medicine to various phases of clinical trials will provide more insights into the bioavailability, biosafety, and efficacy of antimiRs and miRNA mimetics.

REFERENCES

[1] Bartel DP. MicroRNAs: target recognition and regulatory functions. Cell 2009;136(2): 215–33.
[2] Montano M. Detecting disease in blood: what miRNA biomarkers can tell us. In: Science technology webinar series. Science/AAAS Custom Publishing Office; 2012.
[3] Liu X, et al. Involvement of microRNAs in lung cancer biology and therapy. Transl Res 2011;157(4):200–8.
[4] Dorn 2nd GW. MicroRNAs in cardiac disease. Transl Res 2011;157(4):226–35.

[5] Akkina S, Becker BN. MicroRNAs in kidney function and disease. Transl Res 2011; 157(4):236–40.

[6] O'Rourke JR, et al. Essential role for Dicer during skeletal muscle development. Dev Biol 2007;311(2):359–68.

[7] Guay C, et al. Diabetes mellitus, a microRNA-related disease? Transl Res 2011; 157(4):253–64.

[8] Kerr TA, Korenblat KM, Davidson NO. MicroRNAs and liver disease. Transl Res 2011;157(4):241–52.

[9] Long K, Montano M. RNA surveillance-an emerging role for RNA regulatory networks in aging. Ageing Res Rev 2010.

[10] Nishikura K. Functions and regulation of RNA editing by ADAR deaminases. Annu Rev Biochem 2010;79:321–49.

[11] Dai R, Ahmed A. MicroRNA, a new paradigm for understanding immunoregulation, inflammation and autoimmune diseases. Transl Res 2011:157.

[12] Nana-Sinkam SP, Croce CM. MicroRNAs as therapeutic targets in cancer. Transl Res 2011;157(4):216–25.

[13] Zhou T, Garcia JG, Zhang W. Integrating microRNAs into a system biology approach to acute lung injury. Transl Res 2011;157(4):180–90.

[14] Clop A, et al. A mutation creating a potential illegitimate microRNA target site in the myostatin gene affects muscularity in sheep. Nat Genet 2006;38(7):813–8.

[15] Lian JB, et al. MicroRNA control of bone formation and homeostasis. Nat Rev Endocrinol 2012;8(4):212–27.

[16] Townley-Tilson WH, Callis TE, Wang D. MicroRNAs 1, 133, and 206: critical factors of skeletal and cardiac muscle development, function, and disease. Int J Biochem Cell Biol 2010;42(8):1252–5.

[17] Adlakha YK, Saini N. Brain microRNAs and insights into biological functions and therapeutic potential of brain enriched miRNA-128. Mol Cancer 2014;13:33.

[18] Meister B, Herzer S, Silahtaroglu A. MicroRNAs in the hypothalamus. Neuroendocrinology 2013;98(4):243–53.

[19] Li X, et al. A microRNA imparts robustness against environmental fluctuation during development. Cell 2009;137(2):273–82.

[20] Pandit KV, Milosevic J, Kaminski N. MicroRNAs in idiopathic pulmonary fibrosis. Transl Res 2011;157(4):191–9.

[21] Sharma M, et al. Mega roles of microRNAs in regulation of skeletal muscle health and disease. Front Physiol 2014;5:239.

[22] Zacharewicz E, Lamon S, Russell AP. MicroRNAs in skeletal muscle and their regulation with exercise, ageing, and disease. Front Physiol 2013;4:266.

[23] Rezen T, et al. Expression changes in human skeletal muscle miRNAs following 10 days of bed rest in young healthy males. Acta Physiol (Oxf) 2014;210(3):655–66.

[24] Schembri F, et al. MicroRNAs as modulators of smoking-induced gene expression changes in human airway epithelium. Proc Natl Acad Sci USA 2009;106(7):2319–24.

[25] Yendamuri S, Kratzke R. MicroRNA biomarkers in lung cancer: MiRacle or quagMiRe? Transl Res 2011;157(4):209–15.

[26] Novak J, et al. MicroRNAs involved in skeletal muscle development and their roles in rhabdomyosarcoma pathogenesis. Pediatr Blood Cancer 2013;60(11):1739–46.

[27] Dhahbi JM, et al. Deep sequencing identifies circulating mouse miRNAs that are functionally implicated in manifestations of aging and responsive to calorie restriction. Aging (Albany NY) 2013;5(2):130–41.

[28] Liu X, et al. MicroRNA-148/152 impair innate response and antigen presentation of TLR-triggered dendritic cells by targeting CaMKIIalpha. J Immunol 2010; 185(12):7244–51.

[29] Liu Y, et al. Modulation of T cell cytokine production by miR-144* with elevated expression in patients with pulmonary tuberculosis. Mol Immunol 2011;48(9–10): 1084–90.

[30] Saba R, Sorensen DL, Booth SA. MicroRNA-146a: a dominant, negative regulator of the innate immune response. Front Immunol 2014;5:578.

[31] Stittrich AB, et al. The microRNA miR-182 is induced by IL-2 and promotes clonal expansion of activated helper T lymphocytes. Nat Immunol 2010;11(11):1057–62.

[32] Noren Hooten N, et al. Age-related changes in microRNA levels in serum. Aging (Albany NY) 2013;5(10):725–40.

[33] Smith-Vikos T, Slack FJ. MicroRNAs and their roles in aging. J Cell Sci 2012;125(Pt 1): 7–17.

[34] Drummond MJ, et al. Aging and microRNA expression in human skeletal muscle: a microarray and bioinformatics analysis. Physiol Genomics 2011;43(10):595–603.

[35] Jung HJ, Suh Y. MicroRNA in aging: from discovery to biology. Curr Genomics 2012;13(7):548–57.

[36] Rivas DA, et al. Diminished skeletal muscle microRNA expression with aging is associated with attenuated muscle plasticity and inhibition of IGF-1 signaling. FASEB J 2014.

[37] Wang H, et al. Recent progress in microRNA delivery for cancer therapy by non-viral synthetic vectors. Adv Drug Deliv Rev 2015;81:142–60.

[38] van Rooij E, Kauppinen S. Development of microRNA therapeutics is coming of age. EMBO Mol Med 2014;6(7):851–64.

[39] Mizuno H, et al. Identification of muscle-specific microRNAs in serum of muscular dystrophy animal models: promising novel blood-based markers for muscular dystrophy. PLoS One 2011;6(3):e18388.

[40] Miyachi M, et al. Circulating muscle-specific microRNA, miR-206, as a potential diagnostic marker for rhabdomyosarcoma. Biochem Biophys Res Commun 2010;400(1):89–93.

[41] Cortez MA, et al. MicroRNAs in body fluids–the mix of hormones and biomarkers. Nat Rev Clin Oncol 2011;8(8):467–77.

[42] Aoi W, Sakuma K. Does regulation of skeletal muscle function involve circulating microRNAs? Front Physiol 2014;5:39.

[43] Scholer N, et al. Serum microRNAs as a novel class of biomarkers: a comprehensive review of the literature. Exp Hematol 2010;38(12):1126–30.

[44] Pritchard C. Detecting disease in blood: what miRNA biomarkers can tell us. In: Science technology webinar series. Science/AAAS Custom Publishing Office; 2012.

[45] Wang K, et al. Comparing the MicroRNA spectrum between serum and plasma. PLoS One 2012;7(7):e41561.

CHAPTER 2

Clinical and Therapeutic Applications of MicroRNA in Cancer

D.M. Brown, C. Croce, S.P. Nana-Sinkam

Key Concepts

- MicroRNAs (miRNAs) belong to a class of small noncoding RNAs (ncRNAs) that function as guide molecules for posttranscriptional gene regulation.
- The dysregulation of miRNAs is fundamental to the underlying pathogenesis leading to the development and potentiation of many cancers.
- Their relative stability in a wide array of microenvironments has given miRNAs a high potential as noninvasive biomarkers, particularly as an alternative to tumor biopsies.
- miRNA biomarker discovery is generally performed with broad-range profiling methods, such as microarrays, RT-PCR, and next-generation sequencing that can measure the differential expres-sion of up to thousands of ncRNAs.
- Functional miRNA-single-nucleotide polymorphisms in miRNAs may be biomark-ers of disease risk and may predict the clinical outcomes in cancer.
- Dysfunction of the cell cycle–related miRNAs (miR-15a/16 cluster, the miR-17/20 cluster, the miR-221/222 cluster, and the let-7 and miR-34 families) increases cell proliferation, leading to tumor growth promotion.
- miRNA biogenesis is altered, not only at transcriptional levels but also at genomic levels in cancer cells.
- Therapeutic approaches based on miRNA can be divided into two broad categories: miRNA inhibition therapy when the target miRNA is overexpressed and miRNA replacement therapy when the miRNA is repressed.
- In lung cancer, circulating miRNAs have been linked to predisease patterns, disease staging, tumor activity, and metastasis development.
- Expression levels of miR-21, miR-126, miR-199a, and miR-335 in serum have all been associated with key characteristic features of breast cancer, including tumor grade and hormone receptor status.
- A myriad of miRNAs have been implicated as potentially useful in the diagnosis, staging, and treatment of prostate cancer; colorectal cancer; and hepatocellular carcinoma.

Translating MicroRNAs to the Clinic
ISBN 978-0-12-800553-8

- The clinical utility of a number of miRNAs has been well established within the realm of hematologic malignancies, in particular, diffuse large B-cell lymphoma (DLBCL), chronic lymphocytic leukemia (CLL), and acute myeloid leukemia (AML).
- While a plethora of studies have identified altered circulating miRNA in the plasma/sera of cancer patients, there is a lack of standardization among the procedures and methods used in the detection of these ncRNAs.
- Additional studies are needed to aid in the identification and validation of diagnostic, prognos-tic, and therapeutic miRNA biomarkers, via standardized techniques.

1. INTRODUCTION

miRNAs belong to a class of small ncRNAs that function as guide mole-cules for posttranscriptional gene regulation. They are involved in a wide range of biological processes including development, cell proliferation, metabolism, and signal transduction [1]. According to the most recent release of the database on mirbase.org, there have been more than 35,000 miRNA products identified in over 220 species. Once believed to be of little biologic relevance two decades ago, miRNAs have emerged as poten-tial candidates for prognostic biomarkers and therapeutic targets in cancer. In addition to their applications in the setting of cancer, sufficient evidence exists to suggest that miRNAs are dysregulated in nervous system disorders, cardiovascular disorders, muscular disorders, and endocrine disorders such as diabetes mellitus, suggesting that the aberrant expressions of these markers may prove invaluable as diagnostic biomarkers in a wide array of disease states. While the ongoing research is expansive in nature, the focus of much of the recent advances has been inclusive of malignancies. The dysregulation of miRNAs is fundamental to the underlying pathogenesis leading to the development and potentiation of many cancers. Research involving the identification and development of miRNA as biomarkers in cancer has experienced exceptional growth lately as the scientific community has come to realize the versatility of these small ncRNAs to serve as noninva-sive screening assays with the sensitivity and specificity to detect cancer. In fact, miRNA biomolecules are remarkably stable, even in harsh physiologic conditions, including extreme variations in heat and acid/base milieu. Altered expression or dysfunction of miRNAs is associated with tumor formation and progression by influencing the oncogenic pathways that pro-mote tumor progression, such as cell cycle regulation, apoptosis, senescence, metabolism, angiogenesis, and metastasis [2]. The charged miRNAs have small size and low molecular weight making them suitable for packaging into an effective delivery system for utilization as possible clinical cancer therapy effectors. To this end, to achieve effective gene silencing in cancer cells, the development of strategies for efficient in vivo delivery and escape from blood clearance, enzyme degradation, and intracellular trapping, such as an endosome, is required.

2. MICRORNA PROCESSING

The generation of miRNA involves a complex series of well-defined enzymatically modulated steps (Fig. 2.1). miRNA biogenesis begins with transcription by RNA polymerase II of long primary transcripts termed (pri)miRNA, which in turn leads to the formation of primiRNAs. This transcript then undergoes a conformational change, folding on itself to form a double-stranded hairpin structure. An RNase endonuclease III termed Drosha, along with DiGeorge syndrome critical region gene 8 (DGCR8), which functions as a double-stranded RNA-binding domain protein, cleaves this secondary structure resulting in ~70 nt precursor miRNAs (premiRNAs). Under the influence of Exportin 5, a double-stranded binding protein, the premiRNA are subsequently transported from the nucleus to the cytoplasm. Here, they are further converted into double-stranded 18–25 nt RNAs by Dicer, another double-stranded RNA-specific nuclease, which acts to define a cleavage site that results in ~22-nt-long double-stranded dsRNA containing a mature miRNA [1]. One of the strands of mature miRNAs, known as the guide strand, binds to Argonaute (AGO) protein in the RNA-induced silencing complex (RISC). The Ago protein is crucial for the miRNA to form a complex with RISC. Moreover, it allows miRNA maturation even in the absence of Dicer by acquiring the ability to cleave premiRs. This evidence was identified by Cheloufi and colleagues who observed that miR-451 could enter RISC through an alternative biogenesis pathway. After Drosha-mediated cleavage, the premiR was loaded directly into Ago omitting the catalytic activity of Dicer and forming a complex with RISC [2]. Drosha can be substituted by other RNases such as RNase Z [3] or the integrator complex [4] to perform miRNA biogenesis. Some miRNA can even be derived from the tRNA pathway [5]. This works to illustrate the point that miRNA regulation can be augmented in a Drosha/Dicer independent fashion, and tight regulation of this machinery at a molecular level is crucial in maintaining biogenic homeostasis. By the imperfect base pairing between the guide strands of miRNAs and the

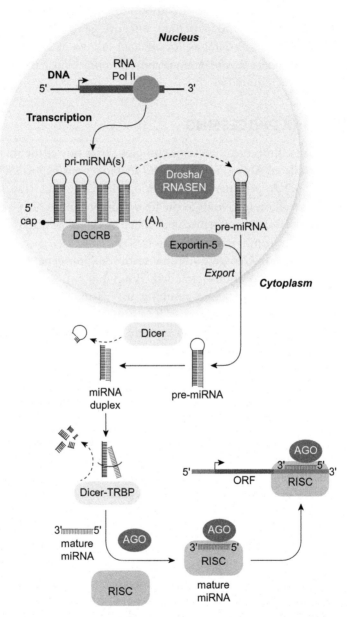

Figure 2.1 MicroRNA processing. *AGO*, Argonaute proteins; DGCR8, DiGeorge syndrome critical region gene 8; ORF, open reading frame; RISC, RNA-induced silencing complex; TRBP, transactivating response RNA-binding protein.

3′ or 5′ untranslated region of the target gene, miRNAs are able to regulate tens to hundreds of messenger RNAs (mRNAs), and induce either RNA degradation of inhibition of translation [6]. The gene suppression mechanism depends on the degree of complementarity sites between the miRNAs and mRNAs [7]. Protected from both cellular and environmental degradation, miRNA are inherently resistant to destruction for multiple reasons. First, miRNAs can be packaged in extracellular vesicles such as microvesicles, exosomes, and apoptotic bodies [8,9]. Second, most miRNAs are bound to RNA-binding proteins such as AGO_2, or lipoprotein complexes such as high density lipoproteins [10,11].

3. CIRCULATING MICRORNA AS BIOMARKERS

Possessing critical regulatory functions, miRNAs are involved in virtually all physiologic processes from cellular development, proliferation, and differentiation to metabolism and homeostasis. Circulating miRNAs were first used as biomarkers in the serum to examine patients with diffused large B-cell lymphoma [12]. An ideal biomarker should, among other properties, be measurable with minimal invasiveness, risk, and discomfort to the patient. Thus, screening biofluids, such as blood, sputum, or urine, is favored over more invasive techniques required to access lung biopsies, bronchial brushings, or bronchoalveolar lavage fluid. Interestingly, miRNAs have been detected in all human fluids, despite the hostile conditions of these environments which generally promote RNA degradation. Their relative stability in a wide array of microenvironments have given miRNAs a high potential as noninvasive biomarkers, particularly as an alternative to tumor biopsies. As this field of research has continued to grow, implications between various miRNA and a wide variety of other human diseases have emerged. miRNA signatures from body fluids are traditionally identified with the use of microarray profiling, real-time PCR array, and next-generation sequencing technologies [13].

4. MICRORNA ISOLATION AND AMPLIFICATION METHODOLOGIES

The choice of starting material is crucial to initial experimental design and the choice of whole blood, peripheral blood mononuclear cells, serum, plasma, or purified exosomes from the same individual will generate very different expression profiles. miRNA biomarker discovery is generally performed with broad-range profiling methods, such as microarrays, RT-PCR, and next-generation sequencing that can measure the differential expression of up to thousands of ncRNAs. With the advent of diagnostic profiling, disease-specific assay panels are now being designed that measure only the differential expression of miRNAs that are clinically relevant to the disease state in question, thereby reducing the cost and complexity of the assay.

A number of companies have started to develop these clinical diagnostic test panels to classify various cancer types [14]. The receptacle used to collect the blood is crucial and should be ethylenediaminetetraacetic acid or citrate-containing, as heparin, a commonly used anticoagulant can inhibit the reverse transcriptase and polymerase enzymes used in PCR [15,16]. Standardized phlebotomy protocols should be utilized to avoid prolonged time from collection to processing and to avoid hemolysis which can lead to augmentation of RNA levels.

5. PACKAGING AND DELIVERY MECHANISMS

The charged miRNAs have small size and low molecular weight making them possible to be formulated into an effective delivery system and become attractive options for clinical cancer therapy development. However, to achieve effective gene silencing in cancer cells, the development of strategies for efficient in vivo delivery and escape from blood clearance, enzyme degradation, and intracellular trapping, such as an endosome, is required. Exosomes are endocytic membrane-derived vesicles that participate in cell-to-cell communication along with protein and RNA delivery [17]. miRNAs have been identified in exosomes, which can be taken up by neighboring or distant cells and subsequently modulate physiology of the recipient cells [18]. Exosomal miRNAs have been detected in the serum of several cell lines, and have been implicated as potential biomarkers for entities like adenocarcinoma and cardiovascular disease.

6. MICRORNA DYSREGULATION IN CANCER

Efficient management of cancer patients depends on early diagnosis and monitoring of treatment. The gold standard for cancer diagnosis remains the histological examination of affected tissue, obtained either by surgical excision, or radiologically guided biopsy. Such procedures, however, are expensive, not without risk to the patient, and require consistent evaluation by expert pathologists. Therefore, there has been great interest in the field of circulating nucleic acids as potential cancer biomarkers. Abnormal expression of miRNAs is associated with all steps of tumorigenesis. Polymorphisms in miRNA genes have been associated with a high risk of developing cancer. In fact, functional miRNA-single-nucleotide polymorphisms in miRNAs may be biomarkers of disease risk and can predict the clinical outcomes in cancer. Numerous studies have conclusively shown altered miRNA profiles in multiple cancer types, including but not limited to lung cancer, breast cancer, prostate cancer, colorectal cancer (CRC), liver cancer, CLL, and AML (Table 2.1). Five groups of

Table 2.1 miRNA in select human malignancies

Condition	Overexpressed	Downregulated
Lung cancer	miR-21, miR-210, miR-182, miR-31, miR-200b, miR-205, miR-183, miR-155, miR-221/222	miR-126-3p, miR-30a, miR-30d, miR-486-5p, miR-451a, miR-126-5p, let-7, miR-143, miR-145, miR-128, miR-29, miR-34
Breast cancer	miR-425, miR-302b, miR-21, miR-126, miR-199a, miR-335, miR-122, miR-10b, miR-34a, miR-195, let-7, miR-155	miR-182, miR-17, miR-34a, miR-30a, miR-126
Prostate cancer	miR-141, miR-375, miR-107, miR-574-3p, miR-18am, miR-221, miR-21	miR-409-3p, let-7
Colorectal cancer	miR-141, miR-29a, miR-92, miR-21, let-7g, miR-200c, miR-155	miR-31, miR-181b, miR-92a, miR-203, miR-34a, let-7
Hepatocellular cancer	miR-1, miR-25, miR-92a, miR-206, miR-375, let-7f, miR-222/223, miR-15b, miR-130b, miR-21, miR-221/222	miR-122, miR-16, miR-199a, miR-1, miR-26a
Chronic lymphocytic leukemia	miR-155	miR 15 a/16-1, miR-29b, miR-181b, miR-34
Acute myeloid leukemia	Let-7b, miR-523, miR-29	Let-7d, miR-150, miR-339, miR-181, miR-204, miR-342

miRNAs, including the miR-15a/16 cluster, the miR-17/20 cluster, the miR-221/222 cluster, and the let-7 and miR-34 families, act by targeting cell cycle regulators to control cell cycle checkpoints and progression [19]. Dysfunction of the cell cycle–related miRNAs increases cell proliferation, leading to tumor growth promotion. Abnormal expression of miRNAs is also correlated with the resistant mechanisms to chemotherapy. Thus, miRNAs can serve as both diagnostic and prognostic biomarkers in cancer [20]. For example, miR-155 overexpression and let-7a downregulation are associated with poor disease outcome in lung cancer [21]. When all of the aforementioned is placed in a clinical context, it is easy to see that miRNA biogenesis is altered, not only at transcriptional levels but also at genomic levels in cancer cells. Although still not completely elucidated, one can conclude that there exists a causal relationship between tumorigenesis and perturbed miRNA biogenesis [22].

7. ROLE OF MICRORNA IN CANCER DIAGNOSIS, PROGNOSTICATION, AND TREATMENT

While the utility of blood-based miRNA as noninvasive biomarkers of disease is an attractive alternative to invasive surgical or bone marrow biopsies in the diagnosis of cancer, they oftentimes lack the sensitivity and specificity for the screening of early stage disease. Unlike the regulatory mRNA expression profiles, tissue miRNA signatures seem to be more reliable in the detection and staging of cancer [23]. Furthermore, the levels of circulating miRNAs can be used to distinguish cancer patients from healthy individuals. The circulating miRNAs exhibit higher stability in body fluids and can be extracted and measured noninvasively. Collectively, circulating miRNAs represent a class of ideal biomarkers for cancer diagnosis and prognosis in the correct clinical context. Therapeutic approaches based on miRNA can be divided into two broad categories: miRNA inhibition therapy when the target miRNA is overexpressed and miRNA replacement therapy when the miRNA is repressed. miRNA antagonists are generated to inhibit endogenous miRNAs that show a gain-of-function in diseased tissues. This technique involves the introduction of a highly chemically modified miRNA passenger strand (antimiR) that binds with high affinity to the active miRNA strand. Since binding is frequently irreversible, the new miRNA duplex is unable to be processed by RISC and/or degraded. By reintroducing a tumor suppressor miRNA, miRNA replacement therapy seeks to restore a loss of function in cancer cells and to reactivate cellular pathways driving a therapeutic response [24]. One example of this relationship can be seen with miR-21, which is overexpressed in multiple malignant phenotypes, and acts to downregulate many tumor suppressor genes, yielding a net downstream effect in proliferation, cell death, metastasis, and chemoresistance. More specifically, miR-21 antagonism reverses epithelial mesenchymal transition phenotype and acts to block angiogenesis in breast cancer via the inactivation of AKT and MAPK pathways [25–27]. Conversely, miRNA mimics, known as miRNA replacement therapy, work to suppress the expression of target genes. Genomic loss of tumor-suppressor miRNAs can be restored by miRNA mimics, which behave like endogenous miRNAs. They are particularly beneficial to cancer cells with low tumor-suppressor miRNA expression. Dysregulated in several epithelial tumors, melanomas, neuroblastomas, leukemias, and sarcomas, the miR-34 family functions as an effector on the p53 pathway [28]. A defect in this tumor-suppressor gene is ultimately responsible for approximately half of all human cancers, and plays an integral role in the suppression of tumor development. Both miRNA antagonists and miRNA mimics are low molecular weight oligonucleotides. Accordingly, they are easier to deliver into the target cells compared with large viral vectors or plasmids normally used for gene therapy. Nevertheless, there is a distinct disconnect between the efficacy of in vitro and in vivo delivery of miRNAs due to factors including poor bioavailability, limited tissue permeability, and instability of delivery vectors. Successful payload delivery remains an obstacle to effective miRNA-based therapeutics. Xue et al. recently investigated the effects of Mir-34a delivery on lung cancer development in a genetically engineered mouse model of lung cancer based on loss of p53 and Kras activation. Tumor-bearing mice were treated with intravenous Mir-34a in a lipid/polymer nanoparticle. They noticed that the therapeutic delivery of Mir-34a delayed tumor progression compared with control animals, suggesting that restoring p53 functions with Mir-34a could be important to obtain an improved response to chemotherapy. In fact, the first clinical trial with miRNA replacement therapy in cancer is ongoing. This phase I study focuses on miR-34 (MRX34) in patients with liver cancer or solid cancer with liver involvement. It is delivered using a liposomal formulation and the first safety data from 26 patients show that it has a manageable safety profile (https://clinicaltrials.gov/ct2/show/NCT01829971).

7.1 Lung Cancer

Second only to prostate and breast cancer in males and females, respectively, lung cancer is the second most common malignancy diagnosed, according to 2015 estimates from the American Cancer Society, yet lung cancer accounts for more deaths than any other cancer in both men and women.

Unfortunately only 15% of lung cancers are diagnosed at a localized stage, which limits treatment modalities and significantly reduces the 1- and 5-year survival rates [29]. Cigarette smoking remains the most important risk factor for the development of lung cancer with risk inversely related to the quantity and duration of tobacco use. Screening with spiral CT has been shown to decrease lung cancer deaths by 16–20% compared to standard X-ray among adults with a 30-pack year smoking history who were current smokers or had quit within 15 years [29]. The National Lung Screening Trial (NLST) using low-dose helical computed tomography (LDCT) in high-risk individuals demonstrated a 20% reduction in lung cancer–specific mortality and a 6.7% reduction in all-cause mortality can be achieved. Nevertheless, the high false-positive rates of NLST, costs, and potential harms from radiation exposure highlight the need for simpler, noninvasive, and more accessible methodologies for effective early cancer detection [30,31]. In lung cancer, circulating miRNAs have been linked to predisease patterns, disease staging, tumor activity, and metastasis development. A recent comprehensive meta-analysis of 20 miRNA expression studies in lung cancer, including a total of 598 tumors and 528 noncancerous control samples, was published. Utilizing a rank aggregation method, the authors identified a statistically significant miRNA metasignature of seven upregulated (miR-21, miR-210, miR-182, miR-31, miR-200b, miR-205, and miR-183) and eight downregulated (miR-126-3p, miR-30a, miR-30d, miR-486-5p, miR-451a, miR-126-5p, miR-143, and miR-145) miRNAs [32]. Although outcomes have varied, due to differences in collection, processing, and analysis of samples, recently published studies have all unanimously agreed that miRNA profiles from tissues may serve as important indicators and tools for classifying lung cancer subtypes and distinguishing primary from metastatic lung tumors. MiR-21 has been described as an miRNA that can differentiate early stage lung cancer patients from healthy individuals. The plasma levels of this miRNA not only serve as a circulating tumor biomarker, but can also determine sensitivity to platinum-based chemotherapy [33,34]. The expression of miR-205 is uniquely suited to aid in the identification of pathologic subtype of lung cancer, proving itself as a valuable tool in distinguishing squamous from nonsquamous NSCLC. Yu et al. showed that in 112 NSCLC patients, low levels of miR-221 and let-7a, and high miR-137, miR-372, and miR-182 can predict overall and disease-free survival outcomes [35]. MiR-21, in combination with miR-210 and miR-486-5p, was shown to be expressed significantly higher in the plasma of patients with malignant solitary pulmonary nodules compared to those with benign nodules and miR-155 with miR-197 and miR-182 was able to distinguish between NSCLC patients and control samples by real-time PCR of plasma [33,36]. Authors involved in the randomized Multicenter Italian Lung Detection trial concluded recently that a specific miRNA signature classifier (MSC) algorithm which included 24 previously identified high value miRNAs was able to risk stratify patients with similar sensitivity and specificity as low dose CT. However, when the MSC algorithm was used in conjunction with the radiographic imaging aforementioned there was a fivefold reduction in the LDCT false-positive rate [37–39].

7.2 Breast Cancer

According to US National Cancer Institute estimates, a woman in the United States has a one in eight lifetime risk of developing breast cancer. Despite the preventative measures currently employed, including routine mammography and MRI in high-risk individuals, the confirmation of diagnosis is inevitably invasive in nature, requiring fine needle aspiration or biopsy to provide definitive tissue diagnosis. One of the first genome-wide studies to examine miRNA plasma expression was performed by Zhao et al. [40]. This study compared the plasma miRNA profiles of 20 early stage breast cancer patients and 20 healthy controls, and found that miR-425 and miR-302b were differentially expressed in early stage breast cancer. Studies involving next-generation sequencing has also yielded valuable results, indicating that circulating miR-21 could provide the means by which normal breast tissue could be separated from invasive carcinomas [41]. In fact, expression levels of miR-21, miR-126, miR-199a, and miR-335 in serum have all been associated with key characteristic features of breast cancer, including tumor grade and hormone receptor status [42]. The serum levels of miR-155 have been shown to discriminate the changes in tumor mass and the effect of chemotherapy, and is more sensitive than routine markers such as carbohydrate antigen 15-3 (CA15-3), carcinoembryonic antigen (CEA) and tissue polypeptide specific antigen. In addition, miR-155 in combination with miR-145 and miR-182 can significantly increase the sensitivity and specificity of breast cancer diagnosis [43]. In another study, circulating blood levels of miR-122, miR-10b, miR-34a, and miR-155 were associated with the presence of overt metastasis, while serum concentrations of these same miRNAs were significantly elevated in patients with ovarian and lung cancer [44,45]. Heneghan et al. found an association between high serum levels of miR-10b and the estrogen receptor status of breast cancer patients. Consequently, it has been suggested that plasma miR-210 levels can be used in assessing the response of breast cancer patients to trastuzumab [46].

7.3 Prostate Cancer

To date, there is no effective early screening test for the diagnosis of prostate cancer. A blood test for PSA, an antigen produced by prostate cells is available but the test is exceedingly nonspecific and of questionable clinical relevance. Mitchell et al. examined a panel of miRNAs in the blood of advanced prostate cancer and showed that miR-141 was highly elevated in the serum. Moreover, miR-141 demonstrated a high correlation with serum PSA levels, and could identify individuals with advanced prostate cancer with 60% sensitivity and 100% specificity [47]. Brase et al. illustrated that circulating miR-375 and miR-141 were the most confirming of the markers for high-risk prostate cancers. Combined, these markers were observed to be significantly upregulated in prostate tissue samples. In yet another recent study, miR-1290 showed the best diagnostic performance in prostate cancer patients compared to a control group made up of healthy individuals, when 735 miRNAs were screened from the serum of affected individuals. Additionally, serum miR-1290 levels could differentiate between controls and patients with early stage pancreatic cancer [48,49]. For physiological and anatomical reasons, urine may prove to be a valuable source of miRNA biomarkers for the detection of urological cancers. Indeed, miR-107 and miR-574-3p were present at significantly higher concentrations in the urine of men with prostate cancer, as compared to healthy controls [50]. A comparative study between serum miR-141 and the serum levels of several standard biomarkers (PSA, circulating tumor cells, and lactate dehydrogenase) indicated that miR-141 has a similar ability to diagnose or predict clinical progression, when compared to other clinically validated biomarkers. More importantly, miR-141 could potentially be used in a test to identify patients with previously undetectable micro-metastases at the time of diagnosis [51].

7.4 Colorectal Cancer

Although preventative screening with the use of colonoscopy and fecal occult blood testing have reduced mortality, CRC continues to contribute greatly to cancer-related deaths in Western countries. While colonoscopy remains the gold standard for early detection and diagnosis, the procedure itself is not without potential risk and inherent complications. Most cancer-related deaths attributed to CRC can be prevented through early detection and surgical excision of the cancer. Thus, the early detection of CRC is very important for reducing the associated mortality rate. Therefore, the development and use of serum biomarkers as a means to diagnose and monitor response to therapy is of particular interest to the research community. In addition to prostate cancer, high levels of plasma miR-141 have also been associated with the presence of distant metastasis and poor prognosis in CRC [52]. Similarly, miR-29a and miR-92 have been found to be elevated in the plasma of early stage of CRC [53,54]. Wang et al. demonstrated that a characteristic miRNA signature (upregulated miR-21 and let-7g, and downregulated miR-31, miR-181b, miR-92a, and miR-203), which also included miR-92a, was able to differentiate CRC samples from cancer-free controls with high sensitivity and specificity. This detection characteristic outperformed any single-factor biomarker currently utilized, including CA19-9 or CEA [55]. Elevated serum levels of miR-200c have been associated with regional and distant metastasis in CRC patients. Additionally, circulating levels of certain miRNA have been linked to chemotherapy responsiveness. A five serum miRNA signature panel (miR-20a, miR-130, miR-145, miR-216, and miR-372) has demonstrated efficacy in detecting chemosensitivity/chemoresistance in CRC patients [56]. Accordingly, an elevated three-panel miRNA signature including miR-27b, miR-148a, and miR-326 has also been linked to the nonresponse of platinum-based chemotherapy in CRC patients [57].

7.5 Hepatocellular Carcinoma

Hepatocellular carcinoma (HCC) represents the fifth most common malignancy worldwide and has poor survival rates. In a study by Li et al., in 2010, six serum miRNAs were found to be upregulated in HCC samples when compared to healthy controls, and included miR-1, miR-25, miR-92a, miR-206, miR-375, and let-7f. Of these six miRNAs, three were clearly shown to be able to differentiate between healthy controls and HCC patients, including miR-25, miR-375, and let-7f [58]. Following a later study, which included a cohort of 20 miRNAs which were selectively upregulated in the setting of HCC, miR-885-5p was identified as being the most notable among a study group that included patients afflicted with not only HCC but also liver cirrhosis, chronic hepatitis B, and healthy controls. Nevertheless, there was no significant difference in the levels of miR-885-5p observed between the control group, cirrhosis group, and HCC group, indicating that this biomarker may prove to be valuable in identifying potential precipitants of liver pathology, such as hepatitis B [59].

While there have been a multitude of studies published pertaining to the potential usefulness of miRNA as a biomarker of HCC, few have been able to demonstrate conclusive evidence that a particular miRNA is able to delineate underlying liver malignancy from that of a chronic hepatitis picture. Moreover, the ability of circulating miRNAs to distinguish HCC from other hepatocellular diseases is absolutely critical for the accurate diagnosis of liver cancer. As prognosis is clearly tied to the early initiation of therapy, an early clinical diagnosis of HCC is of extreme importance. Several circulating miRNAs such as miR-15b, miR-130b, and miR-16 have been shown to hold promise in diagnosing patients with early stage HCC. The combination of miR-15b and miR-130b can increase the sensitivity for liver cancer diagnosis, as compared to α-fetoprotein (AFP) alone. Similarly, serum miR-16 was found to be a more sensitive biomarker for HCC than serum AFP, des-γ-carboxy prothrombin (DCP), and *Lens culinaris* agglutinin-reactive AFP levels [60,61]. Most recently, a study by Wang et al. produced findings that indicated that miR-302b inhibits SMMC-7721 cell invasion and metastasis by targeting AKT2, suggesting that miR-302b might represent a potential therapeutic target for HCC intervention [62]. Considered a tumor suppressor in human cancer, miR-122 expression has been found to be significantly lower in HCC than the level in normal tumor-adjacent tissues. HCC patients with low expression of miR-122 have a poor 3-year survival. Univariate analysis and multivariate Cox regression analysis indicated that miR-122 is an independent prognostic factor in HCC [63].

7.6 Hematologic Malignancies

Originally proposed in 2008, both miR-21 and miR-92 have been validated independently as potentially useful blood biomarkers of DLBCL [64,65]. In this study, miR-21, miR-155, and miR-210 were shown to be significantly elevated in the sera of cancer patients, than in healthy controls. Additional studies further implied that miR-21 was also upregulated in several DLBCL cell lines as well as in tumor tissues, and that in addition to its diagnostic potential, a high miR-21 concentration was also associated with relapse-free survival, suggesting that miR-21 could be a prognostic marker in DLBCL patients. Additionally, the miR-21 concentrations were inversely correlated to DLBCL stages, with higher concentrations noted in early stage disease [65]. In 2002, a seminal study by Calin et al. showed that miR 15 a/16-1 cluster is frequently deleted in CLL, implicating these miR-NAs as tumor suppressors [66]. A recent analysis by Yeh et al. concluded that following analysis of serial plasma samples collected from CLL patients on an ibrutinib clinical trial that the exosome plasma concentration was significantly decreased following ibrutinib therapy. miRNA profiling of plasma-derived exosomes identified a distinct exosome miRNA signature, including miR-29 family, miR-150, miR-155, and miR-223, that has been associated with CLL disease. Interestingly, expression of exosome miR-150 and miR-155 increases with B-cell receptor (BCR) activation [67]. In total, seven plasma miRNAs (miR-150, miR-19b, miR-92a, miR-223, miR-320, miR-484, and miR-17) have been implicated to have diagnostic specificity for CLL. Accordingly, the changes in the circulating miR-195 or miR-20a levels were shown to be the best marker for separating CLL patients from healthy controls [68]. Individuals with low circulating miR-155 expression responded better to standard therapeutics than patients with higher levels of miR-155 [69]. The most common acute leukemia in adults, AML is associated with multiple derangements in the levels of circulating miRNA, including downregulation of let-7d, miR-150, miR-339, and miR-342. Conversely, let-7b and miR-523 levels were found to be upregulated in the AML patients, when compared to healthy controls [70]. Serum levels of miR-181b-5p have also been linked to an increased survival of AML patients, thus implying that miR-181b-5p may act as a predictor of longevity for AML patients [71].

8. MICRORNA IN CLINICAL TRIALS

There are currently two clinical trials in progress, focusing on using miRNA as therapeutic options in the treatment of human malignancies. The first is a multicenter phase I trial, which commenced in 2013 and is estimated to be completed in the fall of 2016 focuses on utilizing a liposomal formulation of miR-34 to treat histologically confirmed unresectable primary liver cancer or advanced metastatic cancer with or without liver metastasis (melanoma, SCLC, NSCLC). For the hematologic malignancy cohorts, patients with ALL, CLL, CML in accelerated or blast phase, lymphoma, MM, or MDS were also included. This dose-escalation study is intended to investigate the safety and pharmacologic principles by administering the drug intravenously for 5 days, followed by an off cycle of 2 weeks duration. The primary outcomes of the study include an estimation of the maximum total tolerated dose for the drug, along with the recommended phase II dose moving forward, and any associated dose-limited toxicities. There have been no study results posted to date (Clinicaltrials.gov, a multicenter phase I study of MRX34, miRNA miR-RX34 liposomal injection). Since its inception in 2014, an additional phase I study has been investigating the effects of an EGFR-targeted miR-16-based miRNA, which is being developed for the treatment of recurrent malignant pleural mesothelioma and non-small cell–lung cancer. Recall that the miR-16 family has long been implicated as a tumor suppressor in a wide range of malignancies. An miR-16 mimic, which is a 23-base-pair synthetic RNA molecule, is packaged into a nanoparticle. These nonliving bacterial minicells are then directed to effector cells via a targeting moiety, which in this specific case, would include an anti-EGFR-specific antibody. Dose-related outcomes and efficacy, along with potential toxicities are important primary outcome measurements of this study. The drug is dosed intravenously once or twice weekly, depending on the dose, with follow ups scheduled at 24 h posttreatment, as well as periodic PET CT assessments. To date, there have been no reported preliminary results. (Clinicaltrials.gov, MesomiR 1: A phase I study of TargomiRs as 2nd or 3rd line treatment for patients with recurrent MPM and NSCLC). A myriad of other clinical trials are currently underway evaluating the potential usefulness of miRNA as biomarkers of a wide array of disease processes. These studies work to underscore the importance of miRNA not only in the identification of disease processes, but also in the prognostication and treatment responsiveness. The aforementioned two studies also highlight the importance of the need for tight regulation of miRNA as downstream modulators of gene expression, and that a loss of function mutation may be 1 day correctable with the in vivo delivery of miRNA mimics.

9. FUTURE DIRECTIONS FOR MICRORNA RESEARCH

As a consequence of the knowledge gained from sequencing the human genome, it is now known that less than 2% of the genome actually encodes proteins, a number much smaller than that previously expected. In light of this fact, the expression of ncRNA is now being investigated with a sense of increased importance. miRNA plays a vital role in the regulation of the genome by exerting an effect on the processes of transcription and translation, which result in protein end products, capable of driving cellular activities. While a plethora of studies have identified altered circulating miRNA in the plasma/sera of cancer patients, there is a lack of standardization among the procedures and methods used in the detection of these ncRNA. There exists great variability in the biologic and technical aspects of previous studies, including the starting materials utilized, technological platforms employed, and statistical methodologies used. Clearly, more studies are required for the robust identification and validation of diagnostic, prognostic, and therapeutic miRNA biomarkers. Additional emphasis placed on improving technique standardization will undoubtedly improve accuracy and enhance the reproducibility of results.

LIST OF ACRONYMS AND ABBREVIATIONS

AGO Argonaute Proteins
AML Acute Myeloid Leukemia
BCR B-cell receptor
CLL Chronic Lymphocytic Leukemia
CRC Colorectal Cancer
DGCR8 DiGeorge Syndrome Critical Region Gene 8
DLBCL Diffuse Large B-cell Lymphoma
HCC Hepatocellular Carcinoma
HDL High density lipoproteins
LDCT Low-dose helical computed tomography
miRNA MicroRNA
miR MicroRNA
ncRNA Noncoding RNA
mRNA Messenger RNA
NSCLC Non-Small Cell–Lung Cancer
ORF Open Reading Frame
RISC RNA-Induced Silencing Complex
TRBP Transactivating Response RNA-binding Protein

REFERENCES

[1] Bartel DP. MicroRNAs: genomics, biogenesis, mechanism, and function. Cell 2004;116:281–2.
[2] Chen Y. In vivo delivery of miRNAs for cancer therapy: challenges and strategies. Adv Drug Deliv Rev 2015;81:128–214.
[3] Bogerd HP, Karnowski HW, Cai X, Shin J, Pohlers M, Cullen BR. A mammalian herpesvirus uses noncanonical expression and processing mechanisms to generate viral microRNAs. Mol Cell 2010;37:135–42.
[4] Cazalla D, Xie M, Steitz JA. A primate herpesvirus uses the integrator complex to generate viral microRNAs. Mol Cell 2011;43:982–92.
[5] Maute RL, Schneider C, Sumazin P, Holmes A, Califano A, Basso K, et al. tRNA-derived microRNA modulates proliferation and the DNA damage response and is down-regulated in B cell lymphoma. Proc Natl Acad Sci USA 2013;110:1404–9.
[6] Denli AM, Tops BB, Plasterk RH, et al. Processing of primary microRNAs by the Microprocessor complex. Nat Rev 2004:231–5.
[7] Cheloufi S, Dos Santos CO, Chong MM, et al. A dicer-independent miRNA biogenesis pathway that requires Ago catalysis. Nature 2010:584–9.

[8] Zernecke A, Bidzhekov K, Noels H, et al. Delivery of microRNA-126 by apoptotic bodies induces CXCL12-dependent vascular protection. Sci Signal 2009;2(100):ra81.

[9] Valadi H, Ekstrom K, Bossios A, Sjostrand M, Lee JJ, Lotvall JO. Exosome-mediated transfer of mRNAs and microRNAs is a novel mechanism of genetic exchange between cells. Nat Cell Biol 2007;9(6):654–9.

[10] Arroyo JD, Chevillet JR, Kroh EM, et al. Argonaute2 complexes carry a population of circulating microRNAs independent of vesicles in human plasma. Proc Natl Acad Sci USA 2011;108(12):5003–8.

[11] Vickers KC, Palmisano BT, Shoucri BM, Shamburek RD, Remaley AT. MicroRNAs are transported in plasma and delivered to recipient cells by high-density lipoproteins. Nat Cell Biol 2011;13(4):423–33.

[12] Lawrie CH, et al. MicroRNA expression distinguishes between germinal center B cell-like and activated B cell-like subtypes of diffuse large B cell lymphoma. Int J Cancer 2007;121(5):1156–61.

[13] Wang J, Zhang KY, Liu SM, Sen S. Tumor-associated circulating microRNAs as biomarkers of cancer. Molecules 2014;19(2):1912–38.

[14] Vencken SF, et al. Non-coding RNA as lung disease biomarkers. Thorax 2015;70: 501–3.

[15] Garcia ME, Blanco JL, Caballero J, et al. Anticoagulants interfere with PCR used to diagnose invasive aspergillosis. J Clin Microbiol 2002;40:1567–8.

[16] Al-Soud WA, Radstrom P. Purification and characterization of PCR-inhibitory components in blood cells. J Clin Microbiol 2001;39:485–93.

[17] Lin J, Li J, Huang B, Liu J, Chen X, Chen XM, et al. Exosomes: novel biomarkers for clinical diagnosis. Sci World J 2015;2015:657086.

[18] Zhang J, Li S, Li L, Li M, Guo C, Yao J, et al. Exosome and exosomal microRNA: trafficking, sorting, and function. Genomics, Proteomics Bioinforma 2015;13(1):17–24.

[19] Yu Z, Baserga R, Chen L, Wang C, Lisanti MP, Pestell RG. microRNA, cell cycle, and human breast cancer. Am J Pathol 2010;176:1058–64.

[20] Iorio MV, Croce CM. MicroRNA dysregulation in cancer: diagnostics, monitoring and therapeutics. A comprehensive review. EMBO Mol Med 2012;4:143–59.

[21] Yanaihara N, Caplen N, Bowman E, Seike M, Kumamoto K, Yi M, et al. Unique microRNA molecular profiles in lung cancer diagnosis and prognosis. Cancer Cell 2006;9:189–98.

[22] Cheng G. Circulating miRNAs: roles in cancer diagnosis, prognosis and therapy. Adv Drug Deliv Rev 2015;81:75–93.

[23] Lee YS, Dutta A. MicroRNAs in cancer. Annu Rev Pathol 2009;4:199–227.

[24] Brighenti M. MicroRNA and MET in lung cancer. Ann Transl Med 2015;3(5):68.

[25] Bader AG, Brown D, Stoudemire J, Lammers P. Developing therapeutic microRNAs for cancer. Gene Ther 2011;18:1121–6.

[26] Han M, Liu M, Wang Y, Chen X, Xu J, Sun Y, et al. Antagonism of miR-21 reverses epithelial-mesenchymal transition and cancer stem cell phenotype through AKT/ERK1/2 inactivation by targeting PTEN. PLoS ONE 2012;7:e39520.

[27] Liu LZ, Li C, Chen Q, Jing Y, Carpenter R, Jiang Y, et al. MiR-21 induced angiogenesis through AKT and ERK activation and HIF-1alpha expression. PLoS ONE 2011;6:e19139.

[28] Wong MY, Yu Y, Walsh WR, Yang JL. microRNA-34 family and treatment of cancers with mutant or wild-type p53, (Review). Int J Oncol 2011;38:1189–95.

[29] American Cancer society. Cancer facts & Figures. Atlanta: American Cancer Society; 2015.

[30] National Lung Screening Trial Research Team, Aberle DR, Adams AM, et al. Reduced lung-cancer mortality with low-dose computed tomographic screening. N Engl J Med 2011;365:395e409.

[31] Hasan N, Kumar R, Kavuru MS. Lung cancer screening beyond low-dose computed tomography: the role of novel biomarkers. Lung 2014:639e–48.

[32] Võsa U, Vooder T, Kolde R, et al. Meta-analysis of microRNA expression in lung cancer. Int J Cancer 2013;132:2884–93.

[33] Shen J, Liu Z, Todd NW, et al. Diagnosis of lung cancer in individuals with solitary pulmonary nodules by plasma microRNA biomarkers. BMC Cancer 2011;11:374.

[34] Wei J, Gao W, Zhu CJ, et al. Identification of plasma microRNA-21 as a biomarker for early detection and chemosensitivity of non-small cell lung cancer. Chin J Cancer 2011;30(6):407–14.

[35] Yu SL, Chen HY, Chang GC, et al. MicroRNA signature predicts survival and relapse in lung cancer. Cancer Cell 2008;13:48–57.

[36] Zheng D, Haddadin S, Wang Y, et al. Plasma microRNAs as novel biomarkers for early detection of lung cancer. Int J Clin Exp Pathol 2011;4:575–86.

[37] Sozzi G, Boeri M, Rossi M, et al. Clinical utility of a plasma based miRNA signature classifier within computed tomography lung cancer screening: a correlative MILD trial study. J Clin Oncol 2014;32:768e73.

[38] Pastorino U, Rossi M, Rosato V, et al. Annual or biennial CT screening versus observation in heavy smokers: 5-year results of the MILD trial. Eur J Cancer Prev 2012;21:308e15.

[39] Boeri M, Verri C, Conte D, et al. MicroRNA signatures in tissues and plasma predict development and prognosis of computed tomography detected lung cancer. Proc Natl Acad Sci USA 2011;108:3713e8.

[40] Zhao H, Shen J, Medico L, Wang D, Ambrosone CB, Liu S. A pilot study of circulating miRNAs as potential biomarkers of early stage breast cancer. PLoS ONE 2010; 5(10):e13735.

[41] Ryu S, Joshi N, McDonnell K, et al. Discovery of novel human breast cancer microRNAs from deep sequencing data by analysis of pri-microRNA secondary structures. PLoS ONE 2011;6(2):e16403.

[42] Wang F, Zheng Z, Guo J, et al. Correlation and quantitation of microRNA aberrant expression in tissues and sera from patients with breast tumor. Gynecol Oncol 2010;119:586–93.

[43] Mar-Aguilar F, Mendoza-Ramirez JA, Malagon-Santiago I, Espino-Silva PK, Santuario-Facio SK, Ruiz-Flores P, et al. Serum circulating microRNA profiling for identification of potential breast cancer biomarkers. Dis Markers 2013;34:163–9.

[44] Roth C, Kasimir-Bauer S, Pantel K, et al. Screening for circulating nucleic acids and caspase activity in the peripheral blood as potential diagnostic tools in lung cancer. Mol Oncol 2011;5:281–91.

[45] Roth C, Kasimir-Bauer S, Heubner M, et al. Increase in circulating microRNA levels in blood of ovarian cancer patients. Adv Exp Med Biol Ed Gahan P Springer 2010: 63–71.

[46] Jung EJ, Santarpia L, Kim J, et al. Plasma microRNA 210 levels correlate with sensitivity to trastuzumab and tumor presence in breast cancer patients. Cancer 2012;118: 2603–14.

[47] Mitchell PS, Parkin RK, Kroh EM, et al. Circulating microRNAs as stable blood-based markers for cancer detection. Proc Natl Acad Sci USA 2008;105(30):10513–8.

[48] Brase JC, Johannes M, Schlomm T, Falth M, Haese A, Steuber T, et al. Circulating miRNAs are correlated with tumor progression in prostate cancer. Int J Cancer 2011;128:608–16.

[49] Li A, Yu J, Kim H, Wolfgang CL, Canto MI, Hruban RH, et al. MicroRNA array analysis finds elevated serum miR-1290 accurately distinguishes patients with low-stage pancreatic cancer from healthy and disease controls. Clin Cancer Res 2013;19:3600–10.

[50] Bryant RJ, Pawlowski T, Catto JW, Marsden G, Vessella RL, Rhees B, et al. Changes in circulating microRNA levels associated with prostate cancer. Br J Cancer 2012;106: 768–74.

[51] Gonzales JC, Fink LM, Goodman Jr OB, Symanowski JT, Vogelzang NJ, Ward DC. Comparison of circulating microRNA 141 to circulating tumor cells, lactate dehydrogenase, and prostate-specific antigen for determining treatment response in patients with metastatic prostate cancer. Clin Genitourin Cancer 2011;9:39–45.

[52] Cheng H, Zhang L, Cogdell DE, et al. Circulating plasma MiR-141 is a novel biomarker for metastatic colon cancer and predicts poor prognosis. PLoS ONE 2011;6:e17745.

[53] Ng EK, Chong WW, Jin H, Lam EK, Shin VY, Yu J, et al. Differential expression of microRNAs in plasma of patients with colorectal cancer: a potential marker for colorectal cancer screening. Gut 2009;58:1375–81.

[54] Huang Z, Huang D, Ni S, Peng Z, Sheng W, Du X. Plasma microRNAs are promising novel biomarkers for early detection of colorectal cancer. Int J Cancer 2010;127:118–26.

[55] Wang J, Huang SK, Zhao M, Yang M, Zhong JL, Gu YY, et al. Identification of a circulating microRNA signature for colorectal cancer detection. PLoS ONE 2014;9:e87451.

[56] Zhang J, Zhang K, Bi M, Jiao X, Zhang D, Dong Q. Circulating microRNA expressions in colorectal cancer as predictors of response to chemotherapy. Anticancer Drugs 2014;25:346–52.

[57] Menendez P, Padilla D, Villarejo P, Palomino T, Nieto P, Menendez JM, et al. Prognostic implications of serum microRNA-21 in colorectal cancer. J Surg Oncol 2013;108:369–73.

[58] Li LM, Hu ZB, Zhou ZX, Chen X, Liu FY, Zhang JF, et al. Serum microRNA profiles serve as novel biomarkers for HBV infection and diagnosis of HBV-positive hepatocarcinoma. Cancer Res 2010;70:9798–807.

[59] Gui J, Tian Y, Wen X, Zhang W, Zhang P, Gao J, et al. Serum microRNA characterization identifies miR-885-5p as a potential marker for detecting liver pathologies. Clin Sci (Lond) 2011;120:183–93.

[60] Liu AM, Yao TJ, Wang W, Wong KF, Lee NP, Fan ST, et al. Circulating miR-15b and miR-130b in serum as potential markers for detecting hepatocellular carcinoma: a retrospective cohort study. BMJ Open 2012;2.

[61] Borel F, Konstantinova P, Jansen PL. Diagnostic and therapeutic potential of miRNA signatures in patients with hepatocellular carcinoma. J Hepatol 2012;56:1371–83.

[62] Wang LL. miR-302b suppresses cell invasion and metastasis by directly targeting AKT2 in human hepatocellular carcinoma cells. Tumor Biol 2015:1010–4283.

[63] Xu QQ. MicroRNA-122 affects cell aggressiveness and apoptosis by targeting PKM2 in human hepatocellular carcinoma. Oncol Rep 2015:1021–335X.

[64] Fang C, Zhu DX, Dong HJ, et al. Serum microRNAs are promising novel biomarkers for diffuse large B cell lymphoma. Ann Hematol 2012;91:553–9.

[65] Chen W, Wang H, Chen H, et al. Clinical significance and detection of microRNA-21 in serum of patients with diffuse large B cell lymphoma in Chinese population. Eur J Haematol 2014;92:407–12.

[66] Calin GA, Dumitru CD, Shimizu M, et al. Frequent deletions and down-regulation of micro-RNA genes miR15 and miR16 at 13q14 in chronic lymphocytic leukemia. Proc Natl Acad Sci USA 2002;99:15524–9.

[67] Yeh Y-Y, Ozer HG, Lehman AM, Maddocks K, Yu L, Johnson AJ, et al. Characterization of CLL exosomes reveals a distinct microRNA signature and enhanced secretion by activation of BCR signaling. Blood 2015:3297–305.

[68] Moussay E, Wang K, Cho JH, van Moer K, Pierson S, Paggetti J, et al. MicroRNA as biomarkers and regulators in B-cell chronic lymphocytic leukemia. Proc Natl Acad Sci USA 2011;108:6573–8.

[69] Ferrajoli A, Shanafelt TD, Ivan C, Shimizu M, Rabe KG, Nouraee N, et al. Prognostic value of miR-155 in individuals with monoclonal B-cell lymphocytosis and patients with B chronic lymphocytic leukemia. Blood 2013;122:1891–9.

[70] Fayyad-Kazan H, Bitar N, Najar M, Lewalle P, Fayyad-Kazan M, Badran R, et al. Circulating miR-150 and miR-342 in plasma are novel potential biomarkers for acute myeloid leukemia. J Transl Med 2013;11:31.

[71] Zhi F, Cao X, Xie X, Wang B, Dong W, Gu W, et al. Identification of circulating microRNAs as potential biomarkers for detecting acute myeloid leukemia. PLoS ONE 2013;8.

CHAPTER 3

MicroRNAs in Kidney Function and Disease

S. Akkina, B.N. Becker

Key Concepts

- Discrete miRNAs are expressed in kidney tissue to a greater degree than other organs but there are noticeably absent miRNAs in kidney tissue as well.
- Organ-specific expression of miRNA provides insight into their functional roles in development and disease in the kidney.
- Deletion of Dicer in various parts of the kidney has shown the importance of miRNA in normal ureter development and glomerular function.
- Differential expression of miRNAs are noted in a variety of renal kidney diseases including acute kidney injury (AKI), renal fibrosis, and diabetic nephropathy.
- Further investigation will provide a more comprehensive understanding of the pathophysiology of kidney disease and may reveal potential diagnostic and therapeutic options.

1. INTRODUCTION

MicroRNAs (miRNAs) are regulatory RNAs that act as posttranscriptional repressors by binding the 3′ untranslated region of target genes [1]. The mammalian genome encodes several hundred miRNAs. Interestingly, the miRNA profile in kidney tissue differs greatly from other tissues and indeed, between different segments of tissue within the kidney [2–5].

As described elsewhere [1,6], miRNA genes, transcribed as long pri-miRNA molecules, have stem-loop structures with imperfect stems. The nuclear RNAse III enzyme Drosha and DiGeorge syndrome critical region gene 8 (DGCR8) complex cleave primiRNA into premiRNA as a hairpin structure with a 3′ overhang. PremiRNA exported into the cytoplasm via exportin-5 is processed by the RNAse III enzyme Dicer into mature miRNA, approximating 19–25 nucleotides in length. A strand of mature miRNA enters the RNA–inducing silencing complex and usually binds to the 3′ untranslated region of target mRNA. This binding reduces expression levels of the target protein via inhibition of translation or elongation, induction of deadenylation, or increased mRNA degradation [7].

Discrete changes in miRNA can have broad effects in kidney tissue. Deletion of Dicer in renin-secreting cells in murine kidneys leads to altered cell number, renin gene expression, lower plasma renin levels, and lower blood pressure levels. Kidneys also develop marked areas of fibrosis and vascular injury [8].

Notably, there is also a Dicer-independent pathway involving the Argonaute protein, AGO_2, that can convert premiRNA to mature miRNA independent of Dicer [9]. This pathway is regulated in cells of kidney origin by Hsp90 [10], and AGO_2 has been often identified as a link in understanding tubular cell injury during AKI [11]. Again, this suggests a direct interplay between an miRNA pathway and an organ-level event in the kidney.

miRNA genes are located throughout the genome including introns of protein-coding genes and noncoding transcription units, exons of noncoding transcription units, or intergenic regions [6,12]. Some miRNA genes are located close together in miRNA clusters and may target the same gene or multiple genes in the same pathway, while other miRNA genes may be duplicated on multiple chromosomes and act on the same target gene [12–14].

2. KIDNEY PHYSIOLOGY

2.1 Kidney Development

miRNAs that are in greater abundance in kidneys compared to other organs include miR-192, -194, -204, -215, and -216 [4]. There are a number of predicted and validated target genes for these miRNAs. Interestingly, there are also a number of miRNAs not present in kidney tissue (Table 3.1). Their absence may permit protein expression at levels necessary for adequate and constant kidney function. What is clear is the miRNAs are integral for normal kidney development. In animal models involving conditional Dicer knockout mice, miRNAs have critical roles in kidney development. Deletion of Dicer1 in developing ureteric bud epithelium led to hydronephrosis and parenchymal cysts [15]. Podocyte-specific Dicer knockout mice exhibited proteinuria by 3 weeks with segmental foot process effacement, and glomerulosclerosis [16,17]. The foot process effacement was associated, in part, with decreased nephrin and podocin expression in the slit diaphragm [18]. More recently, miR-17–92 was found to be essential in nephron development and renal function postnatally [19]. Using a conditional deletion of miR-17–92 in a mouse model, the deletion led to impaired progenitor cell proliferation and the number of developing nephrons were reduced.

Table 3.1 Differential expression of miRNA in human kidney tissue
High levels of expression in kidney

| miR-192 |
| miR-194 |
| miR-204 |
| miR-215 |
| miR-216 |

Low levels of expression in kidney	Highly expressed
miR-133a	Heart/Muscle
miR-133b	Heart/Muscle
miR-1d	Heart/Muscle
miR-296	Heart/Muscle
miR-1a	Heart
miR-122a	Liver
miR-124a	Brain

Postnatally, these mice developed renal disease with albuminuria by 6 weeks and histological changes at 3 months with focal podocyte foot process effacement and glomerulosclerosis. This suggests that individuals with the mutations in miR-17–92 cluster host gene (MIR17HG) such as some with Feingold syndrome may be susceptible to congenital kidney disease [19,20].

Studies in Sprague-Dawley rats suggested differential expression of miRNAs between cortex and medulla as well as reciprocal expression of mRNAs in subsections of kidney tissue [2,21]. Proteomic studies have provided additional confirmation of such differential distribution in the kidney. This, in fact, is logical, given the different functions present with cortical and medullary aspects of kidney tissue. One feature of miRNA expression worthy of additional investigation is the possibility that miRNAs in kidney tissue are linked with the protean degree of transport functions along the nephron. There are preliminary data suggesting that miR-192, an miRNA with markedly greater expression in cortical versus medullary kidney tissue, might be associated with sodium transport. Such links mandate further study [7].

3. BLOOD PRESSURE REGULATION

There is circumstantial evidence in several instances of the potential for miRNA to regulate blood pressure. MiR-155 appears to suppress expression of the type 1 angiotensin II receptor (AT1R) [22] and as such, would affect blood pressure in several ways. Given that AT1R are encoded on human chromosome 21, it made sense for investigators to take a different view of conditions that might affect human chromosome 21 and examine blood pressure regulation. The most common condition associated with low blood pressure and involving human chromosome 21 is trisomy 21, a condition associated with low blood pressure. In fibroblasts from individuals with trisomy 21, miR-155 levels were higher and AT1R levels lower compared to monozygotic twins [23]. Moreover, miR-155 levels were lower in aortic vessels from 16-week-old spontaneously hypertensive rats compared to age-matched Wistar-Kyoto rats [24].

In individuals with biopsy-proven nephrosclerosis, intrarenal expression of miRNAs correlated with severity of disease. Wang et al. studied 34 patients with hypertensive nephrosclerosis and compared intrarenal expression of the miR-200 family, miR-205, and miR-192 to 20 normotensive individuals from nephrectomies due to renal cancer [25]. They found that intrarenal expression of miR-200a, miR-200b, miR-141, miR-429, miR-205, and miR-192 were higher compared to normotensive controls. Proteinuria correlated with the intrarenal expression of each of the overexpressed miRNAs and the estimated glomerular filtration rate correlated with miR-200a and miR-205. In addition, intrarenal expression of ZEB1 was inversely correlated with miR-429, while intrarenal expression of ZEB2 was inversely correlated with miR-429 as well as miR-200a and miR-200b.

There are additional associative data examining polymorphisms in the 3′ untranslated region of the human L-arginine transporter, SCL7A1 [26]. This polymorphism may be linked with a genetic form of hypertension. The long and short variants of this polymorphism can be distinguished in part by an additional miR-122 binding site, though it remains to be determined whether miR-122 potentiates the development of hypertension for individuals with the long variant.

Finally, data from studies in an experimental model of hypertension, the Dahl salt-sensitive rat, raises the possibility that there might be differentially expressed miRNAs in the medullary region of the kidney compared to salt-insensitive, consomic SS-13^{BN} rats [7].

4. KIDNEY DISEASE

4.1 Acute Kidney Injury

AKI is a common problem among hospitalized patients and frequent due to tubular injury from ischemia or drug toxicity. Ischemia-reperfusion is associated with upregulation of miR-24 in renal cells in both mice and humans, while silencing of miR-24 in mice before injury resulted in better survival and kidney function, less apoptosis, less tubular injury, and less inflammatory cell infiltration [27]. In a cisplatin-based mouse model of AKI, miR-34a was found to be upregulated while miR-122 levels were downregulated. A combined miRNA–mRNA analysis shows Foxo3 as a key molecule involved in cisplatin-induced AKI where miR-122 directly suppresses Foxo3 mRNA translation while miR-34a activates Foxo3 via suppression of Sirt1 leading to cell-cycle arrest and resistance to oxidative stress [28,29]. Antagonism of miR-34a increased cell death during cisplatin treatment, suggesting that miR-34a has a protective role for cell survival [30].

Several miRNA have been recognized as potential biomarkers of AKI. Lorenzen et al. looked at miRNA expression among 77 individuals with AKI, 30 healthy controls, and 18 critically ill individuals with an acute myocardial infarction [31]. Using archived blood samples from a prospective study, a pooled sample of RNA from five individuals with AKI and five controls were analyzed via microarray and the top candidates were quantitated with RT-PCR in a validation cohort. In an adjusted analysis, miR-210 was found to be a strong predictor of mortality among critically ill patients with AKI. Similarly, urinary and plasma miR-21 levels were evaluated in those with severe AKI after cardiac surgery [32]. Urinary levels of miR-21 was 100% sensitive and 99% specific in predicting those requiring renal replacement therapy, while urinary and plasma levels were also associated with death, severe AKI, and prolonged hospital and ICU stays.

miRNAs have shown some potential in protecting the kidney from AKI. Hematopoietic miR-126 has been shown to promote vascular integrity within the renal peritubular capillary network and support recovery of the kidney after ischemic injury [33]. MiR-126 facilitates vascular regeneration by modulating mobilization of $Lin^-/Sca-1^+/cKit^+$ hematopoietic stem cells toward the kidney by reducing expression of the chemokine receptor, CXCR4, in the cells in the bone marrow and increasing renal expression of its ligand, stromal cell-derived factor 1. AKI recovery has also been associated with extracellular vesicles derived from mesenchymal stromal cells [34]. In glycerol-induced AKI, gene ontology studies found downregulation of genes in fatty acid metabolism and upregulation of genes associated with inflammation, matrix–receptor interactions, and cell-adhesion molecules. Intravenous administration of mesenchymal stromal cells and extracellular vesicles was found to correct these alterations and induce morphologic and function recovery of AKI, while the Drosha-knockdown model of mesenchymal stromal cells, with global downregulation of miRNA, did not.

4.2 Renal Fibrosis

Interstitial fibrosis is the common endpoint of chronic kidney disease and multiple miRNAs have been associated in the process. Transforming growth factor-β (TGF-β) is a key mediator of fibrosis via Smad2 and Smad3, and has been associated with numerous miRNAs. In fibrosis, miR-21 and miR-214 are inducers of kidney fibrosis, while miR-29 acts as an inhibitor. In the normal kidney, miR-21 expression is typically low but is upregulated significantly in the tubulointerstitial and glomerular areas in obstructive and diabetic nephropathy [35–37]. MiR-21 represses Smad7, an inhibitor of Smad2/3 phosphorylation that induces podocyte apoptosis and inhibits fibrosis in the tubulointerstitium, leading to increased TGF-β signaling in tubular epithelia and activation of the ERK pathway leading to increased renal fibrosis [38,39]. In miR-21 knockout mice, renal fibrosis was significantly reduced with fewer α-smooth muscle actin positive (αSMA+) interstitial myofibroblasts, interstitial collagen I, and fibronectin while lowering mRNA levels encoding TGF-β1 via Smad3 [36]. The effect of miR-214 on fibrosis varies by tissue; in cardiac tissue, miR-214 deletion leads to increased fibrosis after ischemia-reperfusion injury [40] whereas a similar deletion attenuated renal fibrosis in ureteral obstruction [41]. This reduction was confirmed with antimiR therapy prior to unilateral ureteral obstruction fibrosis. Interestingly, miR-214 appears to work independently of Smad2/3 when miR-21 antagonism blocked Smad2/3 activation while miR-214 antagonism did not.

In contrast, miR-29 has been shown to be a downstream inhibitor TGF-β-mediated fibrosis. MiR-29 is highly expressed in the normal kidney [42] but is significantly downregulated in animal models of fibrosis [43]. TGF-β-induced downregulation of miR-29 is mediated by Smad3 which binds to the promoter of miR-29. This is shown in Smad3 knockout mice where expression of miR-29 was increased while renal fibrosis was attenuated after ureteral obstruction. In cardiac tissue, miR-29 downregulation is associated with extracellular matrix deposition while overexpression results in reduced collagen deposition [44]. Similarly in the kidney, knockdown of miR-29 led to TGF-β-induced expression of collagens I and III, which were enhanced while delivery of miR-29b before or after obstructive injury blocked progressive renal fibrosis [43].

Therapeutic interventions are relatively limited in slowing the progression of renal fibrosis. Angiotensin converting enzyme (ACE) inhibitors have shown some success in delaying the onset of end-stage renal disease by reducing angiotensin II which has a direct role in the pathogenesis of glomerular injury. To examine the role of miRNA in spontaneous glomerular disease, miRNA expression profiles were measured in the glomeruli of Munich Wistar Fromter rats. MiR-324-3p was found to be the most upregulated miRNA and was localized to the glomerular podocytes, parietal cells of Bowman's capsule, and most abundantly in the cortical tubules. A predicted target of miR-324-3p was prolyl endopeptidase (Prep) that is involved in the synthesis of N-acetyl-seryl-aspartyl-lysyl-proline (Ac-SDKP), a peptide with antifibrotic and antiinflammatory properties. In this model, increased miR-324-3p was associated with reduced expression of Prep, low urine Ac-SKDP, and increased collagen. The use of an ACE inhibitor, lisinopril, led to lower miR-324-3p levels, higher Prep expression, and increased serum and urine levels of Ac-SDKP and reduced fibrosis. These findings suggest that ACE inhibitors may have some role in renoprotection via miRNAs [38,45].

4.3 Diabetic Kidney Disease

Diabetic nephropathy has yielded interesting findings related to miRNA expression. Work from Natarajan and colleagues identified a key role for miR-192 in diabetic nephropathy. MiR-192 levels are significantly increased in glomeruli obtained from streptozotocin-induced diabetic animals and diabetic db/db mice [46]. MiR-192 in diabetic glomeruli has been shown to be important in mediating TGF-β-induced collagen expression through a mechanism that encompasses TGF-β1-induced downregulation of SIP1/ZEB2 via miR-192 and ZEB1/δEF1 to increase Col1a2 expression through derepression of E-box elements. Inhibition of miR-192 significantly increases ZEB1/2 expression with decreased gene expression of collagen, TGF-β and fibronectin. Proteinuria was also reduced with miR-192 inhibition. Similarly, miR-216a was also found to be upregulated by TGFβ1 in experimental models of diabetic kidney disease.

Some of the same molecules play key roles in epithelial–mesenchymal transition and miRNA has been linked with this process [47]. The miR-200 family (miR-200a, miR-200b, miR-200c, miR-141, and miR-429) prevents expression of ZEB1/δEF1 and SIP1/ZEB2, maintaining an epithelial cell phenotype. To date, this process has been studied in only one kidney cell type, Madin–Darby canine kidney cells stably transfected with the protein tyrosine phosphatase, Pez [48]. Furthermore, miR-192 seems to have effects that may supersede diabetes alone. Deletion of Smad7 promoted miR-192 expression in a model of obstructive kidney disease whereas overexpression of Smad7 reduced miR-192 expression in kidney tissue [49].

Finally, it has become apparent over the last decade that diabetic nephropathy is critically dependent on podocyte biology. As noted earlier, miRNAs also have important roles in development and function of podocytes. Knocking out Dicer specifically in podocytes in mice leads to proteinuria and even death with evident podocyte and glomeruli abnormalities in these mice [16–18]. Two miRNAs, miR-93 and miR-29c, have been found to play important roles in podocytes and glomerular endothelial cells. The expression of miR-93 was lower in glomeruli from diabetic db/db mice compared to their controls and was further supported by similar findings in high glucose-treated podocytes and renal microvascular endothelial cells [50,51]. In contrast, miR-29c expression was increased in glomeruli from diabetic db/db mice. A target of miR-93 was found to be vascular endothelial growth factor-A (VEGF-A) that was confirmed with overexpression of VEGF when miR-93 was inhibited. Increased expression of miR-29c was found to induce podocyte apoptosis leading to reduced levels of Spry1, a direct target of miR-29c. Inhibition of miR-29c by antisense oligonucleotide decreased albuminuria and mesangial matrix accumulation in the db/db mouse model.

4.4 Other Kidney Diseases

Overexpression of miR-17-92 cluster in mice leads to lymphoproliferative disease with concomitant autoimmune injury in kidneys and likely immune complex deposition in glomeruli [52]. Additional studies have documented differential miRNA expression also in lupus nephritis, with 36 upregulated miRNAs, including miR-130b, miR-608, miR-124a, and miR-15b_MM1 and 30 downregulated miRNAs, most notably miR-150 and miR-92b_MM2 [53].

In vitro and in vivo studies suggest that miR-15a is downregulated in liver tissue from patients with autosomal dominant and recessive polycystic kidney disease, congenital hepatic fibrosis, and rats with PKD [54]. Other miRNAs have been shown to be differentially expressed in PKD/Mhm(cy/+) rats that develop PKD and control PKD/Mhm (+/+) rats, including the novel mi-RNAs, miR-31 and miR-217. Signaling molecule interactions, signal transduction, immune system regulation, and cell communication represent more than half of overrepresented miRNA regulatory pathways in this study [55]. Interestingly, a recent report from Li et al.

suggests miR-17 involvement directly targeting PKD2, altering PKD2 expression more so than PKD1 [56]. This suggests a differential effect on cystogenesis mediated by this miRNA.

5. KIDNEY TRANSPLANTATION

miRNAs also appear to correlate with biopsy-proven rejection in kidney transplant recipients. Anglicheau et al. [57] assayed 33 kidney allograft biopsy samples (12 acute rejection samples and 21 normal samples). The studies demonstrated that miRNA expression associated with human peripheral blood mononuclear cells (PBMCs), miR-142-5p, miR-155, miR-223, were increased in acute rejection. A positive association was evident between tubule-specific NKCC-2 mRNA and miR-30a-3p, miR-10b, and let-7c. Similarly, tubule-associated USAG-1 mRNA was positively associated with these transcripts as well. In vitro experiments using PBMCs and phytohemagglutinin noted that miR-155 increased after mitogen stimulation whereas miR-223 and let-7c decreased. Furthermore, in primary human renal epithelial cell cultures, miR-30a-3p, miR-30a-10b, and let-7c were present and cell stimulation decreased miR-30a-3p expression. The changes in PBMC miRNA expression positively correlating with alterations in tubular miRNA levels raises the possibility that the two could be interrelated expanding our understanding of epithelial cell and PBMC activation as they occur in acute cellular rejection.

Interstitial fibrosis/tubular atrophy (IFTA), also described as chronic rejection, has undergone similar studies. Biopsy tissue and urine demonstrate differential expression of miR-142-3p, miR-204, and miR-211 in those with IFTA versus those with normal histology [58]. Further investigations confirmed miR-142-3p in biopsy tissue and found that miR-21, miR-142-5p, another a cluster comprising miR-506 on chromosome X, miR-30b, and miR-30c were also differentially expressed in IFTA [59]. Analysis of 191 urinary cell pellets from 125 deceased donor recipients comparing those with IFTA to those with normal renal function found 22 differentially expressed miRNAs, mostly representing inflammation [60]. Further, these investigators examined 36 individuals with paired biopsies and urine samples and found that differential expression of miR-200b, miR-375, miR-423-5p, miR-193b, and miR-345 in the urine could predict those that would progress to IFTA compared to those that maintained stable renal function [58].

6. BIOMARKERS

The variable expression of miRNA in tissues and diseases make them a valuable tool for understanding, diagnosing, and discovering therapeutic options for diseases. While most studies involving miRNA involve tissue specimens, there are other sources including various body fluids. Serum has been shown to be a reliable source of miRNA biomarkers. miRNA circulates in a stable form, resistant to RNase activity [61], and that levels change with physiological changes such as pregnancy [62]. Urine has also shown some promise as a source for biomarkers. Analysis of miRNAs in urine from bladder of cancer patients and controls show that high ratios of miR-126:miR-152 and miR-182:miR-152 may be able to differentiate these groups [63]. In chronic kidney disease, serum miR-21 levels are high in patients with severe IFTA [64].

There is a strong need for more sensitive biomarkers of diabetic kidney disease as urine microalbumin is notoriously inaccurate, especially in the more common type 2 diabetics. Urine levels of 27 miRNAs that are involved in signaling pathways of renal fibrosis in diabetic kidney disease were associated with different stages of diabetic nephropathy [65]. Analysis of urinary exosomes in type 1 diabetics with early stage renal disease and microalbuminuria show higher levels of miR-130a and miR-145 and lower levels of miR-155 and miR-424 compared to patients with normoalbuminuria [66]. Decreased levels of urinary miR-15 were also seen in biopsy-proven diabetic nephropathy [67].

Urinary exosome analysis in IgA nephropathy, urinary miR-29b, and miR-29c expression are related to proteinuria and renal function, while miR-93 expression is related to glomerular scarring [68]. In lupus nephritis, miRNAs in various biofluids are representative of disease state, including miR-342-3p, miR-223, and miR-20a in plasma, miR-221 and miR-222 in urinary sediment, and miR-371-5p, miR-1224-3p, and miR-423-5p in PBMCs [69]. Autosomal dominant polycystic kidney disease express miR-1 and miR-133b that have been proposed as potential biomarkers to monitor disease progression and response to treatments [70].

7. CLINICAL IMPLICATIONS

Therapeutic options may be focused on manipulating miRNA activity to attenuate disease progression. Mimics consisting of double-stranded synthetic oligonucleotides that are chemically modified for stability and easy uptake into cells can be used to restore levels of miRNA. Inhibitors of miRNA expression include antagomirs which bind directly to miRNA [71], or miRNA sponges which contain tandem repeats of miRNA-binding sites [72]. In animal models of chronic kidney disease, inhibition of miR-21, miR-29c, miR-214, miR-433, and miR-192 and overexpression of miR-29b have been shown to halt progression of renal fibrosis. These inhibitors could be used in diseases such as diabetic nephropathy where increased levels of miR-192 have been found.

8. SUMMARY

These data across various settings demonstrate the burgeoning importance of miRNA in kidney function and disease. The findings to date represent marked strides in identifying unique characteristics about miRNA in kidney tissue and its clinical relevance. Not only are there discrete miRNAs expressed in kidney tissue to a greater degree than other organs but also there are noticeably absent miRNAs in kidney tissue. This organ-specific expression of miRNA provides insight into their functional roles in development and disease in the kidney. Deletion of Dicer in various parts of the kidney has shown the importance of miRNA in normal ureter development and glomerular function. Kidney-specific miRNAs such as miR-192 provide additional insight into the pathogenesis of diabetic nephropathy while the role of miR-155 in AT1R and the miR-17-92 cluster in lymphoproliferative disorders describe indirect methods of injury. Further investigation will provide a more comprehensive understanding of the pathophysiology of kidney disease and may reveal potential therapeutic options.

REFERENCES

[1] Filipowicz W, Bhattacharyya SN, Sonenberg N. Mechanisms of post-transcriptional regulation by microRNAs: are the answers in sight? Nat Rev Genet 2008;9(2):102–14.
[2] Tian Z, Greene AS, Pietrusz JL, Matus IR, Liang M. MicroRNA-target pairs in the rat kidney identified by microRNA microarray, proteomic, and bioinformatic analysis. Genome Res 2008;18(3):404–11.
[3] Liu CG, Calin GA, Meloon B, Gamliel N, Sevignani C, Ferracin M, et al. An oligonucleotide microchip for genome-wide microRNA profiling in human and mouse tissues. Proc Natl Acad Sci USA 2004;101(26):9740–4.
[4] Sun Y, Koo S, White N, Peralta E, Esau C, Dean NM, et al. Development of a microarray to detect human and mouse microRNAs and characterization of expression in human organs. Nucleic Acids Res 2004;32(22):e188.
[5] Shingara J, Keiger K, Shelton J, Laosinchai-Wolf W, Powers P, Conrad R, et al. An optimized isolation and labeling platform for accurate microRNA expression profiling. RNA 2005;11(9):1461–70.

[6] Fazi F, Nervi C. MicroRNA: basic mechanisms and transcriptional regulatory networks for cell fate determination. Cardiovasc Res 2008;79(4):553–61.

[7] Liang M, Liu Y, Mladinov D, Cowley Jr AW, Trivedi H, Fang Y, et al. MicroRNA: a new frontier in kidney and blood pressure research. Am J Physiol Ren Physiol 2009;297(3):F553–8.

[8] Sequeira-Lopez ML, Weatherford ET, Borges GR, Monteagudo MC, Pentz ES, Harfe BD, et al. The microRNA-processing enzyme Dicer maintains juxtaglomerular cells. J Am Soc Nephrol 2010;21(3):460–7.

[9] Cheloufi S, Dos Santos CO, Chong MM, Hannon GJ. A Dicer-independent miRNA biogenesis pathway that requires Ago catalysis. Nature 2010;465(7298):584–9.

[10] Pare JM, Tahbaz N, Lopez-Orozco J, LaPointe P, Lasko P, Hobman TC. Hsp90 regulates the function of argonaute 2 and its recruitment to stress granules and P-bodies. Mol Biol Cell 2009;20(14):3273–84.

[11] Tiemann K, Rossi JJ. RNAi-based therapeutics-current status, challenges and prospects. EMBO Mol Med 2009;1(3):142–51.

[12] Kim VN, Nam JW. Genomics of microRNA. Trends Genet 2006;22(3):165–73.

[13] Mourelatos Z, Dostie J, Paushkin S, Sharma A, Charroux B, Abel L, et al. miRNPs: a novel class of ribonucleoproteins containing numerous microRNAs. Genes Dev 2002;16(6):720–8.

[14] Altuvia Y, Landgraf P, Lithwick G, Elefant N, Pfeffer S, Aravin A, et al. Clustering and conservation patterns of human microRNAs. Nucleic Acids Res 2005; 33(8):2697–706.

[15] Pastorelli LM, Wells S, Fray M, Smith A, Hough T, Harfe BD, et al. Genetic analyses reveal a requirement for Dicer1 in the mouse urogenital tract. Mamm Genome 2009;20(3):140–51.

[16] Shi S, Yu L, Chiu C, Sun Y, Chen J, Khitrov G, et al. Podocyte-selective deletion of Dicer induces proteinuria and glomerulosclerosis. J Am Soc Nephrol 2008; 19(11):2159–69.

[17] Harvey SJ, Jarad G, Cunningham J, Goldberg S, Schermer B, Harfe BD, et al. Podocyte-specific deletion of Dicer alters cytoskeletal dynamics and causes glomerular disease. J Am Soc Nephrol 2008;19(11):2150–8.

[18] Ho J, Ng KH, Rosen S, Dostal A, Gregory RI, Kreidberg JA. Podocyte-specific loss of functional microRNAs leads to rapid glomerular and tubular injury. J Am Soc Nephrol 2008;19(11):2069–75.

[19] Marrone AK, Stolz DB, Bastacky SI, Kostka D, Bodnar AJ, Ho J. MicroRNA-17~92 is required for nephrogenesis and renal function. J Am Soc Nephrol 2014;25(7):1440–52.

[20] de Pontual L, Yao E, Callier P, Faivre L, Drouin V, Cariou S, et al. Germline deletion of the miR-17 approximately 92 cluster causes skeletal and growth defects in humans. Nat Genet 2011;43(10):1026–30.

[21] Tian Z, Greene AS, Usa K, Matus IR, Bauwens J, Pietrusz JL, et al. Renal regional proteomes in young Dahl salt-sensitive rats. Hypertension 2008;51(4):899–904.

[22] Martin MM, Lee EJ, Buckenberger JA, Schmittgen TD, Elton TS. MicroRNA-155 regulates human angiotensin II type 1 receptor expression in fibroblasts. J Biol Chem 2006;281(27):18277–84.

[23] Sethupathy P, Borel C, Gagnebin M, Grant GR, Deutsch S, Elton TS, et al. Human microRNA-155 on chromosome 21 differentially interacts with its polymorphic target in the AGTR1 3' untranslated region: a mechanism for functional single-nucleotide polymorphisms related to phenotypes. Am J Hum Genet 2007;81(2):405–13.

[24] Xu CC, Han WQ, Xiao B, Li NN, Zhu DL, Gao PJ. Differential expression of microRNAs in the aorta of spontaneously hypertensive rats. Sheng Li Xue Bao 2008;60(4):553–60.

[25] Wang G, Kwan BC, Lai FM, Choi PC, Chow KM, Li PK, et al. Intrarenal expression of miRNAs in patients with hypertensive nephrosclerosis. Am J Hypertens 2010;23(1):78–84.

[26] Yang Z, Venardos K, Jones E, Morris BJ, Chin-Dusting J, Kaye DM. Identification of a novel polymorphism in the 3'UTR of the L-arginine transporter gene SLC7A1: contribution to hypertension and endothelial dysfunction. Circulation 2007;115(10):1269–74.

[27] Lorenzen JM, Kaucsar T, Schauerte C, Schmitt R, Rong S, Hubner A, et al. MicroRNA-24 antagonism prevents renal ischemia reperfusion injury. J Am Soc Nephrol 2014;25(12): 2717–29.

[28] Lee CG, Kim JG, Kim HJ, Kwon HK, Cho IJ, Choi DW, et al. Discovery of an integrative network of microRNAs and transcriptomics changes for acute kidney injury. Kidney Int 2014;86(5):943–53.

[29] Brunet A, Sweeney LB, Sturgill JF, Chua KF, Greer PL, Lin Y, et al. Stress-dependent regulation of FOXO transcription factors by the SIRT1 deacetylase. Science 2004;303(5666):2011–5.

[30] Bhatt K, Zhou L, Mi QS, Huang S, She JX, Dong Z. MicroRNA-34a is induced via p53 during cisplatin nephrotoxicity and contributes to cell survival. Mol Med 2010;16(9–10):409–16.

[31] Lorenzen JM, Kielstein JT, Hafer C, Gupta SK, Kumpers P, Faulhaber-Walter R, et al. Circulating miR-210 predicts survival in critically ill patients with acute kidney injury. Clin J Am Soc Nephrol CJASN 2011;6(7):1540–6.

[32] Du J, Cao X, Zou L, Chen Y, Guo J, Chen Z, et al. MicroRNA-21 and risk of severe acute kidney injury and poor outcomes after adult cardiac surgery. PLoS ONE 2013;8(5):e63390.

[33] Bijkerk R, van Solingen C, de Boer HC, van der Pol P, Khairoun M, de Bruin RG, et al. Hematopoietic microRNA-126 protects against renal ischemia/reperfusion injury by promoting vascular integrity. J Am Soc Nephrol 2014;25(8):1710–22.

[34] Collino F, Bruno S, Incarnato D, Dettori D, Neri F, Provero P, et al. Aki recovery induced by mesenchymal stromal cell-derived extracellular vesicles carrying microRNAs. J Am Soc Nephrol 2015.

[35] Zhong X, Chung AC, Chen HY, Dong Y, Meng XM, Li R, et al. miR-21 is a key therapeutic target for renal injury in a mouse model of type 2 diabetes. Diabetologia 2013;56(3):663–74.

[36] Zhong X, Chung AC, Chen HY, Meng XM, Lan HY. Smad3-mediated upregulation of miR-21 promotes renal fibrosis. J Am Soc Nephrol 2011;22(9):1668–81.

[37] Wang J, Gao Y, Ma M, Li M, Zou D, Yang J, et al. Effect of miR-21 on renal fibrosis by regulating MMP-9 and TIMP1 in kk-ay diabetic nephropathy mice. Cell Biochem Biophys 2013;67(2):537–46.

[38] Okada H. Angiotensin converting enzyme inhibitor-modulated microRNAs targeting renal fibrosis. J Am Soc Nephrol 2012;23(9):1441–3.

[39] Lai JY, Luo J, O'Connor C, Jing X, Nair V, Ju W, et al. MicroRNA-21 in glomerular injury. J Am Soc Nephrol 2015;26(4):805–16.

[40] Aurora AB, Mahmoud AI, Luo X, Johnson BA, van Rooij E, Matsuzaki S, et al. MicroRNA-214 protects the mouse heart from ischemic injury by controlling Ca(2) (+) overload and cell death. J Clin Invest 2012;122(4):1222–32.

[41] Denby L, Ramdas V, Lu R, Conway BR, Grant JS, Dickinson B, et al. MicroRNA-214 antagonism protects against renal fibrosis. J Am Soc Nephrol 2014;25(1):65–80.

[42] Kriegel AJ, Liu Y, Fang Y, Ding X, Liang M. The miR-29 family: genomics, cell biology, and relevance to renal and cardiovascular injury. Physiol Genomics 2012;44(4):237–44.

[43] Qin W, Chung AC, Huang XR, Meng XM, Hui DS, Yu CM, et al. TGF-beta/Smad3 signaling promotes renal fibrosis by inhibiting miR-29. J Am Soc Nephrol 2011; 22(8):1462–74.

[44] van Rooij E, Sutherland LB, Thatcher JE, DiMaio JM, Naseem RH, Marshall WS, et al. Dysregulation of microRNAs after myocardial infarction reveals a role of miR-29 in cardiac fibrosis. Proc Natl Acad Sci USA 2008;105(35):13027–32.

[45] Macconi D, Tomasoni S, Romagnani P, Trionfini P, Sangalli F, Mazzinghi B, et al. MicroRNA-324-3p promotes renal fibrosis and is a target of ACE inhibition. J Am Soc Nephrol 2012;23(9):1496–505.

[46] Kato M, Zhang J, Wang M, Lanting L, Yuan H, Rossi JJ, et al. MicroRNA-192 in diabetic kidney glomeruli and its function in TGF-beta-induced collagen expression via inhibition of E-box repressors. Proc Natl Acad Sci USA 2007;104(9):3432–7.

[47] Gregory PA, Bert AG, Paterson EL, Barry SC, Tsykin A, Farshid G, et al. The miR-200 family and miR-205 regulate epithelial to mesenchymal transition by targeting ZEB1 and SIP1. Nat Cell Biol 2008;10(5):593–601.

[48] Bracken CP, Gregory PA, Kolesnikoff N, Bert AG, Wang J, Shannon MF, et al. A double-negative feedback loop between ZEB1-SIP1 and the microRNA-200 family regulates epithelial-mesenchymal transition. Cancer Res 2008;68(19):7846–54.

[49] Chung AC, Huang XR, Meng X, Lan HY. miR-192 mediates TGF-beta/Smad3-driven renal fibrosis. J Am Soc Nephrol 2010;21(8):1317–25.

[50] Long J, Wang Y, Wang W, Chang BH, Danesh FR. MicroRNA-29c is a signature microRNA under high glucose conditions that targets sprouty homolog 1, and its in vivo knockdown prevents progression of diabetic nephropathy. J Biol Chem 2011;286(13):11837–48.

[51] Long J, Wang Y, Wang W, Chang BH, Danesh FR. Identification of microRNA-93 as a novel regulator of vascular endothelial growth factor in hyperglycemic conditions. J Biol Chem 2010;285(30):23457–65.

[52] Xiao C, Srinivasan L, Calado DP, Patterson HC, Zhang B, Wang J, et al. Lymphoproliferative disease and autoimmunity in mice with increased miR-17-92 expression in lymphocytes. Nat Immunol 2008;9(4):405–14.

[53] Dai Y, Sui W, Lan H, Yan Q, Huang H, Huang Y. Comprehensive analysis of microRNA expression patterns in renal biopsies of lupus nephritis patients. Rheumatol Int 2009;29(7):749–54.

[54] Lee SO, Masyuk T, Splinter P, Banales JM, Masyuk A, Stroope A, et al. MicroRNA15a modulates expression of the cell-cycle regulator Cdc25A and affects hepatic cystogenesis in a rat model of polycystic kidney disease. J Clin Invest 2008;118(11):3714–24.

[55] Pandey P, Brors B, Srivastava PK, Bott A, Boehn SN, Groene HJ, et al. Microarray-based approach identifies microRNAs and their target functional patterns in polycystic kidney disease. BMC Genomics 2008;9:624.

[56] Sun H, Li QW, Lv XY, Ai JZ, Yang QT, Duan JJ, et al. MicroRNA-17 post-transcriptionally regulates polycystic kidney disease-2 gene and promotes cell proliferation. Mol Biol Rep 2010;37(6):2951–8.

[57] Anglicheau D, Sharma VK, Ding R, Hummel A, Snopkowski C, Dadhania D, et al. MicroRNA expression profiles predictive of human renal allograft status. Proc Natl Acad Sci USA 2009;106(13):5330–5.

[58] Scian MJ, Maluf DG, David KG, Archer KJ, Suh JL, Wolen AR, et al. MicroRNA profiles in allograft tissues and paired urines associate with chronic allograft dysfunction with IF/TA. Am J Transplant 2011;11(10):2110–22.

[59] Ben-Dov IZ, Muthukumar T, Morozov P, Mueller FB, Tuschl T, Suthanthiran M. MicroRNA sequence profiles of human kidney allografts with or without tubulointerstitial fibrosis. Transplantation 2012;94(11):1086–94.

[60] Maluf DG, Dumur CI, Suh JL, Scian MJ, King AL, Cathro H, et al. The urine microRNA profile may help monitor post-transplant renal graft function. Kidney Int 2014;85(2):439–49.

[61] Mitchell PS, Parkin RK, Kroh EM, Fritz BR, Wyman SK, Pogosova-Agadjanyan EL, et al. Circulating microRNAs as stable blood-based markers for cancer detection. Proc Natl Acad Sci USA 2008;105(30):10513–8.

[62] Gilad S, Meiri E, Yogev Y, Benjamin S, Lebanony D, Yerushalmi N, et al. Serum microRNAs are promising novel biomarkers. PLoS ONE 2008;3(9):e3148.

[63] Hanke M, Hoefig K, Merz H, Feller AC, Kausch I, Jocham D, et al. A robust methodology to study urine microRNA as tumor marker: microRNA-126 and microRNA-182 are related to urinary bladder cancer. Urol Oncol 2010;28(6):655–61.

[64] Glowacki F, Savary G, Gnemmi V, Buob D, Van der Hauwaert C, Lo-Guidice JM, et al. Increased circulating miR-21 levels are associated with kidney fibrosis. PLoS ONE 2013;8(2):e58014.

[65] Argyropoulos C, Wang K, McClarty S, Huang D, Bernardo J, Ellis D, et al. Urinary microRNA profiling in the nephropathy of type 1 diabetes. PLoS ONE 2013;8(1):e54662.

[66] Barutta F, Tricarico M, Corbelli A, Annaratone L, Pinach S, Grimaldi S, et al. Urinary exosomal microRNAs in incipient diabetic nephropathy. PLoS ONE 2013;8(11):e73798.

[67] Szeto CC, Ching-Ha KB, Ka-Bik L, Mac-Moune LF, Cheung-Lung CP, Gang W, et al. Micro-RNA expression in the urinary sediment of patients with chronic kidney diseases. Dis Markers 2012;33(3):137–44.

[68] Wang G, Kwan BC, Lai FM, Chow KM, Li PK, Szeto CC. Urinary miR-21, miR-29, and miR-93: novel biomarkers of fibrosis. Am J Nephrol 2012;36(5):412–8.

[69] Li Y, Fang X, Li QZ. Biomarker profiling for lupus nephritis. Genomics Proteomics Bioinformatics 2013;11(3):158–65.

[70] Ben-Dov IZ, Tan YC, Morozov P, Wilson PD, Rennert H, Blumenfeld JD, et al. Urine microRNA as potential biomarkers of autosomal dominant polycystic kidney disease progression: description of miRNA profiles at baseline. PLoS ONE 2014;9(1):e86856.

[71] Krutzfeldt J, Rajewsky N, Braich R, Rajeev KG, Tuschl T, Manoharan M, et al. Silencing of microRNAs in vivo with 'antagomirs'. Nature 2005;438(7068):685–9.

[72] Ebert MS, Neilson JR, Sharp PA. MicroRNA sponges: competitive inhibitors of small RNAs in mammalian cells. Nat Methods 2007;4(9):721–6.

CHAPTER 4

MicroRNAs in the Pathogenesis, Diagnosis, and Treatment of Liver Disease

B.B. Madison, T.A. Kerr

Key Concepts

Significant advances have been made in the understanding of microRNAs (miRNAs) and liver disease, and the liver will likely be the first successful organ for tissue-specific disease treatment via miRNA manipulation. Because of its unique vasculature the liver efficiently and rapidly accumulates exogenously administered small RNAs, and thus liver disease will likely be at the forefront of efforts to implement miRNA-/siRNA-targeted treatments. Some trepidation must be taken with the therapeutic targeting of miRNAs since a single miRNA may have many targets; broad alterations in these targets may have unintended consequences. However, miRNAs as modulators of gene expression (controlling expression like a rheostat) can be partially inhibited or enhanced to temper a biological response below a pathological threshold, while possibly avoiding side effects associated with more potent manipulations. Manipulation of miR-122 is a key example. MiR-122 may be an effective target for ameliorating hepatic steatosis, where titrating the dosage and level of inhibition appear critical. Headway has already been made in clinical trials with the targeting of miR-122 in hepatitis C through production of neutralizing antagomirs. Such drugs may also be effective for fatty liver disease. However, miR-122 appears to be tumor suppressive and an important antiviral miRNA in hepatitis B virus (HBV) infection, so efforts to target hepatitis C virus (HCV) must be made with caution in individuals at risk for HBV or hepatocellular carcinoma (HCC). Approaches for HBV have been more direct, with the development of siRNAs to directly target viral mRNAs, an approach that may allow for the design of highly specific therapies. The HBV-X protein appears to be a high value target that modulates many host miRNA genes. For either HBV or HCV the risk of developing HCC is greatly increased with chronic infection. Targeting or preventing HCV or HBV infection with miRNA therapeutics will likely help reduce HCC risk and incidence. However, treatment of HCC must still be pursued, and targeting miRNAs and their pathways hold promise. Many oncogenic (eg, miR-21 and miR-221/miR-222) and tumor-suppressive (eg, miR-122 and miR-199a) miRNAs have already been identified in HCC. Lastly, miRNA biomarkers and therapeutic miRNAs for the diagnosis and treatment of acute liver injury may soon be effective because of salient biological profiles that are emerging. All of these concepts will be explored in greater detail in this chapter.

Translating MicroRNAs to the Clinic
ISBN 978-0-12-800553-8

1. INTRODUCTION

RNA interference (RNAi), discovered by Mello and Fire in the early 1990s [1], provides a common mechanism by which endogenously or exogenously encoded RNAs target messenger RNA (mRNA) transcripts for attenuated translation, deadenylation, and degradation, thereby modulating gene expression. The requisite cell machinery, conserved throughout eukaryotic cells including hepatocytes [2], facilitates posttranscriptional mRNA targeting by endogenously or virally encoded miRNAs. Within the liver, the physiological importance of miRNAs has been demonstrated in metabolism [3], immunity [4], viral hepatitis, and oncogenesis. This chapter will describe the biogenesis of miRNAs and the role of microarray technology in detecting miRNAs, with a primary focus on developments in miRNA research as it relates to the pathogenesis of nonalcoholic fatty liver disease (NAFLD), viral hepatitis (C and B), and HCC. In addition, the role of miRNAs as biomarkers of liver injury and HCC will be discussed. New technologies provide a bountiful platform for the production of new therapeutic compounds; one can now easily produce antagonists and mimics of miRNAs in the form of small RNAs, antagomirs and gapmers, all of which can be synthesized with modified sugar backbones, including locked nucleic acids (LNAs), peptide nucleic acids, with phosphorothioate linkages. These modifications provide enhanced stability and resistance to degradation and/ or increased permeability through a phospholipid bilayer. Combined with the natural tendency of these molecules to accumulate in the liver [5], the therapeutic manipulation of miRNAs in liver disease is likely to provide effective treatments.

2. MICRORNA BIOGENESIS

MiRNAs are transcribed in mono or polycistronic form as single-stranded RNA transcripts from genomic, viral, or plasmid DNA. The resultant transcript, termed pri-miRNA (genomically encoded) or shRNA (viral or plasmid encoded) is cleaved in the nucleus by the RNase Drosha to a 60–90 bp hairpin-configured pre-miRNA. The pre-miRNA is exported from the nucleus via a GTP-dependent Ran/Exportin 5 complex. In the cytoplasm, the premiRNA undergoes further processing by the Dicer complex to a mature 20–22 base miRNA. The "guide" strand is loaded onto the RNA-induced silencing complex (RISC) [6,7], where it variably directs target transcript cleavage, degradation, or P-body sequestration, based upon the degree of complementarity with its target(s) (reviewed in Ref. [2]). Each miRNA, by targeting a range of targets (up to hundreds), may broadly modify the cellular transcriptome and maintain balanced cellular physiology. Aside from miRNAs, competing endogenous miRNAs can sponge miRNAs away from other targets, and can be protein-coding mRNAs, pseudogene transcripts, long noncoding RNAs, or circular RNAs. The competing endogenous RNA (ceRNA) phenomenon is reviewed by Salmena et al. [8]. Over or underexpression of certain miRNAs or ceRNAs in pathologic conditions (described later) may significantly alter cellular metabolism and other processes resulting in disease.

3. MICRORNA ANALYSIS AND BIOINFORMATICS

The first evidence for miRNA function was observed in 1993 [9], and has increased understanding for the importance of miRNAs in physiology in the last 13 years. To date, there have been 28,645 miRNAs described in 223 species, with nearly ~2600 described in humans (miRBase release 21 [10]). As each miRNA can regulate hundreds of target mRNA transcripts, developments in microarray, deep-sequencing technologies, and bioinformatics have been central to understanding miRNA function. A common approach to investigate the role of miRNAs in disease processes is to profile miRNA expression patterns between disease and control tissue (neoplastic vs nonneoplastic tissue or metastatic vs nonmetastatic cancer). To facilitate nonbiased miRNA expression profiling, several platforms have been developed, including deep sequencing, direct multiplexed measurement of gene expression with color-coded probe pairs (eg, NanoString) and traditional microarrays.

Once candidate miRNAs for disease processes are identified, the challenge remains to identify and validate physiologic target mRNAs. Advances in bioinformatics and RNA arrays have improved the accuracy of miRNA target identification (reviewed in Ref. [11]). Computerized algorithms based on miRNA seed pairing and conservation of miRNA recognition elements (MREs) (targetScan.org, pictar.mdc-berlin.de, microRNA.org, and others) allow the identification of candidate transcripts but the large number of candidates identified may create challenges in identifying physiologically relevant targets. In vitro miRNA overexpression or antagonism followed by functional and/or cellular mRNA analysis may identify miRNA-targeted genes. These results can then be narrowed to include only those transcripts with the appropriate MRE. Using these techniques, it may be difficult to distinguish primary (direct targets) versus secondary (compensatory changes) of miRNA activity. A more direct approach to identifying miRNA targets is achieved via immunoprecipitation of RISC-associated proteins (like AGO2) and deep-sequencing of coprecipitated miRNA fragments. This is often achieved through a cross-linking step, usually with ultraviolet light, to preserve the endogenous complex in a procedure called cross-linking, immunoprecipitation, and high-throughput sequencing (CLIP-seq). Even with stringent criteria and biochemical identification, each miRNA may target up to hundreds of mRNA transcripts, necessitating gene ontogeny and interactome analyses to facilitate the identification of pathways preferentially targeted by a given miRNA.

4. HEPATIC MICRORNAs AS METABOLIC MODULATORS AND THEIR IMPORTANCE IN NAFLD

NAFLD is the most common liver disease in the Western hemisphere and at one end of the spectrum includes a severe form of the disease called non-alcoholic steatohepatitis (NASH). MiRNAs have been implicated in regulating key hepatic metabolic functions [3] and over the last few years some of the relevant pathways have been selectively interrogated. In keeping with their emerging importance as pleiotropic modulators of key cellular metabolic functions, there was considerable interest in the findings from liver-specific elimination of mature miRNA expression in mice [12,13] particularly in relation to a metabolic phenotype. This illustrates the importance of miRNAs in liver metabolism. Liver-specific reduction of mature miRNAs was achieved using conditional *Dicer1* deletion (germ line deletion is embryonic lethal [14]) to disrupt cleavage of pre-miRNAs into their mature processed form. Studies using an Albumin-Cre transgene (Dicer-LKO[alb-Cre]) demonstrated efficient, progressive postnatal *Dicer1* deletion with a striking metabolic phenotype at 3 weeks of age that included hepatic steatosis with increased triglyceride and cholesterol ester accumulation and impaired regulation of blood glucose, with fasting mice becoming rapidly hypoglycemic [12]. This striking phenotype in Dicer-LKO[alb-Cre] mice, however, contrasts with other findings in which conditional *Dicer1* deletion was driven by an alfa-fetoprotein-albumin fusion Cre (Dicer-LKO[alflb-Cre]), where Dicer expression was decreased at embryonic day 18 and almost completely downregulated at birth [13]. These latter Dicer-LKO[alflb-Cre] livers exhibited no gross metabolic abnormalities and no changes in serum glucose or cholesterol levels. The dramatic differences in the phenotypes presumably reside in the timing for *Dicer* deletion rather than the extent of knockdown of the target since hepatocytes from both lines demonstrated effective downregulation of Dicer expression and of mature miRs, including miR-122 [12,13]. Studies in Dicer-LKO[alb-Cre] mice demonstrated that approximately one-third of mice older than 6 months developed hepatocellular cancers in which there was variable degrees of hepatic steatosis [12].

The molecular pathways underlying the hepatic steatosis phenotype in Dicer-LKO[alb-Cre] liver is yet to be fully explained but there was decreased expression of miR-122 target genes including those involved in cholesterol synthesis [12]. Nevertheless, as emphasized in studies summarized earlier, the dramatic hepatic steatosis associated with *Dicer1* deletion presumably reflects changes other than miR-122-dependent pathways since the effects of miR-122 knockdown alone appeared to attenuate hepatic steatosis. Future work will be required to clarify the extent to which miR-122-dependent and miR-122-independent pathways modulate hepatic lipid metabolism in vivo. Other work has suggested that miR-335 may represent a biomarker for hepatic lipid accumulation in mice, since increased accumulation was noted in genetically obese mice (both ob/ob and db/db) in association with hepatic steatosis [15]. This and other recent work have laid the groundwork for future high throughput screens of miRNAs that might be useful predictors of lipid droplet formation and metabolic liver disease in humans [16].

With regards to specific miRNAs, several have been identified as regulators of metabolism in the liver, including miR-122, miR-33, miR-34a, miR-21, and miR-378, among others (see Table 4.1). Initial studies in mice used a loss-of-function approach with either specific antagomirs [17], or by antisense oligonucleotide (ASO), to mediate knockdown of miR-122 [18], one of the most abundant miRNAs in adult liver [19]. Either targeting strategy effectively decreased hepatic miR-122 expression in mice, leading to decreased serum cholesterol levels and also decreased expression and/or activity of hepatic HMG-CoA reductase (HMGCR) [17,18]. ASO-mediated knockdown of miR-122 decreased hepatic lipogenesis and afforded mice protection against high fat diet–induced hepatic steatosis and a trend toward reduced serum transaminase levels, raising the possibility that therapeutic targeting of miR-122 might be a consideration for patients with metabolic syndrome [18,20]. Antagonism of miR-122 in mouse liver was associated with significant changes (>1.4-fold up- or downregulated) in mRNA expression of a large number of transcripts (>300 in each direction), with enrichment for those mRNAs in which there was at least one copy of the

Table 4.1 MicroRNAs (miRNAs) involved in liver disease

Liver disease	Mitigating miRNAs (for possible augmentation or supplementation)	Exacerbating miRNAs (for possible targeted inhibition)	miRNAs with ambivalent or unknown roles
Nonalcoholic fatty liver disease or steatohepatitis	miR-21 miR-29a miR-451 miR-185 miR-378 miR-205 miR-206 miR-613 miR-130a	miR-122 miR-34a miR-103/107 miR-17 miR-24 miR-27	miR-122 miR-21 miR-33a/b miR-155
Hepatitis C virus infection	miR-155 miR-221 miR-181a miR-130a miR-199a miR-17 miR-27a miR-194 miR-196 miR-29a,b,c miR-181c Let-7	miR-122 miR-21 miR-146a miR-141 miR-373 miR-192/miR-215 miR-491	miR-126 miR-192 miR-198 miR-345
Hepatitis B virus infection	miR-122 miR-101 miR-192 miR-92a miR-375 miR-125a/b miR-141 miR-205	miR-21 miR-27a miR-1 miR-15b miR-372/373	miR-223
Hepatocellular carcinoma	miR-122, miR-199a, Let-7, miR-146a, miR-34a, miR-101, miR-125b, miR-26a, miR-16, miR-148a, miR-145, miR-223, miR-195, miR-141, miR-200, miR-126, miR-124, miR-375, miR-1, miR-29, miR-15a	miR-21 miR-221 miR-224 miR-222 miR-155 miR-195 miR-141 miR-18a miR-181 miR-130b miR-92a	miR-233

Table 4.1 MicroRNAs (miRNAs) involved in liver disease—cont'd

Liver disease	Mitigating miRNAs (for possible augmentation or supplementation)	Exacerbating miRNAs (for possible targeted inhibition)	miRNAs with ambivalent or unknown roles
Liver injury	miR-21 miR-146a miR-29a/b/c miR-200a miR-29	miR-155 miR-34a miR-199a	miR-122 miR-223

miR-122 seed sequence (CACTCC) within the 3′ untranslated region (UTR). These findings were extended in studies using LNA antagomirs to knock down miR-122 expression, where 199 hepatic mRNA transcripts were observed to be upregulated within 24 h of LNA administration [21]. Consistent with these LNA and ASO studies, inactivation of miR-122 in mice ($Mir122^{-/-}$ mice) causes a decrease in serum cholesterol and serum lipoprotein levels, but mice also develop steatohepatitis, fibrosis, and HCC [22,23]. Partial inhibition of miR-122, rather than complete ablation, may be more amenable to potential therapies. Serum levels of ApoB-100 and ApoB-48 appeared normal in $Mir122^{-/-}$ mice, but they exhibited decreased levels of ApoE, along with decreased expression of many lipogenesis genes [22]. Administration of the microsomal triglyceride transfer protein (MTTP) to mice significantly reduced the fatty liver phenotype and returned serum cholesterol and triglyceride levels to normal, suggesting that lipid transport into VLDL is a primary defect in these mice [22]. While several target genes were predicted and confirmed in this model, silencing of the target and transcription factor Klf6 with an shRNA was sufficient to reduce fibrosis [22]. However, no other relevant miR-122 targets were identified. Other studies have identified possibly relevant miR-122 direct targets, such as CD320 (transcobalamin receptor) and Bckdk (branched-chain α-ketoacid dehydrogenase kinase), but miR-122-dependent functional roles have yet to be dissected for these or other targets [21]. These observations establish a role for miR-122 in hepatic lipid metabolism but illustrate an intrinsic difficulty in assigning pathways and mechanisms resulting from changes in a single miRNA, because targeting of multiple transcripts makes it difficult to assign function to any one target.

In vitro studies have helped illuminate some of the relevant targets of miR-122 (Fig. 4.1). The functional consequences of miR-122 downregulation in HepG2 cells revealed increased mRNA abundance of sterol

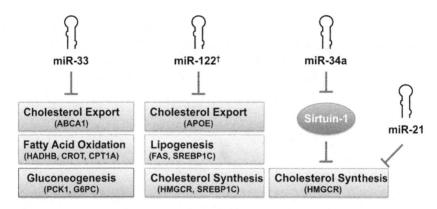

Figure 4.1 *miRNA regulatory pathways in liver metabolism, as relevant to nonalcoholic fatty liver disease (NAFLD).* Left panel: miR-33 represses cholesterol export, fatty acid oxidation, and gluconeogenesis, perhaps via direct targets (in parentheses). Middle panel: miR-122 represses cholesterol export, lipogenesis, and cholesterol synthesis, perhaps via direct targets (in parentheses). †However, miR-122 has roles garnered from in vivo mouse studies that conflict with this paradigm (see main text). Both miR-33 and miR-122 appear to regulate NAFLD in a context-dependent manner with conflicting roles observed with several models. Right panel: MiR-34a appears to promote cholesterol synthesis whereas miR-21 appears to inhibit, likely in part through regulation of HMGCR. *ABCA1*, adenosine triphosphate binding cassette A1; *APOE*, apolipoprotein E; *CPT1A*, carnitine palmitoyltransferase 1A; *CROT*, carnitine O-octanyltransferase; *FAS*, fatty acid synthase; *G6PC*, glucose-6-phosphatase; *HADHB*, hydroxyacyl-CoA dehydrogenase-3-ketoacyl-CoA; *HMGCR*, HMG-CoA reductase; *PCK1*, phosphoenolpyruvate carboxykinase; *SREBP1C*, sterol regulatory element binding transcription factor 1.

regulatory element binding protein 1-c (encoded by the *SREBF1* gene), fatty acid synthase (FAS), and HMGCR, in keeping with the associations found in subjects with NASH. Silencing of miR-122 in HepG2 cells produced a corresponding increase in protein expression for these targets, suggesting that miR-122 silencing mediates effects on key transcriptional regulators of hepatic lipid metabolism [24], but appears to conflict with results from antagomir and ASO studies described earlier in mice. Two studies have also identified PKM2 as a target of miR-122, and via this enzyme miR-122 represses glycolysis and increases oxidative metabolism [25,26]. Other identified targets, such as protein kinase interferon-inducible double-stranded RNA-dependent activator (PRKRA) or N-myc downstream-regulated gene 3 (NDRG3) may also affect lipid metabolism, although mechanisms involving these targets appear elusive for now. Clearly, further studies are needed to dissect the mechanisms of miR-122 regulation of hepatic lipid and cholesterol metabolism.

Several publications have collectively illustrated features of the homeo-static regulation of hepatic cholesterol through the coordinated transcription of *SREBF2* and miR-33 [27–30]. There are two isoforms, miR-33a and miR-33b, which are located within introns of the *SREBF2* and *SREBF1* genes, respectively. This arrangement is also true in mice, where the *Srebf2* gene and the miR-33 gene are both coexpressed [27–30]. The miR-33 miRNAs repress genes involved in cholesterol export and fatty acid oxidation, with targets including adenosine triphosphate binding cassette A1 (ABCA1), carnitine O-octanyltransferase, carnitine palmitoyltransferase 1A, hydroxyacyl-CoA dehydrogenase-3-ketoacyl-CoA, AMP-activated protein kinase, phosphoenolpyruvate carboxykinase, and glucose-6-phosphatase (Fig. 4.1). The functional consequences of miR-33 expression include decreased expression of the cholesterol export pump, ABCA1 [27,28,30], and decreased expression of genes involved in fatty acid oxidation [29]. The net effects and integrated response to cellular cholesterol depletion thus includes a regulated program in which increased *SREBF2* transcription upregulates sterol synthesis while miR-33 induction decreases cholesterol export (via decreased ABCA1 expression) and attenuates degradation of intracellular fatty acids [27–30]. Specific repression of miR-33 with antagomirs in mice oddly has disparate outcomes, depending on whether mice are treated for short-term or long-term periods with anti–miR-33. Short-term treatment with anti–miR-33 increases circulating HDL and enhances reverse cholesterol transport, with the beneficial effect of reducing the pathogenesis of atherosclerosis [31–33]. Long-term treatment of mice on a high fat diet, however, increased circulating triglycerides and caused lipid accumulation in the liver [34]. Consistent with this, inactivation of miR-33a in knockout mice (*Mir33a*$^{-/-}$ mice) causes a significant enhance-ment of high fat diet–induced obesity and steatosis [35]. This study went on to show that this was dependent on the miR-33 target Srebf1; the pheno-type of *Mir33a*$^{-/-}$ mice was alleviated by inactivation of one allele of *Srebf1* (*Mir33a*$^{-/-}$/*Srebf1*$^{+/-}$ mice) [35].

Natural products and nutraceuticals (ie, bioactive compounds purified from food products) may prove pleiotropic in their ability to modulate some of these relevant miRNAs. Of potential therapeutic value, it has been observed that polyphenols and grape seed proanthocyanidins can modulate miR-33 and miR-122 levels, and have been effective at reducing dyslipid-emia in rodent models [36,37]. Further investigations into possible mecha-nisms by ^{1}H NMR spectroscopy found that resveratrol and epigallocatechin gallate bind directly to miR-33a and miR-122 [36], and thus it is possible that these compounds affect processing and maturation of these miRNAs. Indirect modulation of miRNAs in this fashion may possibly avoid any toxicity associated with direct miRNA modulation, considering that differing outcomes are seen with short-term versus long-term or complete versus partial ablation of target miRNAs.

The related question of whether miRNA expression profiles are associated with NAFLD and NASH has also been explored. Sanyal et al. reported findings in two groups of subjects, including a group with metabolic syndrome and NASH and a control group matched for body mass index with features of metabolic syndrome but without liver enzyme, ultrasound, or histologic evidence of NASH [24]. These investigators found 23 miRNAs to be upregulated and 23 miRNAs to be downregulated, with further analysis indicating significant increases in miR-34a and miR-146b and decreased expression of miR-122 in NASH subjects. The findings in human subjects with NASH showing decreased miR-122 expression might appear somewhat at odds with the findings alluded to above in mice treated by therapeutic targeting of miR-122 in which there was *protection* against high fat–induced hepatic steatosis, increased fatty acid oxidation, lower plasma cholesterol levels, and lower transaminases levels than mice receiving a control ASO [17,18,21]. It is important to bear in mind, however, that a protective effect from preemptive knockdown of miR-122 in high fat–fed wild-type mice bears only indirect physiological comparison to steady-state cross-sectional observations in obese human subjects with fatty liver disease. The results emphasize the complexity of dissecting cause and effect relationships in hepatic miR-122 expression and metabolic liver disease. Yet another nuance to this complexity has emerged from recent findings in which the role of miR-122 was examined in relation to the known circadian rhythm of hepatic metabolic functions. MiR-122 mRNA was expressed at a relatively constant level throughout the day, consistent with its long half-life (>24 h), but ASO-mediated knockdown revealed either induction or suppression of hundreds of candidate mRNAs of which circadian transcripts were highly enriched among miR-122 targets [38]. The findings thus strongly imply that miR-122 plays a role in the circadian regulation of hepatic metabolic function although the targets involved are yet to be cataloged [38].

In regard to specific metabolic pathways that may be regulated by miRNAs, recent work has implicated miR-122 and miR-422a in the posttranscriptional regulation of CYP7A1, the rate-limiting enzyme controlling bile acid synthesis in human hepatocytes [39]. Chiang et al. demonstrated that both miR-122 and miR-422a decreased reporter activity of a chimeric luciferase construct containing selected 3′ UTR cassettes from CYP7A1 mRNA [39]. It is well established that CYP7A1 mRNA exhibits rapid turnover in hepatocytes and the 3′ UTR is enriched in A+U sequences along with the canonical AUUUAUUA instability motif, suggesting that posttranscriptional regulation of bile acid synthesis may be an attractive model in which to study the role of miRNAs [40]. Specific metabolic pathway modification in the liver is also seen with miR-378, which inhibits the PI3K subunit p110α, a downstream component of intracellular insulin signaling. Overexpression of miR-378 in mice causes hepatic insulin resistance, whereas inactivation of miR-378 causes hypoglycemia, increased liver triglycerides, and augmented insulin sensitivity, with a critical role for p110α [41]. However, the role of miR-378 in human hepatocyte metabolism remains to be adequately explored, but may prove to be a viable target considering the intriguing results with mouse models.

The importance of combinatorial interactions among miRNAs was further illustrated in recent work in which alterations in hepatic steatosis was produced by administration of an adenovirus encoding a dominant negative c-Jun followed by an examination of the associated changes in miRNA expression [42]. These authors found nine miRNAs (including miR-122 and miR-370) to be differentially expressed in the livers of the adenovirus-treated mice and demonstrated that increased abundance of miR-370 was associated with increased expression of hepatic lipogenic target mRNAs (including SREBP-1c, FAS, and DGAT2). The authors went on to demonstrate that transfection of miR-370 itself induced the expression of miR-122 and that knockdown of miR-122 attenuated the effects of miR-370 overexpression [42]. Those findings together suggest that miR-370 modulates hepatic lipogenic genes indirectly through pathways that include miR-122 targets. Feeding wild-type mice a high fat diet for a period of up to 8 weeks also resulted in increased expression of miR-122 (but not miR-370) in liver, suggesting that dietary modulation of miRNA expression is a relevant consideration [42].

Various studies have attempted to link serum levels of specific miRNAs to NAFLD disease. Greatly disparate results have been found for miR-21 and NAFLD, with some reporting elevation and others reporting depletion of miR-21 in serum of patients with NAFLD [43–45]. Two studies both found that levels of miR-34a and miR-122 were higher in patients with NAFLD [43,44]. Both miR-34a and miR-21 appear to converge on HMGCR (Fig. 4.1). This enzyme catalyzes the rate-limiting step of mevalonic acid synthesis and the ultimate production of cholesterol and other isoprenoids. In NAFLD miR-34a is elevated and represses sirtuin-1, resulting in dephoshorylation and activation of HMGCR [46]. One study found reduction of miR-21 in NAFLD, coincident with elevation of HMGCR with additional findings that miR-21 directly targets HMGCR, triggering both translational inhibition and mRNA degradation [45] (Fig. 4.1). In HepG2 cells treated with palmitic acid and oleic acid transfection with miR-21 caused a decrease in levels of triglycerides and cholesterol, which was dependent on HMGCR repression [45]. Thus, miR-34a, which is elevated in NAFLD contributes to HMGCR activity, whereas miR-21 depletion in NAFLD contributes to derepression of HMGCR. One caveat with any attempts at therapeutic enhancement of miR-21 levels is that miR-21 is a well-established pro-oncogeneic miRNA in HCC and other cancers, which will be discussed later.

As discussed earlier, antagonism of miR-122 also causes a decrease in HMGCR activity. Synergy or cooperation between elevated miR-34a and/or miR-122 and depleted miR-21 is possible, with the need for future studies to examine possible convergence on HMGCR. Another consideration in regard to the role of miRNAs in NAFLD is highlighted by a report examining visceral adipose tissue profiles in a small cohort (12) of subjects that revealed alterations in miRNAs targeting adipokines and cytokines [47]. The miRNAs identified in this study potentially targeted cytokines such as CCL3, IL6, and ghrelin/obestatin. Pro-inflammatory cytokines such as IL-6 and TNFα are linked to the pathogenesis of NAFLD with roles in driving insulin resistance and inflammation [48]. The identification of miR-NAs or siRNAs that antagonize these cytokines may prove beneficial to the discovery of new therapies.

5. MICRORNAs AND HEPATITIS C VIRUS

Experience with RNAi in plants and invertebrates would argue for a conserved role for miRNAs in the innate response to viral infections. However, discoveries in human viral infections have revealed unexpected findings that have enlarged our understanding of miRNA function within mammalian cells.

The role of miRNAs in modulating the response to hepatotrophic virus infection has been most extensively studied in the setting of HCV infection, the most common etiologic agent underlying chronic hepatitis in the

United States. Exposure to HCV leads to chronic infection in the majority of subjects and, as a consequence of infection, typically ranging from two to four decades, individuals are at risk for the development of cirrhosis and hepatocellular cancer [49]. The prevalence of HCV infection in the United States is 1.6% and is the most common indication for liver transplantation. The HCV virus is a positive sense, single-stranded RNA virus of 9.6 kB [50] whose genome includes a 5′ noncoding region (NCR) containing four conserved structural domains and an internal ribosomal entry site (IRES) that permits cap-independent translation of viral RNA with minimal requirement for canonical translation factors [51]. The resulting polyprotein consists of four structural and six nonstructural proteins that undergo further proteolysis by viral and host enzymes.

Robust, sustainable cell-culture models of HCV infection first became available in 1999 with the advent of subgenomic replicon systems [52]. A curious feature of these early replicon systems was that efficient replication could be sustained in the Huh7 but not HepG2 cell line, even though both transformed cell lines have their origin in human hepatocellular cancers. The biologic basis for this efficiency was first delineated in 2005 when Jopling et al. demonstrated that miR-122 was detectable in Huh7 but not HepG2 [53]. Further, the HCV genome contained recognition sites for miR-122's seed sequence within its NCR. In cells stably transfected with an HCV replicon, sequestration of miR-122 with chemically modified ASOs resulted in an 80% decrease in accumulation of replicon RNA (Fig. 4.2). The viral elements that interact with miR-122 have been mapped to two conserved sites in the 5′ NCR between stem-loop I and II complementary to the seed sequence of miR-122 [54].

The finding was all the more surprising because it seemed counterintuitive to the traditional notion of RNAi as an innate antiviral response, such as in plants and invertebrates [55]. Indeed, siRNA targeting of DICER1, DROSHA, DGCR8, and the RISC effector complex appears to inhibit HCV replication [56]. While the precise mechanism underlying HCV's interaction with miR-122 is incompletely understood, the position of the miR-122's binding site within the 5′ NCR is critical. Translocation of the binding site to the 3′NCR in a luciferase reporter mRNA upregulated reporter activity when miR-122 levels were diminished [54]. MiR-122 has been postulated to increase both RNA replication and translation, the latter independent of viral replication [57]. Upregulation of translation by miR-122 has been observed in reporter constructs and also in constructs carrying full-length HCV genomes [58].

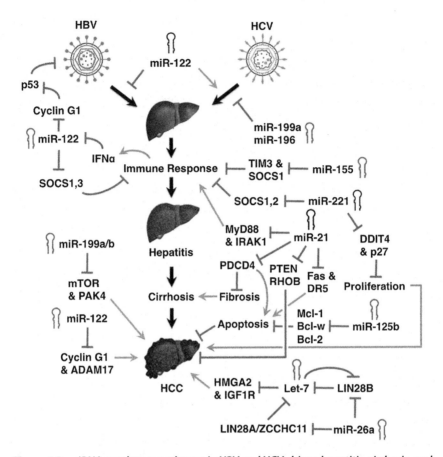

Figure 4.2 *miRNA regulatory pathways in HBV and HCV-driven hepatitis, cirrhosis, and HCC.* Chronic HBV and HCV infection can lead to hepatitis and cirrhosis, which greatly increases the risk of HCC. MiR-122 has differential roles on HBV and HCV infection. MiR-122 represses HBV infection via regulation of cyclin G1 and SOCS1/3, which regulates viral transcription via p53 and the immune response, respectively. Conversely, miR-122 directly binds to the 5′ noncoding region of HCV to promote viral RNA replication and protein translation. MiR-199a and miR-196 also appear to bind to HCV RNA, although they appear to repress HCV RNA replication. MiR-155 and miR-221 modulate HCV viral clearance by repressing several immune-repressive factors (TIM3, SOCS1, and SOCS2), whereas miR-21 represses immunostimulatory factors, MyD88 and IRAK1. MiR-221, which may have antiviral effects relevant to HCV, unfortunately has pro-oncogenic effects via repression of the anti-proliferative factors p27 and DDIT4. Other pro-oncogenic (oncomiRs) include miR-21; targeted repression of miR-21 may help by promoting an immure response against HCV and by repressing cellular transformation in HCC. MiR-21 regulates cellular apoptosis and metastasis via FAS, DR5, PTEN, and PDCD4. Unlike these oncomiRs, there appear to be numerous tumor suppressive (anti-oncomiRs) such as miR-125b, miR-199a/b, miR-122, miR-26a, and Let-7, which target the indicated mRNAs. Inhibitory effects are illustrated with perpendicular lines while positive effects are illustrated with *arrows*. *ADAM17*, a disintegrin and metalloprotease 17; *DDIT4*, DNA-damage inducible transcript 4; *HBV*, hepatitis B virus; *HCC*, hepatocellular carcinoma; *HCV*, hepatitis C virus; *IRAK1*, interleukin-1 receptor-associated kinase 1; *PAK4*, p21 (RAC1) activated kinase 4; *PDCD4*, programmed cell death protein 4; *PTEN*, phosphatase and tensin homolog; *SOCS*, suppressor of cytokine signaling.

In a separate experiment, Jangra et al. [59] studied mutations in full-length HCV constructs capable of creating infectious virions in vitro. Non-overlapping mutations were introduced into either the IRES or miR-122's binding site in separate constructs. In constructs harboring IRES mutants the production of infective virus was downregulated by 28-fold compared to greater than 3000-fold reduction in constructs with disruption of the miR-122 binding site [59].

The importance of these observations is that a complete description of miR-122's role in HCV infection requires looking beyond HCV replication, stability, and translation to study other mechanism including posttranslational targets or RNA targets relevant to HCV biology. One such target is heme oxygenase-1 (HO-1), the enzyme that catalyzes the degradation of heme to biliverdin. HO-1 is an inducible enzyme upregulated in conditions of oxidative stress. Incubation of biliverdin with cell lines carrying HCV replicons reduced HCV replication by induction of antiviral interferons [60]. HO-1 is transcriptionally repressed by heterodimers comprised of the transcription factor BACH1 and proteins of the Maf family. The 3′ UTR of BACH1 mRNA contains binding sites for miR-122, whose importance was confirmed by silencing miR-122, which then increased HO-1 mRNA levels twofold. Further, silencing of BACH1 by siRNA or chemical means with cobalt protoporphyrin or heme decreased HCV RNA [61].

Despite these findings, miR-122 is not required for HCV RNA replication. Later generation replicons, including those cloned from other HCV genotypes, have been shown to successfully replicate in HepG2 cells [62], the human cervical cancer–derived HeLa cell [63], and mouse liver cells (hepa1-6) [64]. Further, the cell culture findings have yet to be completely correlated with clinical outcomes of infection. For example, there was no correlation between HCV RNA viral load and levels of miR-122 in liver tissue of HCV-infected subjects [65]. Additionally, nonresponders to antiviral therapy tended to have lower pretreatment liver tissue miR-122 levels than responders [65]. In other studies, there was an inverse correlation between hepatic miR-122 expression and severity of hepatic fibrosis [66]. Even considering those findings, there is still compelling evidence that targeting miR-122 may emerge as an effective strategy in the treatment of HCV infection. Lanford et al. treated chimpanzees chronically infected with HCV with an LNA-modified oligonucleotide complementary to miR-122 [67]. HCV RNA levels fell by 2.6 orders of magnitude in the primate receiving the highest dose of the agent and showed histologic improvement in liver biopsy specimens. Furthermore, deep sequencing of the 5′ NCR showed no evidence for selection of adaptive mutations to the miR-122 recognition site. Similar results were also seen in a human phase II clinical trial testing an anti-miR-122 antagomir (Miraversen), an LNA-modified phosphorothioate antisense DNA oligonucleotide [68]. These results were very promising, where a subcutaneous injection of Miraversen gave a dose-wise reduction of serum HCV RNA levels, and up to a 3-log reduction over 10–18 weeks [68]. Given the long-term reduction, infrequent (monthly) injections may be sufficient for Miraversen effectiveness [68]. The synthesis of small molecule inhibitors of miR-122 also raises the possibility for new avenues of treatment of HCV infection [69].

Though miR-122 is the best studied of the miRNAs to interact with HCV, it is not unique. MiR-199a also recognizes sequences in the 5′ NCR of HCV and downregulates HCV RNA replication [70] (Fig. 4.2). MiR-196 also contains within its seed sequence a region complementary to sequences in HCV and both inhibits HCV expression and downregulates BACH1 [71] (Fig. 4.2). MiR-196 is also one of eight miRNAs upregulated in response to interferon signaling [72]. Another important immune-regulating miRNA is miR-155, which is induced by TLR3/4 signaling and affects HCV replication by potentiating antiviral pathways activated by TLR3 [73]. MiR-155 levels are elevated in patients with HCV, and appear inversely proportional to HCV serum loads, suggesting relevant antiviral effects of this miRNA. One possible mechanism involves miR-155 repression of Tim-3 (HAVCR2), a negative regulator of immune signaling that is upregulated on NK cells in patients with chronic HCV infection. Effects of miR-155 in NK cells appear key, where overexpression of miR-155 (or inhibition of Tim-3) in NK cells from HCV patients causes an increase in interferon-γ (IFNγ) production [74]. MiR-155 also has antiviral effects against other viruses; in vitro studies of HIV infection demonstrated that induction of miR-155 downstream of TLR3/TLR4 signaling prevented infection of macrophages, and overexpression of miR-155 alone repressed HIV infection of macrophages and other cells. Like miR-155, miR-221 is also induced by HCV infection, but in an nuclear factor-κB (NFκB)-dependent manner, and appears to enhance IFNα-mediated antiviral effects against HCV [75]. These effects were reported to be due to miR-221-mediated direct inhibition of suppressor of cytokine signaling 1 and 2 (SOCS1 and SOCS2, respectively), negative regulators of IFN/JAK/STAT signaling [76]. Lastly, miR-21 is induced by HCV infection like miR-155 and miR-221, but miR-21 induction helps HCV evade an immune response by directly repressing myeloid differentiation factor 88 (MyD88) and interleukin-1 receptor-associated kinase 1 (IRAK1) [77]. These factors are both integral and necessary for mediating an HCV-induced type I interferon response against HCV. Therefore, *augmentation* of miR-155 and miR-221 levels (or activity) and/or *inhibition* of miR-21 may prove beneficial in the treatment of HCV infections, and possibly other viral infections.

There are frequently potential downsides to targeting an abundant endogenous miRNA such as miR-122. It is worth noting that miR-122-associated suppression of HO-1 was associated with decreased replication of HBV [78]. In other words, while targeting miR-122 expression may become a relevant strategy to attenuate HCV replication, the data suggest that such a strategy would increase HBV replication. This would be an important consideration in coinfected individuals. In other work (expanded later) findings from subjects with HBV and HCC suggest that miR-152 is frequently inversely correlated with the expression of DNA methyltransferase I [79]. The findings also indicated alterations in global methylation profiles, suggesting that the epigenetic changes associated with alterations in miR-152 expression may be useful predictors of HCC in patients with chronic HBV infection.

6. MICRORNAs AND HEPATITIS B VIRUS

Although US and European incidence of HBV infection is relatively low, at about 1% of the population, the worldwide incidence of HBV infection is much higher, affecting over 200 million people, with sub-Saharan Africa and East Asia having the highest rate of infection, at around 5–10% (World Health Organization). This hepatotropic virus, of the genus *Orthohepadnavirus*, is the major cause of HCC and causes cellular damage to the liver via T cell–mediated responses to infected hepatocytes. Thus, miRNAs that modulate the immune response to HBV will likely affect the pathogenesis and severity of HBV infections. As mentioned earlier, miR-122 appears to have opposing effects on HBV versus HCV, and is involved in the interferon response against HBV. Several studies have noted the antiviral effects of miR-122 on HBV. Oddly, interferon-alpha (IFNα) treatment triggers depletion of miR-122, which appears to blunt the efficacy of IFNα antiviral treatment [80] (Fig. 4.2). Consistent with this, individuals who have had successful treatment (ie, sustained virological responders) or have an inactive infection have significantly lower levels of miR-122, which could reflect successful treatment or endogenous responses by IFNα [81]. However, sustaining miR-122 expression in the context of IFNα stimulation may prove synergistic for treatment regimens, as it appears to enhance the antiviral effects of IFNα, in vitro [80]. It is possible that miR-122 sustains the interferon response by directly inhibiting SOCS1 and SOCS3 [82,83], which are negative regulators of IFNα signaling via IFNα receptor 1 (IFNAR1), have miR-122 seed sequences in their 3′ UTRs, and are downregulated by miR-122. In addition to effects on interferon signaling, miR-122 directly inhibits NDRG3, a member of the NDRG family, and cyclin G1. The latter appears to negatively affect recruitment of p53 to the HBV promoter and ensuing inhibition of transcription; ie, miR-122 blocks the inhibitory effect of cyclin G1 on p53-mediated repression of HBV [84]. Other endogenous miRNAs that interfere or represses HBV, and could be exploited for treatment of HBV or HBV-associated HCC, include miR-125a/b [85], miR-101, miR-192, miR-92a, miR-141, miR-148a, miR-193b, and miR-205 (see also Table 4.1). However, miRNAs such as miR-92a, which is induced by c-Myc, has oncogenic roles in HCC, so precautions should be taken to prevent any exacerbation of HCC risk [86]. A clear understanding of endogenous miRNAs is required before considering therapeutic utility, especially in light of conflicting studies, such as those focused on miR-15b, which has been reported to both enhance and repress HBV replication [87,88].

Although expanded inoculation with the HBV vaccine is the most effective method to prevent HBV infection, there are currently no treatments that can eradicate the virus in infected individuals, who must continue drug regimens for the rest of their lives. Treatment of HBV with siRNAs against viral mRNAs does appear effective, from multiple studies, mostly in vitro. One exciting new approach that exploits RNAi against HBV also combined this strategy with an additional small RNA-mediated silencing method called U1 small nuclear RNA-snRNA-interference (U1i) [89]. U1i consists of a modified U1 snRNA that targets a specific mRNA and blocks polyadenylation of that transcript. By combining U1i and RNAi, the inhibitory effects on HBV transcription were significantly greater than either treatment alone [89]. This approach could likely be used against other viruses that rely on transcription of genes from a DNA genome. In addition to directly targeting the HBV mRNA, approaches that target cellular miRNAs that are supportive of HBV replication may also be effective. Such miRNAs, include miR-1, miR-373, and miR-501, which have been shown to enhance HBV infection or replication.

HBV is the most significant cause of HCC in East Asia and Africa, and this cancer will be discussed in more detail later. However, one viral protein is sufficient to modulate HCC-relevant miRNAs. The HBV-X protein affects both oncomiRs and anti-oncomiRs (tumor suppressor miRNAs). Possible anti-oncomiRs that are repressed by the HBV-X protein include miR-15a/16, miR-101, miR-122, miR-132, miR-136, miR-148a, miR-152, miR-192, miR-205, miR-338-3p, miR-375, and Let-7a. Possible oncomiRs that are induced by HBV-X protein include miR-21, miR-29a, and miR-181. Blocking the effects of this HBV-X protein therefore appears likely to ameliorate transformation and HCC induced by HBV. Unfortunately, as an underserved disease affecting poorer parts of the world, HBV may not benefit from new advances in miRNA and RNAi-mediated treatments, especially if such treatments are costly to develop or manufacture.

7. MICRORNAs AND HEPATOCELLULAR CARCINOMA

Liver cancer is the second leading cause of cancer-related deaths in males, worldwide [90]. HCC is the most common type of liver cancer, accounting for about 75% of all hepatic malignancies. In Western Africa, Central Africa, the Indochinese Peninsula (mainland Southeast Asia), and Mongolia (and developing countries, overall), liver cancer is the most common and deadliest cancer among men. Data from the US and Europe indicates that liver cancer is also the second most aggressive cancer (after pancreatic cancer), having a 5-year survival rate of about 10%. The high rate of liver cancer in developing countries is undoubtedly connected to infectious disease incidence, with HBV and HCV as causative agents of chronic inflammation and cirrhosis. Second, food spoilage and the fungal-derived carcinogen aflatoxin is also a major contributor to HCC. Viral hepatitis and aflatoxin exposure may also interact, with some evidence that aflatoxin is immunosuppressive [91] and may synergize with HBV in the induction of liver cancer [92]. In the United States, HCC incidence is rising, likely due to rising obesity and the associated risk of NAFLD, NASH, and cirrhosis [93]. As next-generation sequencing is applied to the examination of miRNA abundance in HCC, we will be learning a lot more about miRNAs and HCC in the coming years.

While global loss of miRNAs appears to promote tumorigenesis in HCC and other cancers, suggesting a tumor-suppressive effect, miRNAs can also contribute to oncogenesis by targeting the expression of tumor suppressor genes (a role that has earned the name "oncomiRs"). In aggregate, miRNAs tend to repress tumorigenesis and promote differentiation in multiple tissues, with the liver appearing no different. The tumor-suppressive function of miRNAs is most evident in studies examining the consequences of genetic deletion of mouse *Dicer1*, which is required for the final cleavage and processing of almost all miRNAs in the cytoplasm. In mouse models, haploinsufficiency for *Dicer1* was sufficient to greatly accelerate tumorigenesis in a mutant *Kras* model of lung cancer and a *p107/Rb* model of retinoblastoma [94,95]. Defects in DROSHA/DGCR8 processing in the nucleus are also associated with carcinogenesis [96], with some limited reports claiming that DROSHA is downregulated in HBV-infected hepatocytes at the transcriptional level [97], and others reporting that DROSHA is upregulated in HBV-associated HCC [98]. More evidence tends to indicate that miRNA-biogenesis machinery is compromised in HCC, with decreased levels of DROSHA and microprocessor accessory proteins (like DGCR8, p68, and p72) and DICER1 in HCC samples, especially in samples with no evidence of HBV or HCV infection [99]. Decreased levels of p68, DGCR8, and DICER1 were also associated with more aggressive tumor pathology and significant reductions in patient survival [99]. Evidence for a role of Dicer is especially convincing given that mouse models of liver-specific *Dicer1* inactivation develop spontaneous HCC, albeit after 6–12 months [12]. Tumors arise from a small population of *Dicer1*-null hepatocytes, even after the liver is repopulated by "escaping" *Dicer1*-wildtype hepatocytes [12]. This strongly suggests a cell-autonomous role for miRNAs in hepatocyte transformation. MiRNA-encoding genes are frequently located at sites of DNA deletion or amplification in malignancy [100], and while an association of miRNAs with cancer was earlier demonstrated in the setting of loss of miR-15 and miR-16 expression in chronic lymphocytic leukemia [101], altered miRNA expression has been associated with numerous cancers including lymphoma [102], breast, prostate, colorectal cancer [103], and others. As described either in functional or descriptive studies of HCC etiology and progression, candidate anti-oncomiRs include miR-122, miR-199a, Let-7, miR-146a, miR-34a, miR-101, and miR-26a, while possible oncomiRs include miR-21, miR-221, mir-222, and miR-224, some of which will be discussed in more detail later (also see Table 4.1).

One recently described and fascinating phenomenon involves a type of RNA called ceRNAs, which can affect miRNA targeting and potency, without actually affecting miRNA abundance. Instead, ceRNAs operate by sequestering miRNAs away from other targets, and can involve mRNAs, pseudogene RNAs, lncRNAs, and other noncoding RNAs. Several endogenous ceRNAs, including a PTEN pseudogene, can modulate carcinogenesis by regulating PTEN levels, a key tumor suppressor in the PI3K-AKT signaling cascade [104,105]. In the liver, one veritable ceRNA discovered is the lncRNA-activated by TGF-β (lncRNA-ATB), which is overexpressed in HCC and can sequester the pro-epithelial (and anti-metastatic) miR-200 family of miRNAs [106]. Via sequestration of miR-200 miRNAs, upregulation of ZEB1 and ZEB2 (miR-200 targets) caused increased invasion and EMT of HCC cell lines [106]. Further investigations of potential ceRNAs and their roles in HCC are certainly needed. The ceRNA phenomenon is reviewed by Salmena et al. [8].

The ceRNA phenomenon is also likely at play in the interaction between the anti-oncomiR miR-122 and the polycistronic HCV transcript. The translation of this long HCV mRNA depends on binding of miR-122 to sites in the 5′UTR, and is also required for viral replication [107,108]. It appears likely that the HCV mRNA could act as a ceRNA by sequestering miR-122, and compromising its tumor-suppressive properties. Studies in a rat model of HCC revealed 23 upregulated and 4 downregulated miRNAs, with miR-122 being the most consistently downregulated miRNA in HCC tissue [109]. These authors also examined human HCC miR-122 expression, revealing significantly decreased miR-122 expression in 10 out of 20 HCC tumors and similar downregulation of miR-122 in hepatoma cell lines (HepG2, Hep3B, and H-7 cells) compared to normal liver tissue. These data suggest that HCC tissue may bear characteristic miRNA expression patterns useful in tissue or serum-based diagnostic approaches and may implicate miR-122 as among the most characteristically altered species.

7.1 Tumor-Suppressive Anti-oncomiRs and Hepatocellular Carcinoma

The identification of aberrantly expressed miRNAs in HCC has led to miRNA target identification and an increased understanding of the molecular basis of HCC. In a microarray-based comparison of miRNA expression between cirrhotic and HCC tissue, 35 different miRNAs were identified, including several implicated in other human malignancies [110]. These include members of the let-7 family, miR-221, miR-145, and the normal liver-enriched miR-122a (see Table 4.1 and Fig. 4.2). The most convincing evidence of the tumor-suppressive properties of miR-122 come from in vivo studies in mice where the liver-specific inactivation of a floxed *Mir122* gene triggered hepatosteatosis, cirrhosis, and HCC [23]. Furthermore, this study demonstrated that tumorigenesis in an inducible model of HCC via expression of the *Myc* oncogene could be largely rescued through AAV-mediated delivery of a miR-122-expressing transgene [23].

MiR-122a targets in liver predicted *in silico* using miRanda, TargetScan, and PicTar algorithms implicated cyclin G1 (*CCNG1*), which was then formally evaluated as a miR-122a target. Transfection of miR-122 into HEP3B hepatoma cells decreased cyclin G1 expression and further analysis revealed an inverse correlation between cyclin G1 protein (by western blot) and miR-122a expression (comparing HCC and cirrhotic liver) suggesting that decreased miR-122a may allow overexpression of genes involved in cell-cycle progression and increased risk of malignant transformation. MiR-122 was significantly downregulated in HCC tissue compared to nontumor adjacent tissue [111]. To identify potential miR-122 target genes, computational models identified 32 transcripts as miR-122 targets revealing targets enriched for genes regulating cell movement, morphology, signaling, and transcription. ADAM17 (a disintegrin and metalloprotease 17) was validated as a negatively regulated target of miR-122 and shown to regulate cell migration and invasiveness of two HCC cell lines (SK-Hep1 and Mahlavu). Rescue of miR-122 expression or RNAi-mediated suppression of ADAM17 in Mahlavu cells prior to injection into nude mice significantly decreased tumor growth and angiogenesis. These data suggested that restoration of physiologic miRNA targeting in hepatocytes could decrease the oncogenic properties of hepatoma cells.

Other potential tumor-suppressive anti-oncomiRs include mir-199a/b (5p and 3p), Let-7, mir-26a, mir-125b, miR-146a, miR-34a, and miR-145. Both mir-199a-5p and mir-199a/b-3p appear to have anti-oncomiR function (Fig. 4.2). Mir-199a/b-3p is one of the most abundant miRNAs in the liver, with expression downregulated in HCC and decreased levels associated with decreased survival in patients [112]. Reduced expression of mir-199a/b in HCC was associated with a significantly decreased time to recurrence in patients who underwent surgical resection. It has been shown that mir-199a/b is decreased in a variety of malignancies including HCC and through bioinformatic approaches, mammalian target of rapamycin (mTOR), a regulator of cell proliferation, was identified as a potential target of mir-199a/b. In three HCC cell lines, mir-199a/b expression was inversely related to mTOR expression. In vitro restoration of miR-199a-3p expression in HCC cells resulted in cell cycle arrest, decreased invasion, and increased sensitivity of the cells to doxorubicin challenge. Reduced expression of miR-199a-3p in HCC was associated with a significantly decreased time to recurrence in patients who underwent surgical resection. This miRNA also decreased HCC growth in tumor xenografts [112]. One study identified PAK4 as a relevant target of miR-199a/b, and its necessity in activation of the RAF/MEK/ERK pathway in HCC cells [112]. Other reported targets of miR-199a-5p include FZD7, E2F3, and clathrin heavy chain (CLTC) while miR-199a/b-3p appears to inhibit HIF-1α (HIF1A) and caveolin-2 (CAV2).

Let-7 miRNAs (especially Let-7a, Let-7b, and Let-7c) are also very abundant in normal liver, and like many cancers, loss of Let-7 is associated with cancer, disease progression, and poorer prognosis [113,114]. The activity of Let-7 is likely mediated via repression of a multitude of oncofetal mRNAs and other pro-proliferative and/or pro-metastatic targets, such as HMGA2, IGF2BP1, IGF2BP2, and NR6A1 [115–117]. Let-7 biogenesis is tightly controlled, revealed by the discovery of several factors that regulate processing by DGCR8/DROSHA in the nucleus, and by DICER1 in the cytoplasm. Most notable are LIN28A and LIN28B, which are RNA-binding proteins that directly bind to Let-7 pre- and pri-miRNAs and block their processing [118,119]. In HCC, loss of Let-7 or gain of LIN28B expression has been reported. LIN28B expression in circulating HCC cells is also associated with reduced patient survival [120], and may be an important biomarker for diagnosis and treatment strategies (Fig. 4.2). LIN28B expression has been documented in HCC tumors and it is necessary and sufficient to drive tumorigenesis of xenografts using HCC36 and HA22T HCC cell lines, respectively [121]. In these studies LIN28B expression appeared to be responsible for loss of Let-7, which was shown to negatively regulate HMGA2 and IGF1R expression (Fig. 4.2), although the relative importance of these targets in HCC remains to be explored. MiRNA expression profiling in hepatoma cells compared with human hepatocytes found 26 miRNAs, including members of the Let-7 family, to be downregulated in hepatoma cells [122]. This was associated with upregulated expression of Bcl-XL, an anti-apoptotic protein found increased in HCC tissue. Restoration of Let-7c and Let-7g led to decreased Bcl-XL expression in Huh7 hepatoma cells, and decreased expression of a Bcl-XL 3′-UTR-containing reporter mRNA transcript in a target sequence-specific fashion. Restoration of Let-7c in Huh7 hepatoma cells led to increased sensitivity to staurosporine-induced apoptosis and increased sensitivity to sorafenib, an inhibitor that affects PDGFR, VEGFR, and RAF tyrosine kinases. Conversely, overexpression of Let-7c in normal hepatocytes had no effect on sensitivity to staurosporine-induced apoptosis.

Recent analysis of miRNA expression data, candidate target analysis, and clinical information suggests that miRNA expression may predict prognosis and response to chemotherapy. In one study, 78 matched cancerous and noncancerous tissues from HCC patients were evaluated by miRNA profiling [123]. From this, eight differentially expressed miRNAs were identified and validated by RT-PCR. In vitro analysis demonstrated that overexpression of miR-125b in HepG2 cells impaired cell growth, possibly via modulation of AKT signaling pathways. Kaplan–Meier survival analysis demonstrated that high levels of miR-125b correlated with improved survival in HCC patients. While miR-125b appears to be oncogenic in other tissues, miR-125b appears to exhibit anti-oncomiR properties in HCC, which is likely associated with its ability to repress the anti-apoptotic factors MCL1, BCL-W (BCL2L2), BCL-2, and interleukin-6 receptor (IL6R) [124,125] (Fig. 4.2). Indeed, in functional studies in multiple cell lines miR-125b could induce apoptosis, and correlations of apoptosis in patient-derived HCC tissues indicated a positive association between miR-125b expression and levels of apoptosis [124].

A larger study of 241 patients correlating miRNA expression changes with survival [126] revealed increased miR-26 expression in females than in males and significantly decreased expression in HCC tissue. Transcriptomic analysis of HCC tissue with low miR-26 levels suggested that altered NFκB and IL-6 signaling might play a role in hepatic oncogenesis. Kaplan–Meier analysis revealed that increased miR-26 levels were associated with increased patient survival. Conversely, although decreased HCC miR-26 expression predicted poor survival, this pattern of reduced miR-26 expression was associated with a favorable response to interferon therapy. These observations were validated in 214 additional patients with similar findings. Intersection with a Let-7 signature has also been observed. MiR-26a expression correlated with Let-7 and could induce the expression of multiple Let-7 family miRNAs in tumor cell lines, including HepG2 [127], with further data showing that the inhibitory effects of miR-26a on cellular proliferation could be rescued with a Let-7 antagomir. This group found that miR-26a directly targeted and inhibited both LIN28B and ZCCHC11 mRNAs, which are negative regulators of Let-7 biogenesis [127] (see Fig. 4.2).

7.2 Oncogenic OncomiRs and Hepatocellular Carcinoma

In contrast to the tumor suppressor qualities of miR-122, mir-199, mir-26a, and Let-7, miR-221/miR-222, miR-21, and miR-130b appear to act as oncomiRs in hepatocytes. MiRNA expression profiling in 104 HCC and 90 adjacent cirrhotic liver tissue samples revealed 12 miRNAs linked to progression to HCC [128]. The most upregulated of these, miR-221/222 was transfected into HepG2 cells leading to increased cell proliferation.

Decreased proliferation was observed when cells were treated with an antagomir directed toward miR-221. Injection of miR-221 overexpressing immortalized liver progenitor cells into irradiated nu/nu mice led to decreased tumor latency compared to control immortalized liver progenitor cells alone. Analysis of mRNA expression in HCC tissue revealed 15 transcripts containing putative miR-221 binding sites. In vitro miR-221 targeting assays demonstrated potent suppression of DNA-damage inducible transcript 4 (DDIT4) and p27 (CDKN1B), targets involved in limiting cell proliferation. These findings suggest that overexpression of miR-221 appears to have pro-oncogenic consequences (Fig. 4.2).

Identification of stem cell-like cells within HCC tissue has allowed investigation into the role of miRNA in this population. Cells with a CD133[+] surface phenotype display cancer stem cell characteristics including long-term self-renewal, tumor initiation, and resistance to chemotherapy [129,130]. Recent data suggest that enrichment with CD133[+] cells is associated with decreased disease-free and overall patient survival [131]. Quantitative PCR miRNA analysis of CD133[+] cells derived from HCC tissue and comparison to hepatoma cell lines revealed eight candidate miRNAs that were differentially regulated [131]. Of these, miR-130b most closely correlated with CD133 expression. MiR-130b was preferentially expressed in CD133[+] cells in resected HCC specimens.

MiR-21 is a key oncomiR with potent effects acting to promote tumorigenesis in a variety of tissues. MiR-21 is induced by the HBV protein, HBx, and the DNA-binding factor HMGB1 [132], with several groups showing that this oncomiR targets PDCD4 and PTEN [133–135], which are known to have key tumor-suppressive properties in HCC and other cancers. MiR-21 has also been found to target and inhibit expression of RHOB, RECK, TIMP3, and mitogen-activated protein kinase-kinase 3 (MAP2K3) [136]. Inhibition of RHOB, PTEN, and PDCD4 may lead insight into one salient attribute of this miRNA: the ability of miR-21 to drive metastasis and invasive phenotypes of HCC cells (Fig. 4.2). In multiple cancer types, these factors appear critical to repress metastasis. Intersection and interaction of PTEN and RHOB also appears to enhance effects, as RHOB and PI3-kinase (PI3K) mutually repress each other [137]. MiR-21-mediated repression of both PTEN and RHOB could therefore cooperatively (or synergistically) activate the PI3K pathway. This has yet to be explored in HCC.

In addition to miRNA-mediated regulation of oncogenes and tumor suppressor genes, mutations or modifications to the templated miRNA sequence may dramatically alter miRNA maturation or targeting efficiency, resulting in oncogenesis. Single nucleotide polymorphisms (SNPs) in miRNA stem-loop structures revealed three SNPs with high allelic frequency (>40%) in the Han Chinese population [138]. One of these, miR-146a was previously reported to be overexpressed in HCC [139]. Genotype distribution analysis comparing HCC with control subjects revealed that the GG genotype was associated with a twofold increased risk for HCC compared with the CC genotype. In vitro studies demonstrate that the GG genotype was associated with higher levels of miR-146a production and promotion of cell proliferation in the NIH-3T3 immortalized cell line. These data show that mutations in miRNA encoding loci themselves can increase oncogenic risk.

7.3 MicroRNAs in Diagnostics and Treatment of Hepatocellular Carcinoma

Cirrhosis, regardless of the cause, is a significant risk factor for HCC formation and the early detection of tumors is an important challenge. Current recommendations include ultrasound imaging every 6–12 months, which carries significant cost, has imperfect sensitivity and specificity, and are not available to all patients. Recent data suggest that serum miRNA analysis may be effective for detection of HCC. Using a murine *Myc*-induced HCC model, serum miRNA analysis revealed altered patterns of miRNA in mice with HCC compared to control mice [140]. Regression of HCC in these mice was accompanied by normalization of miRNA expression patterns. A recent study in humans reported serum miRNA profiles in patients with HCC compared to healthy controls and patients with chronic hepatitis B [141]. The findings revealed elevations in miR-21, miR-122, and miR-233 in patients with HCC and hepatitis B compared to healthy control patients. Serum miR-21 and miR-122 were also significantly higher in patients with chronic hepatitis B compared with subjects with HCC. Elevations in these miRNAs were interpreted to represent a consequence of liver injury rather than tumorigenesis itself but no comparison of miRNA profiles was undertaken in serum from patients with HCC and those with cirrhosis. These data suggest that serum-based miRNA analysis may complement and extend current HCC screening strategies, and may increase availability of HCC screening in high-risk populations.

MiRNA expression profiles may also predict HCC clinical behavior. In a study examining 482 cancerous and noncancerous resection specimens from 241 patients, and a cohort of 131 patients, a unique 20-miRNA signature (including miR-122a) was predictive of HCC venous invasion versus nonmetastatic HCC [142]. Prospective validation of this miRNA signature in 110 additional cases demonstrated that miRNA analysis could significantly and independently predict survival and relapse. Though candidate target analysis was not performed in this study, many predicted targets of these 20 miRNAs were previously shown to be included in a 153-mRNA metastasis signature from hepatic tumors [143].

Introduction of miR-130b into CD-133$^-$ cells resulted in enhanced proliferation, resistance to chemotherapy, and the ability to be passaged from one generation to another [131]. PicTar and Targetscan miR-130b target prediction revealed 289 potential downstream targets. When combined with miRNA microarray analysis of miR-130b transfected cells, three putative miR-130b targets were found, including the tumor suppressor gene *TP53INP1*. In vitro luciferase reporter assays using the *TP53INP1* 3' UTR validated this transcript as a target of miR-130b. These data support the direct role of miR-130b in hepatic neoplasia and suggest a potential role for miR-130b antagonism in HCC therapy. As described earlier, miR-21 also appears to be a very relevant oncomiR, with promising new studies indicating that antagomirs to miR-21 hold promise, with blockage of miR-21 reducing HCC cellular growth, in vitro and in xenograft tumor models.

Identification of altered miRNAs in HCC may have value in predicting the response to pharmacotherapy. These data strongly suggest that miRNA profiling of HCC tissue may predict response to chemotherapy, and that restoration or rescue of certain downregulated miRNAs may increase pharmacologic efficacy in HCC treatment. However, targeting miRNAs with compounds such as Miraversen, a miR-122 antagonist in clinical trials for HCV treatment, may bear some long-term risk. For example, antagonism of miR-122, a likely HCC-relevant tumor suppressor, may increase risk of HCC.

In vivo evidence in murine HCC models further suggests that therapeutic restoration of miRNA deficient in HCC tissue may have anti-tumor effects. Using a liver-specific tetracycline-repressible *MYC* transgene in which mice form HCC-like lesions on withdrawal of doxycycline [144], researchers found miR-26 to be dramatically downregulated, consistent with human HCC expression patterns. Expression of miR-26 in HepG2 cells resulted in cell cycle arrest via posttranscriptional repression of Cyclin D2 (CCND2) and Cyclin E2 (CCNE2). Therapeutic adenoviral delivery of miR-26 in mice with HCC resulted in a dramatic protection from HCC formation by reducing cancer cell proliferation and increasing tumor-specific apoptosis [144].These results strongly suggest that miRNA-based therapeutics may be efficacious in medical treatment of HCC and in neoadjuvant therapy prior to resection or transplantation. In an orthotopic tumor model, administration of cholesterol-conjugated Let-7a was able to reduce tumor size in nude (*Foxn1^nu*) mice with liver tumors created by injection of the HCC cell line HepG2 [145].This cholesterol modification enhanced delivery of Let-7a to the liver, and therefore may show promise for Let-7a and other miRNA-targeted therapies for HCC, especially considering the associated benefit of reduced side effects with enhanced specificity.

8. MICRORNAs AS BIOMARKERS OF LIVER INJURY

Serum levels of alanine aminotransferase (ALT) along with aspartate aminotransferase (AST) are the primary serum biomarker of parenchymal liver injury in a variety of clinical scenarios [146]. However, there are significant limitations to the use of aminotransferases as biomarkers of liver injury. First, elevations in serum aminotransferases can reflect nonhepatic injury (particularly skeletal muscle injury), and thus complicate noninvasive assessment of hepatic injury. Second, in situations such as acute acetaminophen toxicity elevations in serum aminotransferases may occur after a critical therapeutic window.Third, serum transaminase concentrations generally do not effectively discriminate between etiologies of liver injury.

In light of these limitations, recent findings highlight the potential for serum transcriptome analysis as biomarkers of both acute and chronic liver injury. Hepatocyte-specific mRNA profiles were analyzed in a rat model of acute chemical (D-galactosamine and acetaminophen)-induced liver [147], demonstrating albumin and a1-microglobulin/bikunin in peripheral blood after liver injury. Notably, albumin mRNA was detected in serum 2 hours after liver injury, prior to elevations in serum ALT or AST. These transcripts were not elevated after bupivicane-HCl-induced skeletal muscle damage suggesting that serum mRNA profiles may effectively differentiate hepatocyte injury from alternative sources transaminases and other studies have confirmed this pattern of hepatocyte-specific mRNA release in experimentally induced liver but not muscle injury [148]. Furthermore, transcriptomic profiling revealed DGAL and accidental acetaminophen overdose (APAP)-specific patterns of serum mRNA detection suggesting that RNA patterns in serum might provide clues to the etiology of liver injury, and perhaps the offending agent in drug-induced liver injury.

MiRNAs are also potentially useful biomarkers because of their abundance and stability. In a mouse model of acetaminophen-induced liver injury, the utility of miRNAs were assessed [149]. Acetaminophen-induced liver injury resulted in a significant increase in microarray-determined serum concentration of hepatocyte-specific miRNAs including miR-122 and miR-192. Increased abundance of these miRNAs in serum was dose dependent and occurred within 1 hour after acetaminophen exposure, prior to increases in serum transaminase concentration. Studies in humans following APAP also identified the liver-specific miRNAs, miR-122 and miR-192, as significantly elevated in serum, as identified by deep sequencing of small RNAs [150]. In addition, this study also identified elevated levels of the liver-specific miRNAs, miR-483, miR-210, and miR-194, in serum from APAP patients [150]. MiR-194, along with miR-122 and miR-147a, has also been reported to be highly elevated serum following liver transplantation, and appears to be associated with peritransplant ischemic liver injury [151]. Little else is known regarding these miRNAs (ie, miR-483, miR-210, and miR-194) and liver injury. Concordance between mouse and human studies, however, suggest that miR-122 and miR-192 may prove to be reliable markers for acetaminophen-induced liver damage, and perhaps other liver injuries.

A similar study investigating the utility of miRNAs as biomarkers of either liver, muscle, or brain injury using a rat model revealed that miR-122 and miR-133a were specific serum markers of liver and muscle injury respectively, whereas AST and ALT were elevated in experimentally induced injury to either tissue [152]. Though there are a limited number of studies to date, the data described earlier highlight the strong potential for serum RNA analysis, particularly miRNA, in detection and investigation of liver injury.

Frequently, the same miRNAs are involved in multiple different liver pathologies. As discussed earlier, miR-21 is deregulated in NAFLD, and also acts as a potent oncomiR in HCC. Consistent with this latter role, rodent injury models and regeneration models have found that miR-21 is induced by alcohol and represses apoptosis in hepatocytes and hepatic stellate cells (HSCs) [153,154]. One study implicated IL-6/Stat3-mediated induction of miR-21, which repressed apoptosis by targeting and repressing Fas ligand (FASLG) and death receptor 5 (DR5) [154]. Examinations of liver fibrosis in mouse models and human patients have also noted induction of miR-21 by thioacetamide and carbon tetrachloride in HSCs

[155]. In HSCs miR-21 represses apoptosis, but also significantly acceler-ates fibrosis, which was dependent on miR-21 repression of programmed cell death protein 4 (PDCD4), a pro-apoptotic protein [155] (Fig. 4.2). The effects on HSCs and fibrosis were associated with miR-21-mediated enhancement of TGF-β signaling, a known enhancer of fibrosis, and direct repression of Smad-7, an inhibitory co-Smad that represses downstream activation of the TGF-β pathway. Further studies are needed to explore the effects of miR-21 on fibrosis, while miR-21 does appear to have potent anti-apoptotic effects in various contexts, although any benefits from therapeutic manipulation of miR-21 remain to be explored in injury models. In contrast, repression of miR-155, another modulator of apop-tosis, may help attenuate liver injury. This immune-modulating antiviral miRNA, discussed earlier, also appears to exacerbate liver injury in an ischemia-reperfusion injury (IRI) rodent model and a Fas-induced liver injury model [156,157]. The effects of miR-155 in the IRI model also appear to be due to immune-modulating function; miR-155 directly inhibits SOCS1, a negative regulator of JAK/STAT signaling that represses development of Th17 cells and M2 macrophages. Thus, miR-155 represses an anti-inflammatory pathway that could otherwise be augmented for therapeutic benefits in IRI. Repression of miR-155 has certainly been demonstrated to be effective in mouse models of lymphoma and new antagomirs appear very promising for cancer treatment in these models and could be applied to liver injury [5,158].

LIST OF ACRONYMS AND ABBREVIATIONS

ADAM17 A disintegrin and metalloprotease 17
ALT Alanine aminotransferase
ASO Antisense oligonucleotide
AST Aspartate aminotransferase
FAS Fatty acid synthase
HCC Hepatocellular carcinoma
HCV Hepatitis C virus
HMG-CoA 3-Hydroxy-3-methyl-glutaryl-CoA
HO-1 Heme oxygenase-1
IRES Internal ribosomal entry site
LNA Locked nucleic acid
miRNA MicroRNA
MRE miRNA recognition element
mTOR Mammalian target of rapamycin
NAFLD Nonalcoholic fatty liver disease
NASH Nonalcoholic steatohepatitis

NCR Noncoding region
PCR Polymerase chain reaction
RISC RNA-induced silencing complex
RNAi RNA interference
RT-PCR Reverse transcription PCR
SNP Single nucleotide polymorphism
SREBP-1c Sterol regulatory element binding protein 1-c
UTR Untranslated region

ACKNOWLEDGMENTS

We apologize to colleagues whose works were unable to cite due to space limitations. The authors have no potential conflicts of interest to declare.

REFERENCES

[1] Fire A, et al. Potent and specific genetic interference by double-stranded RNA in *Caenorhabditis elegans*. Nature 1998;391(6669):806–11.
[2] Bartel DP. MicroRNAs: genomics, biogenesis, mechanism, and function. Cell 2004; 116(2):281–97.
[3] Krutzfeldt J, Stoffel M. MicroRNAs: a new class of regulatory genes affecting metabolism. Cell Metab 2006;4(1):9–12.
[4] Lu LF, Liston A. MicroRNA in the immune system, microRNA as an immune system. Immunology 2009;127(3):291–8.
[5] Cheng CJ, et al. MicroRNA silencing for cancer therapy targeted to the tumour microenvironment. Nature 2015;518(7537):107–10.
[6] Kawamata T, Tomari Y. Making RISC. Trends Biochem Sci 2010;35(7):368–76.
[7] Ding L, Han M. GW182 family proteins are crucial for microRNA-mediated gene silencing. Trends Cell Biol 2007;17(8):411–6.
[8] Salmena L, et al. A ceRNA hypothesis: the Rosetta Stone of a hidden RNA language? Cell 2011;146(3):353–8.
[9] Lee RC, Feinbaum RL, Ambros V. The *C. elegans* heterochronic gene lin-4 encodes small RNAs with antisense complementarity to lin-14. Cell 1993;75(5): 843–54.
[10] Griffiths-Jones S. The microRNA registry. Nucleic Acids Res 2004;32(Database issue):D109–11.
[11] Thomas M, Lieberman J, Lal A. Desperately seeking microRNA targets. Nat Struct Mol Biol 2010;17(10):1169–74.
[12] Sekine S, et al. Disruption of Dicer1 induces dysregulated fetal gene expression and promotes hepatocarcinogenesis. Gastroenterology 2009;136(7):2304–15. e1–4.
[13] Hand NJ, et al. Hepatic function is preserved in the absence of mature microRNAs. Hepatology 2009;49(2):618–26.
[14] Bernstein E, et al. Dicer is essential for mouse development. Nat Genet 2003;35(3): 215–7.
[15] Nakanishi N, et al. The up-regulation of microRNA-335 is associated with lipid metabolism in liver and white adipose tissue of genetically obese mice. Biochem Biophys Res Commun 2009;385(4):492–6.
[16] Whittaker R, et al. Identification of MicroRNAs that control lipid droplet formation and growth in hepatocytes via high-content screening. J Biomol Screen 2010;15(7):798–805.

[17] Krutzfeldt J, et al. Silencing of microRNAs in vivo with 'antagomirs'. Nature 2005;438(7068):685–9.

[18] Esau C, et al. miR-122 regulation of lipid metabolism revealed by in vivo antisense targeting. Cell Metab 2006;3(2):87–98.

[19] Chang J, et al. miR-122, a mammalian liver-specific microRNA, is processed from hcr mRNA and may downregulate the high affinity cationic amino acid transporter CAT-1. RNA Biol 2004;1(2):106–13.

[20] Girard M, et al. miR-122, a paradigm for the role of microRNAs in the liver. J Hepatol 2008;48(4):648–56.

[21] Elmen J, et al. Antagonism of microRNA-122 in mice by systemically administered LNA-antimiR leads to up-regulation of a large set of predicted target mRNAs in the liver. Nucleic Acids Res 2008;36(4):1153–62.

[22] Tsai WC, et al. MicroRNA-122 plays a critical role in liver homeostasis and hepatocarcinogenesis. J Clin Invest 2012;122(8):2884–97.

[23] Hsu SH, et al. Essential metabolic, anti-inflammatory, and anti-tumorigenic functions of miR-122 in liver. J Clin Invest 2012;122(8):2871–83.

[24] Cheung O, et al. Nonalcoholic steatohepatitis is associated with altered hepatic MicroRNA expression. Hepatology 2008;48(6):1810–20.

[25] Liu AM, et al. miR-122 targets pyruvate kinase M2 and affects metabolism of hepatocellular carcinoma. PLoS ONE 2014;9(1):e86872.

[26] Wang X, et al. MicroRNA-122 targets genes related to liver metabolism in chickens. Comp Biochem Physiol B Biochem Mol Biol 2015;184:29–35.

[27] Najafi-Shoushtari SH, et al. MicroRNA-33 and the SREBP host genes cooperate to control cholesterol homeostasis. Science 2010;328(5985):1566–9.

[28] Rayner KJ, et al. MiR-33 contributes to the regulation of cholesterol homeostasis. Science 2010;328(5985):1570–3.

[29] Gerin I, et al. Expression of miR-33 from an SREBP2 intron inhibits cholesterol export and fatty acid oxidation. J Biol Chem 2010;285(44):33652–61.

[30] Marquart TJ, et al. miR-33 links SREBP-2 induction to repression of sterol transporters. Proc Natl Acad Sci USA 2010;107(27):12228–32.

[31] Rayner KJ, et al. Antagonism of miR-33 in mice promotes reverse cholesterol transport and regression of atherosclerosis. J Clin Invest 2011;121(7):2921–31.

[32] Marquart TJ, et al. Anti-miR-33 therapy does not alter the progression of atherosclerosis in low-density lipoprotein receptor-deficient mice. Arterioscler Thromb Vasc Biol 2013;33(3):455–8.

[33] Rotllan N, et al. Therapeutic silencing of microRNA-33 inhibits the progression of atherosclerosis in Ldlr-/- mice–brief report. Arterioscler Thromb Vasc Biol 2013;33(8):1973–7.

[34] Goedeke L, et al. Long-term therapeutic silencing of miR-33 increases circulating triglyceride levels and hepatic lipid accumulation in mice. EMBO Mol Med 2014;6(9):1133–41.

[35] Horie T, et al. MicroRNA-33 regulates sterol regulatory element-binding protein 1 expression in mice. Nat Commun 2013;4:2883.

[36] Baselga-Escudero L, et al. Resveratrol and EGCG bind directly and distinctively to miR-33a and miR-122 and modulate divergently their levels in hepatic cells. Nucleic Acids Res 2014;42(2):882–92.

[37] Baselga-Escudero L, et al. Grape seed proanthocyanidins repress the hepatic lipid regulators miR-33 and miR-122 in rats. Mol Nutr Food Res 2012;56(11):1636–46.

[38] Gatfield D, et al. Integration of microRNA miR-122 in hepatic circadian gene expression. Genes Dev 2009;23(11):1313–26.

[39] Song KH, et al. A putative role of micro RNA in regulation of cholesterol 7alpha-hydroxylase expression in human hepatocytes. J Lipid Res 2010;51(8):2223–33.

[40] Baker DM, et al. One or more labile proteins regulate the stability of chimeric mRNAs containing the 3′-untranslated region of cholesterol-7alpha -hydroxylase mRNA. J Biol Chem 2000;275(26):19985–91.

[41] Liu W, et al. Hepatic miR-378 targets p110alpha and controls glucose and lipid homeostasis by modulating hepatic insulin signalling. Nat Commun 2014;5:5684.

[42] Iliopoulos D, et al. MicroRNA-370 controls the expression of microRNA-122 and Cpt1alpha and affects lipid metabolism. J Lipid Res 2010;51(6):1513–23.

[43] Yamada H, et al. Associations between circulating microRNAs (miR-21, miR-34a, miR-122 and miR-451) and non-alcoholic fatty liver. Clin Chim Acta 2013;424: 99–103.

[44] Cermelli S, et al. Circulating microRNAs in patients with chronic hepatitis C and non-alcoholic fatty liver disease. PLoS ONE 2011;6(8):e23937.

[45] Sun C, et al. miR-21 regulates triglyceride and cholesterol metabolism in non-alcoholic fatty liver disease by targeting HMGCR. Int J Mol Med 2015;35(3):847–53.

[46] Min HK, et al. Increased hepatic synthesis and dysregulation of cholesterol metabolism is associated with the severity of nonalcoholic fatty liver disease. Cell Metab 2012;15(5):665–74.

[47] Estep M, et al. Differential expression of miRNAs in the visceral adipose tissue of patients with non-alcoholic fatty liver disease. Aliment Pharmacol Ther 2010;32(3): 487–97.

[48] Tilg H. The role of cytokines in non-alcoholic fatty liver disease. Dig Dis 2010; 28(1):179–85.

[49] Ghany MG, et al. Diagnosis, management, and treatment of hepatitis C: an update. Hepatology 2009;49(4):1335–74.

[50] Suzuki T, et al. Molecular biology of hepatitis C virus. J Gastroenterol 2007;42(6): 411–23.

[51] Pestova TV, et al. A prokaryotic-like mode of cytoplasmic eukaryotic ribosome binding to the initiation codon during internal translation initiation of hepatitis C and classical swine fever virus RNAs. Genes Dev 1998;12(1):67–83.

[52] Lohmann V, et al. Replication of subgenomic hepatitis C virus RNAs in a hepatoma cell line. Science 1999;285(5424):110–3.

[53] Jopling CL, et al. Modulation of hepatitis C virus RNA abundance by a liver-specific MicroRNA. Science 2005;309(5740):1577–81.

[54] Jopling CL, Schutz S, Sarnow P. Position-dependent function for a tandem microRNA miR-122-binding site located in the hepatitis C virus RNA genome. Cell Host Microbe 2008;4(1):77–85.

[55] Hamilton AJ, Baulcombe DC. A species of small antisense RNA in posttranscriptional gene silencing in plants. Science 1999;286(5441):950–2.

[56] Randall G, et al. Cellular cofactors affecting hepatitis C virus infection and replication. Proc Natl Acad Sci USA 2007;104(31):12884–9.

[57] Henke JI, et al. microRNA-122 stimulates translation of hepatitis C virus RNA. EMBO J 2008;27(24):3300–10.

[58] Chang J, et al. Liver-specific microRNA miR-122 enhances the replication of hepatitis C virus in nonhepatic cells. J Virol 2008;82(16):8215–23.

[59] Jangra RK, Yi M, Lemon SM. Regulation of hepatitis C virus translation and infectious virus production by the microRNA miR-122. J Virol 2010;84(13):6615–25.

[60] Lehmann E, et al. The heme oxygenase 1 product biliverdin interferes with hepatitis C virus replication by increasing antiviral interferon response. Hepatology 2010;51(2):398–404.

[61] Shan Y, et al. Reciprocal effects of micro-RNA-122 on expression of heme oxygenase-1 and hepatitis C virus genes in human hepatocytes. Gastroenterology 2007;133(4):1166–74.

[62] Date T, et al. Genotype 2a hepatitis C virus subgenomic replicon can replicate in HepG2 and IMY-N9 cells. J Biol Chem 2004;279(21):22371–6.

[63] Kato T, et al. Nonhepatic cell lines HeLa and 293 support efficient replication of the hepatitis C virus genotype 2a subgenomic replicon. J Virol 2005;79(1):592–6.

[64] Zhu Q, Guo JT, Seeger C. Replication of hepatitis C virus subgenomes in nonhepatic epithelial and mouse hepatoma cells. J Virol 2003;77(17):9204–10.

[65] Sarasin-Filipowicz M, et al. Decreased levels of microRNA miR-122 in individuals with hepatitis C responding poorly to interferon therapy. Nat Med 2009;15(1): 31–3.

[66] Marquez RT, et al. Correlation between microRNA expression levels and clinical parameters associated with chronic hepatitis C viral infection in humans. Lab Invest 2010;90(12):1727–36.

[67] Lanford RE, et al. Therapeutic silencing of microRNA-122 in primates with chronic hepatitis C virus infection. Science 2010;327(5962):198–201.

[68] Janssen HL, et al. Treatment of HCV infection by targeting microRNA. N Engl J Med 2013;368(18):1685–94.

[69] Young DD, et al. Small molecule modifiers of microRNA miR-122 function for the treatment of hepatitis C virus infection and hepatocellular carcinoma. J Am Chem Soc 2010;132(23):7976–81.

[70] Murakami Y, et al. Regulation of the hepatitis C virus genome replication by miR-199a. J Hepatol 2009;50(3):453–60.

[71] Hou W, et al. MicroRNA-196 represses Bach1 protein and hepatitis C virus gene expression in human hepatoma cells expressing hepatitis C viral proteins. Hepatology 2010;51(5):1494–504.

[72] Pedersen IM, et al. Interferon modulation of cellular microRNAs as an antiviral mechanism. Nature 2007;449(7164):919–22.

[73] Jiang M, et al. MicroRNA-155 controls Toll-like receptor 3- and hepatitis C virus-induced immune responses in the liver. J Viral Hepat 2014;21(2):99–110.

[74] Cheng YQ, et al. MicroRNA-155 regulates interferon-gamma production in natural killer cells via Tim-3 signalling in chronic hepatitis C virus infection. Immunology 2015;145.

[75] Ding CL, et al. HCV infection induces the upregulation of miR-221 in NF-kappaB dependent manner. Virus Res 2015;196:135–9.

[76] Xu G, et al. MiR-221 accentuates IFNs anti-HCV effect by downregulating SOCS1 and SOCS3. Virology 2014;462–463:343–50.

[77] Chen Y, et al. HCV-induced miR-21 contributes to evasion of host immune system by targeting MyD88 and IRAK1. PLoS Pathog 2013;9(4):e1003248.

[78] Qiu L, et al. miR-122-induced down-regulation of HO-1 negatively affects miR-122-mediated suppression of HBV. Biochem Biophys Res Commun 2010;398(4):771–7.

[79] Huang J, et al. Down-regulated microRNA-152 induces aberrant DNA methylation in hepatitis B virus-related hepatocellular carcinoma by targeting DNA methyltransferase 1. Hepatology 2010;52(1):60–70.

[80] Hao J, et al. Inhibition of alpha interferon (IFN-alpha)-induced microRNA-122 negatively affects the anti-hepatitis B virus efficiency of IFN-alpha. J Virol 2013;87(1): 137–47.

[81] Brunetto MR, et al. A serum microRNA signature is associated with the immune control of chronic hepatitis B virus infection. PLoS ONE 2014;9(10):e110782.

[82] Gao D, et al. Down-regulation of suppressor of cytokine signaling 3 by miR-122 enhances interferon-mediated suppression of hepatitis B virus. Antiviral Res 2015;118:20–8.

[83] Li A, et al. MiR-122 modulates type I interferon expression through blocking suppressor of cytokine signaling 1. Int J Biochem Cell Biol 2013;45(4):858–65.

[84] Wang S, et al. Loss of microRNA 122 expression in patients with hepatitis B enhances hepatitis B virus replication through cyclin G(1) -modulated P53 activity. Hepatology 2012;55(3):730–41.

[85] Potenza N, et al. Human microRNA hsa-miR-125a-5p interferes with expression of hepatitis B virus surface antigen. Nucleic Acids Res 2011;39(12):5157–63.

[86] Jung YJ, et al. c-Myc-mediated overexpression of miR-17-92 suppresses replication of hepatitis B virus in human hepatoma cells. J Med Virol 2013;85(6):969–78.

[87] Dai X, et al. Modulation of HBV replication by microRNA-15b through targeting hepatocyte nuclear factor 1alpha. Nucleic Acids Res 2014;42(10):6578–90.

[88] Wu CS, et al. Downregulation of microRNA-15b by hepatitis B virus X enhances hepatocellular carcinoma proliferation via fucosyltransferase 2-induced Globo H expression. Int J Cancer 2014;134(7):1638–47.

[89] Blazquez L, et al. Increased in vivo inhibition of gene expression by combining RNA interference and U1 inhibition. Nucleic Acids Res 2012;40(1):e8.

[90] American Cancer Society. Global cancer facts & figures. 2nd ed. 2011.

[91] Bondy GS, Pestka JJ. Immunomodulation by fungal toxins. J Toxicol Environ Health B Crit Rev 2000;3(2):109–43.

[92] Kew MC. Aflatoxins as a cause of hepatocellular carcinoma. J Gastrointestin Liver Dis 2013;22(3):305–10.

[93] Michelotti GA, Machado MV, Diehl AM. NAFLD, NASH and liver cancer. Nat Rev Gastroenterol Hepatol 2013;10(11):656–65.

[94] Kumar MS, et al. Dicer1 functions as a haploinsufficient tumor suppressor. Genes Dev 2009;23(23):2700–4.

[95] Lambertz I, et al. Monoallelic but not biallelic loss of Dicer1 promotes tumorigenesis in vivo. Cell Death Differ 2010;17(4):633–41.

[96] Thomson JM, et al. Extensive post-transcriptional regulation of microRNAs and its implications for cancer. Genes Dev 2006;20(16):2202–7.

[97] Ren M, et al. Correlation between hepatitis B virus protein and microRNA processor Drosha in cells expressing HBV. Antiviral Res 2012;94(3):225–31.

[98] Liu AM, et al. Global regulation on microRNA in hepatitis B virus-associated hepatocellular carcinoma. OMICS 2011;15(3):187–91.

[99] Kitagawa N, et al. Downregulation of the microRNA biogenesis components and its association with poor prognosis in hepatocellular carcinoma. Cancer Sci 2013;104(5):543–51.

[100] Calin GA, et al. Human microRNA genes are frequently located at fragile sites and genomic regions involved in cancers. Proc Natl Acad Sci USA 2004;101(9): 2999–3004.

[101] Calin GA, et al. Frequent deletions and down-regulation of micro- RNA genes miR15 and miR16 at 13q14 in chronic lymphocytic leukemia. Proc Natl Acad Sci USA 2002;99(24):15524–9.

[102] van den Berg A, et al. High expression of B-cell receptor inducible gene BIC in all subtypes of Hodgkin lymphoma. Genes Chromosomes Cancer 2003;37(1):20–8.

[103] Michael MZ, et al. Reduced accumulation of specific microRNAs in colorectal neoplasia. Mol Cancer Res 2003;1(12):882–91.

[104] Tay Y, et al. Coding-independent regulation of the tumor suppressor PTEN by competing endogenous mRNAs. Cell 2011;147(2):344–57.

[105] Poliseno L, et al. A coding-independent function of gene and pseudogene mRNAs regulates tumour biology. Nature 2010;465(7301):1033–8.

[106] Yuan JH, et al. A long noncoding RNA activated by TGF-beta promotes the invasion-metastasis cascade in hepatocellular carcinoma. Cancer Cell 2014;25(5):666–81.

[107] Niepmann M. Activation of hepatitis C virus translation by a liver-specific microRNA. Cell Cycle 2009;8(10):1473–7.

[108] Fukuhara T, et al. Expression of microRNA miR-122 facilitates an efficient replication in nonhepatic cells upon infection with hepatitis C virus. J Virol 2012;86(15): 7918–33.

[109] Kutay H, et al. Downregulation of miR-122 in the rodent and human hepatocellular carcinomas. J Cell Biochem 2006;99(3):671–8.

[110] Gramantieri L, et al. Cyclin G1 is a target of miR-122a, a microRNA frequently down-regulated in human hepatocellular carcinoma. Cancer Res 2007;67(13): 6092–9.

[111] Tsai WC, et al. MicroRNA-122, a tumor suppressor microRNA that regulates intrahepatic metastasis of hepatocellular carcinoma. Hepatology 2009;49(5):1571–82.

[112] Hou J, et al. Identification of miRNomes in human liver and hepatocellular carcinoma reveals miR-199a/b-3p as therapeutic target for hepatocellular carcinoma. Cancer Cell 2011;19(2):232–43.

[113] Bussing I, Slack FJ, Grosshans H. let-7 microRNAs in development, stem cells and cancer. Trends Mol Med 2008;14(9):400–9.

[114] Boyerinas B, et al. The role of let-7 in cell differentiation and cancer. Endocr Relat Cancer 2010;17(1):F19–36.

[115] Lee YS, Dutta A. The tumor suppressor microRNA let-7 represses the HMGA2 oncogene. Genes Dev 2007;21(9):1025–30.

[116] Boyerinas B, et al. Identification of let-7-regulated oncofetal genes. Cancer Res 2008;68(8):2587–91.

[117] Gurtan AM, et al. Let-7 represses Nr6a1 and a mid-gestation developmental program in adult fibroblasts. Genes Dev 2013;27(8):941–54.

[118] Piskounova E, et al. Determinants of microRNA processing inhibition by the developmentally regulated RNA-binding protein Lin28. J Biol Chem 2008;283(31): 21310–4.

[119] Viswanathan SR, Daley GQ, Gregory RI. Selective blockade of microRNA processing by Lin28. Science 2008;320(5872):97–100.

[120] Cheng SW, et al. Lin28B is an oncofetal circulating cancer stem cell-like marker associated with recurrence of hepatocellular carcinoma. PLoS ONE 2013;8(11):e80053.

[121] Wang YC, et al. Lin-28B expression promotes transformation and invasion in human hepatocellular carcinoma. Carcinogenesis 2010;31(9):1516–22.

[122] Shimizu S, et al. The let-7 family of microRNAs inhibits Bcl-xL expression and potentiates sorafenib-induced apoptosis in human hepatocellular carcinoma. J Hepatol 2010;52(5):698–704.

[123] Li W, et al. Diagnostic and prognostic implications of microRNAs in human hepatocellular carcinoma. Int J Cancer 2008;123(7):1616–22.

[124] Gong J, et al. MicroRNA-125b promotes apoptosis by regulating the expression of Mcl-1, Bcl-w and IL-6R. Oncogene 2013;32(25):3071–9.

[125] Zhao A, et al. MicroRNA-125b induces cancer cell apoptosis through suppression of Bcl-2 expression. J Genet Genomics 2012;39(1):29–35.

[126] Ji J, et al. MicroRNA expression, survival, and response to interferon in liver cancer. N Engl J Med 2009;361(15):1437–47.

[127] Fu X, et al. miR-26a enhances miRNA biogenesis by targeting Lin28B and Zcchc11 to suppress tumor growth and metastasis. Oncogene 2014;33(34):4296–306.

[128] Pineau P, et al. miR-221 overexpression contributes to liver tumorigenesis. Proc Natl Acad Sci USA 2010;107(1):264–9.

[129] Ma S, et al. Identification and characterization of tumorigenic liver cancer stem/progenitor cells. Gastroenterology 2007;132(7):2542–56.

[130] Ma S, et al. CD133+ HCC cancer stem cells confer chemoresistance by preferential expression of the Akt/PKB survival pathway. Oncogene 2008;27(12):1749–58.

[131] Ma S, et al. miR-130b Promotes CD133(+) liver tumor-initiating cell growth and self-renewal via tumor protein 53-induced nuclear protein 1. Cell Stem Cell 2010;7(6):694–707.

[132] Chen M, et al. High-mobility group box 1 promotes hepatocellular carcinoma progression through mir-21-mediated matrix metalloproteinase activity. Cancer Res 2015;75(8):1645–56.

[133] Damania P, et al. Hepatitis B virus induces cell proliferation via HBx-induced microRNA-21 in hepatocellular carcinoma by targeting programmed cell death protein4 (PDCD4) and phosphatase and tensin homologue (PTEN). PLoS ONE 2014;9(3):e91745.

[134] Qiu X, et al. HBx-mediated miR-21 upregulation represses tumor-suppressor function of PDCD4 in hepatocellular carcinoma. Oncogene 2013;32(27):3296–305.

[135] Meng F, et al. MicroRNA-21 regulates expression of the PTEN tumor suppressor gene in human hepatocellular cancer. Gastroenterology 2007;133(2):647–58.

[136] Xu G, et al. MicroRNA-21 promotes hepatocellular carcinoma HepG2 cell proliferation through repression of mitogen-activated protein kinase-kinase 3. BMC Cancer 2013;13:469.

[137] Jiang K, et al. Akt mediates Ras downregulation of RhoB, a suppressor of transformation, invasion, and metastasis. Mol Cell Biol 2004;24(12):5565–76.

[138] Xu T, et al. A functional polymorphism in the miR-146a gene is associated with the risk for hepatocellular carcinoma. Carcinogenesis 2008;29(11):2126–31.

[139] Huang YS, et al. Microarray analysis of microRNA expression in hepatocellular carcinoma and non-tumorous tissues without viral hepatitis. J Gastroenterol Hepatol 2008;23(1):87–94.

[140] Maller Schulman BR, et al. The let-7 microRNA target gene, Mlin41/Trim71 is required for mouse embryonic survival and neural tube closure. Cell Cycle 2008;7(24):3935–42.

[141] Xu J, et al. Circulating MicroRNAs, miR-21, miR-122, and miR-223, in patients with hepatocellular carcinoma or chronic hepatitis. Mol Carcinog 2010;50.

[142] Budhu A, et al. Identification of metastasis-related microRNAs in hepatocellular carcinoma. Hepatology 2008;47(3):897–907.

[143] Ye QH, et al. Predicting hepatitis B virus-positive metastatic hepatocellular carcinomas using gene expression profiling and supervised machine learning. Nat Med 2003;9(4):416–23.

[144] Kota J, et al. Therapeutic microRNA delivery suppresses tumorigenesis in a murine liver cancer model. Cell 2009;137(6):1005–17.

[145] Liu YM, et al. Cholesterol-conjugated let-7a mimics: antitumor efficacy on hepatocellular carcinoma in vitro and in a preclinical orthotopic xenograft model of systemic therapy. BMC Cancer 2014;14:889.

[146] Ozer J, et al. The current state of serum biomarkers of hepatotoxicity. Toxicology 2008;245(3):194–205.

[147] Miyamoto M, et al. Detection of cell-free, liver-specific mRNAs in peripheral blood from rats with hepatotoxicity: a potential toxicological biomarker for safety evaluation. Toxicol Sci 2008;106(2):538–45.

[148] Wetmore BA, et al. Quantitative analyses and transcriptomic profiling of circulating messenger RNAs as biomarkers of rat liver injury. Hepatology 2010;51(6):2127–39.

[149] Wang K, et al. Circulating microRNAs, potential biomarkers for drug-induced liver injury. Proc Natl Acad Sci USA 2009;106(11):4402–7.

[150] Krauskopf J, et al. Application of high-throughput sequencing to circulating microRNAs reveals novel biomarkers for drug-induced liver injury. Toxicol Sci 2015;143(2):268–76.

[151] Farid WR, et al. Hepatocyte-derived microRNAs as serum biomarkers of hepatic injury and rejection after liver transplantation. Liver Transpl 2012;18(3):290–7.

[152] Laterza OF, et al. Plasma MicroRNAs as sensitive and specific biomarkers of tissue injury. Clin Chem 2009;55(11):1977–83.

[153] Dippold RP, et al. Chronic ethanol feeding enhances miR-21 induction during liver regeneration while inhibiting proliferation in rats. Am J Physiol Gastrointest Liver Physiol 2012;303(6):G733–43.

[154] Francis H, et al. Regulation of the extrinsic apoptotic pathway by microRNA-21 in alcoholic liver injury. J Biol Chem 2014;289(40):27526–39.

[155] Zhang Z, et al. The autoregulatory feedback loop of microRNA-21/programmed cell death protein 4/activation protein-1 (MiR-21/PDCD4/AP-1) as a driving force for hepatic fibrosis development. J Biol Chem 2013;288(52):37082–93.

[156] Tang B, et al. MicroRNA-155 deficiency attenuates ischemia-reperfusion injury after liver transplantation in mice. Transpl Int 2015;28(6):751–60.

[157] Chen W, et al. Deletion of Mir155 prevents Fas-induced liver injury through up-regulation of Mcl-1. Am J Pathol 2015;185(4):1033–44.

[158] Babar IA, et al. Nanoparticle-based therapy in an in vivo microRNA-155 (miR-155)-dependent mouse model of lymphoma. Proc Natl Acad Sci USA 2012;109(26): E1695–704.

CHAPTER 5

Clinical Application of MicroRNAs in Liver Diseases

G. Song, H. Vollbrecht

Key Concepts

Since liver diseases are lethal, we are reviewing the potential of microRNAs as a novel diagnostic and therapeutic approaches for liver diseases.

1. INTRODUCTION

Liver disease is a major cause of morbidity and mortality worldwide. Despite the great progress in the management of liver disease, the effective diagnostic and therapeutic approaches are lacking and there is an urgent need for non-invasive markers to improve diagnostics and prognostic ability in liver pathology and develop novel therapeutic agents for liver disorders. MicroR-NAs (miRNA) are naturally-occurring small noncoding RNAs that play important roles in lipid metabolism, apoptosis, proliferation, stem cell differentiation, and organ regeneration. They function primarily by binding to the 3′-untranslated regions (3′-UTR) of specific messenger RNAs (mRNAs), leading to mRNA cleavage or translational repression [1]. More and more evidence has shown that miRNAs play important roles in the pathogenesis of many types of human diseases, including liver diseases [1–6]. The introduction of specific miRNA mimics or their inhibitors into diseased cells and tissues has been shown to induce favorable therapeutic responses [3]. Given the ability to generate therapeutic inhibitors and mimics for miRNAs and relative efficiency for delivery to the liver, miRNAs are an attractive therapeutic target for the treatment of liver disease. In addition, miRNAs are naturally-occurring small noncoding RNAs in humans and our preliminary data and clinical trials from others show that injection of miRNA mimics or inhibitors exhibits negligible liver toxicity, indicating that these small molecules might represent a potentially unique type of "natural therapeutic agent" with low toxicity. Indeed, the first miRNA therapeutic agent for hepatitis C (HCV) has entered into a phase III clinical trial.

Translating MicroRNAs to the Clinic
ISBN 978-0-12-800553-8

Moreover, miRNAs have been identified as biomarkers that can often be detected in the systemic circulation. We review the roles of miRNAs in liver physiology and pathophysiology and their potential as new therapeutic and diagnostic approaches, focusing on non-alcoholic fatty liver disease (NAFLD), hepatocellular carcinoma (HCC), viral hepatitis, liver fibrosis and cirrhosis, and liver failure. We also discuss miRNAs as diagnostic and prognostic markers and miRNA-based therapeutic approaches for liver disease.

2. ROLE OF MICRORNAs IN LIVER PATHOLOGY AND PATHOPHYSIOLOGY

2.1 Non-alcoholic Fatty Liver Disease

NAFLD is the most common cause of chronic liver disease among populations of Western countries [7]. Literature defines the disease by accumulation of fat in hepatocytes exceeding 5–10%, with no significant alcohol intake or suspected viral etiology. The spectrum of NAFLD ranges from simple fatty liver with benign prognosis (hepatosteatosis) to a potentially progressive form non-alcoholic steatohepatitis (NASH), which dramatically increases the risks of fibrosis, cirrhosis, liver failure, and HCC, resulting in increased morbidity and mortality [8,9]. The pathogenesis of NAFLD is complex and widely considered to reflect the metabolic syndrome including type 2 diabetes, insulin resistance, and obesity [10]. Many hypotheses have been developed to explain the pathogenesis of NAFLD. "Two-hit hypothesis" is the most widespread and prevailing model to explain the progression of simple steatosis to more severe NASH. The "first hit" involves lipid accumulation in the hepatocytes that is the prerequisite for hepatocyte injury to develop [11]; whereas cytokines, adipokines, bacterial endotoxin, mitochondrial dysfunction, and/or endoplasmic reticulum stress represent the second hit for the progression of simple steatosis to NASH [12]. Recently, the traditional "two-hit" pathophysiological theory has been challenged as knowledge of the interplay between insulin resistance, adipokines, adipose tissue inflammation, and other less-recognized pathogenic factors have increased over the last year [13]. In particular, it has been suggested that hepatic steatosis may represent an epiphenomenon of several distinct injurious mechanisms rather than a true "first hit" [14]. In the "multiple-hit" model, the first hit is insulin resistance and its associated metabolic disturbances. After the initial hepatic infiltration, the liver becomes extremely vulnerable to a series of hits that may follow, leading to hepatocyte injury and finally progressing from simple steatosis to NASH/fibrosis.

Such multiple pathogenic factors may include oxidative damage, dysregulated hepatocyte apoptosis, activation of the profibrogenic transforming growth factor (TGF)-β pathway, dysregulation of multiple adipokines, and hepatic stellate cell (HSC) activation. While it might be possible to reverse the early stages of hepatic steatosis by aggressive management of weight, diet, and exercise, there is no definitive treatment to stop the progression of NASH largely because there is still a lack of comprehensive understanding of its underlying pathogenesis. Growing evidence show that miRNAs are critical regulators of metabolic syndrome by simultaneously modulating expression of genes controlling both lipid and glucose homeostasis. As described earlier, NAFLD is involved in multiple hits, highlighting miRNAs as important molecular regulators controlling NAFLD pathogenesis by simultaneously targeting several genes controlling different types of processes, in addition to their potential for the treatment of NAFLD.

Several studies have investigated profiles of miRNA expression from livers of human NAFLD patients and observed altered miRNA expression [15]. Further, miRNA expression profiles from livers of mice with induced simple steatosis and/or NASH have also helped identify the numerous miRNAs involved in the molecular pathology of NAFLD [16–21]. Dysregulated miRNAs are important regulators of hepatic lipid accumulation and the progression of simple steatosis to NASH. miR-122 is a critical and abundant miRNA specifically expressed in the liver that regulates lipid and fatty acid metabolism, as well as cholesterol accumulation [22,23]. Reduced expression of miR-122 is thought to be critical to the NAFLD/NASH pathogenesis. miRNA profiles of liver biopsies from NASH patients show a significant decrease in miR-122 expression [15]. Mice with deleted miR-122 develop hepatosteatosis, steatohepatitis, fibrosis, and subsequent development of HCC [22,24]. All these evidence indicate the critical role of miR-122 in the pathogenesis of both human and mouse NAFLD. Mechanistically, miR-122 overexpression leads to a decreases in mRNA levels of *Srebp1c* (sterol regulatory element-binding protein 1c), *Srebp2*, HMGCR, (3-Hydroxy-3-Methylglutaryl-CoA Reductase) and FAS (fatty acid synthase) [15]. Of note, SREBPs are transcription factors that activate transcription of genes encoding enzymes involved in triglyceride and cholesterol synthesis [25]. SREBP1c is a master controller of fatty acid and triglyceride synthesis, while SPREB2 carries out cholesterol synthesis [26]. HMGCR is the enzyme generally considered to catalyze the rate-limiting step in cholesterol synthesis [27]. FAS catalyzes de novo synthesis of fatty acids [28]. Thus, silencing of miR-122 expression in liver could contribute to increased levels of these proteins that control lipid synthesis and subsequent fat accumulation in hepatocytes [15].The progression of simple steatosis to NASH is a critical step of malignancy transformation of HCC. A very recent publication shows that silencing of miR-122 is an early event during hepatocarcinogenesis from NASH. HCC tumors from livers of NASH-model mice show significantly lower miR-122 expression compared to normal liver tissues. Clinical samples from human NASH patients show similar results [29]. All these findings indicate that miR-122 could be a useful molecular marker for NAFLD/NASH, liver fibrosis, and cancer, in addition to its potential as therapeutic targets for these metabolic diseases. miR-185 underexpression has been implicated in contributing to the NAFLD/NASH development. Treatment of hepatocytes with fatty acids leads to downregulation of miR-185. Further, miR-185 overexpression in mice maintained on a high fat diet reduces liver steatosis and improves insulin sensitivity compared to a control group [30]. Insulin resistance is critical to NAFLD pathogenesis and often occurs concurrently with the development of NAFLD [31,32]. Research has implicated miR-29a as an important repressor of lipoprotein lipase (LPL). LPL is responsible for cellular uptake of lipids from the circulation. miR-29a repression in mice livers caused LPL-dependent uptake of lipids into the liver and led to steatosis.This suggests that miR-29a may be an important therapeutic target for NAFLD by modeling lipid uptake from circulation system [33]. Together, miRNAs induces hepatosteatosis by modifying insulin signaling pathway, lipid and fatty acid synthesis, and uptake of lipids from the circulation.

Contrary to the inhibitory effects of miR-122 and miR-185 on simple steatosis and its transition to NASH, miR-24 is a positive regulator of hepatic lipid accumulation. miR-24 is elevated in livers of patients with NAFLD/NASH [15]. A high-fat diet treatment of mice significantly induces hepatic expression of miR-24 and subsequently inhibits expression of its target gene, *Insig1* (insulin induced protein 1) [34]. Insig1 is a membrane protein in the endoplasmic reticulum that regulates lipid synthesis by retaining SREBPs in the ER and preventing their activation in the Golgi apparatus [35–37]. It is well-established that overexpression of *Insig1* in the liver inhibits lipogenesis and *Insig1* knockdown increases hepatic and plasma triglycerides [38,39], indicating that downregulation of *Insig1* due to miR-24 overexpression is likely important in the initial stage of NAFLD development by modulating Srebps maturation. miR-10b is another miRNA that is upregulated in steatotic hepatocytes and its overexpression increased intracellular lipid content by inhibiting *PPARα* (peroxisome proliferator-activated receptor alpha) [40]. PPARα is a very important regulator of lipogenesis and NASH progression. Therefore, the crosstalk between miR-10b and *PPARα* is an important pathway for the modulation of NAFLD pathogenesis and liver damage.

The transition of simple steatosis to NASH is a critical step of malignant transformation of HCC. Hepatocyte apoptosis plays an important role in this process. Recent findings reveal that deregulation of several miRNAs plays an important role in this apoptosis. Upregulation of miR-34a is correlated with increased NASH severity and hepatocyte apoptosis [41,42]. miR-34a functions by directly targeting *SIRT1* (Sirtuin 1), an NAD-dependent deacetylase that regulates apoptosis in response to oxidative stress [42,43]. Repression of *SIRT1*, which occurs with miR-34a overexpression, increases *p53* expression [44]. p53 acts in a positive feedback loop with miR-34a, as p53 itself activates transcription of miR-34a [45]. *p53* overexpression often occurs with NASH due to inflammation and also contributes to the hepatocyte apoptosis characteristic of NASH [46–48], suggesting that miR-34a is able to promote the progression of NASH by inducing SIRT1-p53 axis-mediated apoptosis. miR-34a expression is significantly correlated with NAFLD severity in human NAFLD patients, confirming its usefulness as a novel and noninvasive biomarker. miR-296-5p is another miRNA that is thought to contribute to hepatocyte apoptosis [49]. Liver biopsies from NASH patients exhibit decreased accumulation of miR-296-5p. Lipoapoptosis occurs when free fatty acids accumulate in cells and is modulated by PUMA (proapoptotic BH3-only protein), a potent proapoptotic protein. miR-296-5p is thought to inhibit apoptotic damage in human fatty liver diseases by directly inhibiting *PUMA* expression [49]. Many more miRNAs that modulate NASH progression are being identified with the development of miRNA omics, which will further contribute to the explanation of NASH pathogenesis. For example, miR-155, miR199a/b-3p, and let-7 are another three miRNAs that are involved in the transition of hepatosteatosis to NASH [16,50]. miR-155 and let-7 expression are increased during the progression of liver inflammation from steatosis, but miR-199a/b-3p is reduced during this process. However, the mechanism by which miR-155 and let-7 regulates the progression of NASH are unknown.

Dicer1, Drosha and DGCR8 (DiGeorge Syndrome Critical Region Gene 8) are three critical enzymes for miRNA biogenesis that are essential for cleaving the primary transcript of miRNA into a mature miRNA molecule [51]. Dysregulation of miRNA processing machinery has also been shown to occur in NAFLD. One study looked at the expression of genes that process miRNAs in visceral adipose tissue (VAT) [51]. VAT is thought to contribute to pathogenesis of NASH progression by releasing proinflammatory molecules like adipokines and cytokines, which then act in an autocrine manner on hepatocytes to promote NAFLD progression [52]. Significant increases in expression of *Dicer1, Drosha,* and *DGCR8* were found in VAT of NASH patients. All these findings indicate that miRNAs as a group has an effect on the pathogenesis of NASH.

As described earlier, miRNAs are critical regulator of NAFLD and their dysregulation has been observed in both mouse and human fatty liver tissues and blood, suggesting the potential of miRNAs as new diagnostic and therapeutic approaches for NAFLD. Biopsy is the gold standard for NAFLD diagnosis. However, a liver biopsy is not routinely done when simple fatty liver or NASH is the likely diagnosis, as there is some risk involved when doing a liver biopsy. A liver biopsy is mainly done if the diagnosis is in doubt, or if there is concern that cirrhosis has developed. Therefore, the identification of reliable and specific biomarkers is required. Unfortunately, only a few biomarkers have a potential diagnostic and prognostic relevance, and most of them can only be evaluated through invasive methods. miRNA is stable in serum of human patients and levels of several miRNAs is significantly correlated with the grade of NAFLD. Several recent works showed a positive correlation between miR-122 serum levels and the severity of steatosis [53,54]. Recently, plasma levels of miR-34a together with miR-122, miR-181a, miR-192, and miR-200b were shown to be significantly associated with the severity of NAFLD-in mice [55]. Circulating levels of miR-16, together with miR-34a and miR-122, may potentially be useful biomarkers for the assessment of disease stage in NAFLD subjects [56].

2.2 Hepatocellular Carcinoma

HCC is one of the most prevalent cancers worldwide. Understanding the molecular mechanisms of tumorigenicity, invasion, and metastasis is important for developing new and targeted treatments and identifying useful prognostic and diagnostic markers for the disease. Many studies have implicated miRNAs as important regulators of hepatocarcinogenesis.

miR-21 is a well-studied miRNA that is highly expressed in many types of human cancers, such as prostate, breast, colon, pancreas, lung, liver, and stomach cancers [57]. High expression of miR-21 is associated with tumor stage and poor prognosis of HCC [58]. Serum levels of miR-21 are also elevated in HCC patients, and there is a strong correlation between miR-21 levels in serum and in HCC tumor [59]. miR-21 can promote cell cycle progression, reduce cell death, and favor angiogenesis and invasion [60]. A recent study indicated that miR-21 downregulates expression of *HEPN1* (HCC, downregulated 1), a gene frequently silenced in HCC. *HEPN1* silencing due to miR-21 overexpression promotes cell proliferation [61]. miR-21 has also been shown to act by suppressing the expression of heparin-degrading endosulfatase (hSulf-1) and PTEN, leading to activation of the AKT/ERK pathways and subsequently the epithelial–mesenchymal transition, ultimately enhancing HCC proliferation, invasion, and metastasis [62]. Suppression of miR-146a is associated with HCC progression. Methylation within the miR-146a promoter often occurs in HCC tissue and is correlated with downregulation of this miRNA [63,64]. Research suggests that miR-146a functions by indirectly repressing VEGF (vascular endothelial growth factor) through a pathway that includes direct upregulation of APC (adenomatous polyposis coli) and suppression of HAb18G, effectively promoting expression of nuclear factor kappa B (NF-κB). Thus, when miR-145a is suppressed in HCC, *VEGF* expression increases, which subsequently promotes angiogenesis, migration, and invasion [64]. miR-221 and miR-222 are encoded together and contain similar sequence, allowing them to target many of the same genes [59]. They are both found to be significantly upregulated in HCC, and their expressions increase with poor prognosis [65]. Research suggests that these miRNAs modulate proteins associated with the cell cycle and apoptosis. miR-221/222 have been shown to downregulate p27Kip1, a member of the Cip/Kip family of cyclin-dependent kinase (CDK) inhibitors that negatively control cell cycle expression [66]. miR-221 has also been shown to downregulate CDKN1C/p57, another cell cycle inhibitor. Low CDKN1C/p57 expression is associated with advanced stage HCC, invasion, and proliferation [67]. Thus, miR-221/222 overexpression may result in dysregulation of the cell cycle in HCC. miR-221/222 also contribute to HCC by inhibiting apoptosis. miR-221 represses *Bmf* expression. Bmf is a pro-apoptotic protein that triggers caspase activation [68]. In addition, miR-221 seems to associate with histone deacetylase 6 (HDAC6), resulting in deregulation of this HDAC. *HDAC6* is a known tumor suppressor gene that mediates autophagic cell death in HCC [69]. miRNA profiling of HCC tumors have identified many more miRNAs that have increased expression in HCC [70]. However, their function needs to be investigated further in HCC pathogenesis.

In contrary, some miRNAs have reduced expression in HCC and function as tumor suppressors. miR-122 is the most abundant miRNA in the liver, accounting for about 70% of all the miRNA [71]. The loss of miR-122 is strongly associated with invasion, metastasis, and poor prognosis of HCC. Mice with deleted miR-122 exhibit NASH, fibrosis, and HCC, which can be offset through re-introduction of miR-122 [24]. It has been found that *c-MYC* (v-myc avian myelocytomatosis viral oncogene homolog), a well-known oncogene, represses transcription of miR-122 in HCC by reducing the expression of HNF-3β, a transcription factor that can activate transcription of miR-122 in livers. Conversely, miR-122 has been shown to inhibit *c-MYC* by downregulating expression of E2F1 and TFDP2, two transcription activators of cMYC [72]. Thus, miR-122 reduction due to *c-MYC* overexpression might be an important step in HCC progression. miR-122 repression also contributes to invasion and metastasis. Loss of miR-122 expression is associated with loss of the hepatocyte apoptosis and gain of invasive properties [73]. miR-122 has been shown to trigger the mesenchymal to epithelial transition, suppressing cell motility and invasion by targeting *RhoA* (ras homolog family member A) [74]. RhoA enhances cell adhesion and cell junctions, effectively decreasing cell motility [74]. miR-122 expression is activated by HNF4α [75], an established tumor-suppressor gene [76]. Thus, it is possible that invasion and metastasis of HCC may progress due to miR-122 repression. miR-122 may be an attractive therapeutic target for HCC, due to its well-accounted involvement in the pathogenesis of HCC. miR-135a upregulation has been implicated in portal vein tumor thrombus (PVTT) in HCC patients. PVTT has been correlated with poor prognosis of HCC, with 50–80% of HCC patients having portal or hepatic vein invasion of the cancer. Overexpression of miR-135a seems to promote PVTT, while silencing significantly decreases PVTT. Further, miR-135a overexpression is correlated with poor prognosis and survival of patients with PVTT [77].

miRNAs also can contribute to hepatocarcinogenesis by impairing chromosome structure. miR-151 is associated with chromosomal aberrations that are frequently observed in HCC, and is often found to be amplified [78]. miR-151 precursor can generate two mature miRNAs: miR-151-3p and miR-151-5p. While both miRNAs are increased in HCC compared to normal liver tissue, only miR-151-5p seems to promote HCC metastasis. The miR-151 transcript is located within an intron of focal adhesion kinase gene (FAK), which is often overexpressed in HCC and promotes tumor migration and invasion [78]. A recent study explored the potential of tumor suppressor miRNAs to suppress liver tumors induced by c-Myc of AKT/Ras oncogenes in mice, and found that miR-101 exhibited the greatest inhibition of liver tumor development. miR-101 may be a potential target for HCC therapeutics [79].

In addition to miRNAs that are dysregulated, changes in expression of the enzymes controlling miRNA biogenesis have also been implicated in HCC. A study of HCV-related HCC tumor tissue revealed that the genes encoding miRNA biogenesis enzymes including *Drosha, DGCR8, AGO1,* and *AGO2* were significantly overexpressed while *Dicer1* and *AGO3* were significantly downregulated in HCC tumor tissue. Drosha was the most differentially expressed in HCC tumors. This enzyme, controls the processing of precursor miRNAs, and thus its overexpression may lead to an increase in expression of certain miRNAs in HCC tumors [80]. The major causes of HCC include hepatitis B virus (HBV), HCV, obesity, and chemical carcinogen, suggesting the complexity of HCC pathogenesis. Although more and more miRNAs have been identified to be related to HCC [60], no single miRNA or a combination of several miRNAs can be used to explain the underlying mechanism for this malignancy. To develop new therapeutic approach for HCC, more detailed mechanism by which miRNAs regulate carcinogenesis is needed.

2.3 Hepatitis B

Hepatitis B is a serious liver infection caused by HBV infection, a leading risk factor for HCC. Data from Hepatitis B Foundation shows that more than 2 billion people are infected by HBV worldwide; 400 million people are chronically infected; and an estimated 1 million people die each year from hepatitis B and its complications. HBV is a small enveloped DNA virus that belongs to the *Hepadnaviridae* family. HBV infect hepatocytes by integrating the host genome. For some people, hepatitis B infection becomes chronic, meaning it lasts more than 6 months sometime for the whole life. Having chronic hepatitis B increases your risk of developing liver failure, fibrosis, cirrhosis, and HCC. Although HBV-related liver diseases have been studied extensively, the prognosis of HBV-related HCC is poor, particularly for advanced stage diagnosis, and the effective therapeutic approach is lacking. Therefore, there is an urgent need for an optimal early biomarker for the screening purpose and an effective and low-adverse effects therapeutic target for HBV infection. Accumulating evidence suggests that miRNAs are involved in the HBV life cycle and that HBV can modulate the expression of endogenous cellular miRNAs to create a favorable environment for its replication and survival [81–83]. More recently, miRNAs in serum has been confirmed as a potential diagnostic biomarker [82,84].

Mechanistically, miRNAs can promote or inhibit HBV replication. Hepatic injury is commonly seen in patients infected with HBV, due to immune reaction by the host. miR-146a is a miRNA related to innate host immunity whose expression is found to be differentially expressed in hepatocytes infected with HBV. Research suggests that the HBV X protein enhances expression of miR-146a through NF-κB-mediated activation of the miR-146a promoter. miR-146a downregulates complement factor H, resulting in inhibition of complement activation and tissue inflammation associated with hepatic injury [85]. The precise modulation of HBV gene expression is essential for replication of the virus. HBV sequences are transcribed under the control of the preC/pregenomic, S1, S2, and X promoters [86]. Recent finding shows that increased miR-155 level could help to reduce HBV viral load. The findings show that miR-155 expression is significantly reduced during HBV infection. Further, ectopic expression of miR-155 in vitro reduces HBV load as evidenced from reduced viral DNA, mRNA, and subsequently reduced level of secreted viral antigens (HBsAg and HBeAg). Mechanistically, miR-155 is able to inhibit expression of CCAAT/enhancer-binding protein-β gene (C/EBP-β), a positive regulator of HBV transcription. Taken together, miR-155 level could help to reduce HBV viral load by targeting *C/EBP-β* [87]. miR-130a and miR-26b are another two potential inhibitors of HBV replication [88,89]. miR-122 is highly abundant in the liver and an inhibitor of NAFLD as described earlier. Interestingly, miR-122 is also an important host factor for the HCV and promotes HCV replication. In contrast to HCV, miR-122 inhibits replication of the HBV [90]. Further, miR-1 and miR-372 along with miR-373 can enhance HBV replication by activating nuclear receptor farnesoid X receptor-alpha (FXRα) and nuclear factor I/B, respectively [91,92]. miR-21 is a major promoter of HBV-related HCC [93–95]. Interestingly, some laboratories showed that miR-15b can promote viral replication [96], but others confirmed that downregulation of miR-15b reduces virus load and allow persistent HBV infection [97,98]. These contradictory results further explain the complexity on the role of miRNA in HBV infection. It is expected that more miRNAs that control HBV infection will be identified, which will not only contribute to the ultimate elucidation of molecular mechanism of hepatitis B but also provide more therapeutic targets for this disorder. In addition, serum miRNAs have expression alteration after HBV infection, which can serve as novel biomarkers for HBV infection and diagnosis of HBV-positive HCC. For example, serum miR-210 level is correlated with the severity of hepatitis due to HBV infection [99]. miR-125b is increased and miR-223 is reduced significantly in the serum of HBV-positive HCC patients [100]. In summary, many miRNAs in livers and serum infected with HBV have dysregulated expression. Based on the profiling data, it is hard to use one single miRNA as a biomarker for the diagnosis of HBV infection. A combination of several dysregulated miRNAs are a good choice for the development of diagnostic approach for HBV in the future.

2.4 Hepatitis C

Hepatitis C infection is caused by the HCV. HCV infects more than 185 million individuals worldwide [101]. About 15% of cases end up being acute hepatitis C, in which the immune system is able to completely destroy the virus. For about 75–85% of infected people, however, the immune system is not able to completely get rid of the HCV, and they develop chronic infection and end up having a long-term liver infection. Twenty percent of patients chronically infected with HCV progress to cirrhosis, and ultimately HCC. HCV is an enveloped single-stranded RNA virus of positive polarity which is the sole member of the genus Hepacivirus within the family Flaviviridae. HCV have a very high mutation rate. As a result, development of an effective vaccine against HCV has long been defined as a difficult challenge and the observation that convalescent humans and chimpanzees could be reinfected after reexposure [102]. On the other hand, progress in the understanding of antiviral immune responses in patients with viral clearance has elucidated key mechanisms playing a role for control of viral infection. In 2014 two oral drug regimens *VIEKIRA PAK and Harvoni* were approved by FDA for the treatment of chronic hepatitis C genotype 1 infection in adults. Viekira Pak contains three new drugs—ombitasvir, paritaprevir, and dasabuvir—that work together to inhibit the growth of HCV. Unfortunately, HCV has extremely high mutation rate and these drugs are not effective for some genotypes. Therefore, early diagnosis and new and effective therapeutic agent for hepatitis C are needed.

Some publications have revealed that many miRNAs are dysregulated in HCV-infected patients. These miRNAs can modulate viral stability, replication, and protein translation that is required for the virus life cycle. miR-122 is a liver-specific miRNA that binds to two sites (S1 and S2) on the 5′ UTR (untranslated region of mRNA) of the HCV genome and promotes the viral life cycle. This liver-abundant miRNA positively affects viral RNA stability, translation, and replication [103]. A therapeutic agent based on miR-122-ASO (anti-sense oligonucleotide) has entered into clinical for HCV. miR-196b is able to inhibit growth of HCV [104]. Overexpression of miR-199a* inhibited HCV genome replication in two cells bearing replicons (replicon cell) HCV-1b or -2a [105]. miR-155 is overexpressed and miR-196b is reduced in HCV-infected patients. Recent data indicate that miR-155 induces an inflammatory state and promotes virus replication and persistence even after the completion of antiviral treatment, while miR-196b inhibits HCV replication with cytoprotective, anti-inflammatory, and antioxidant properties [106,107]. In addition to dysregulated miRNAs in HCV-infected livers, some serum miRNAs have altered expression after HCV infection. Upregulated miR-21 is a biomarker for necroinflammation in hepatitis C patients with and without HCC [108]. All these findings indicate the diagnostic and therapeutic potential of miRNAs for the treatment of HCV infection.

2.5 Liver Failure

Acute liver failure (ALF) is attributed to many etiologies and can occur due to toxic overdose, viral hepatitis, inflammation, and cancer. miRNAs have been implicated in ALF caused by these factors; however, the study of miRNAs in ALF has been complicated by the various etiologies associated with ALF. Fortunately, different miRNAs has dysregulated expression in ALF due to different factors, which makes us easy to study ALF and develop personalized medicine for this fatal disease. In general, miRNAs have effects on ALF by modulating apoptosis or proliferation that is required for hepatocyte replication.

Acetaminophen (APAP) is the most common cause of ALF in the United States, and is reported to be responsible for approximately 50% of cases of ALF. miR-122 and miR-192 are often found to be upregulated in ALF patients due to APAP in both liver tissue and serum, and amount of upregulation of these miRNAs correlates with the degree of liver degeneration. Further findings show that the elevation of these miRNAs occurs prior to rises in alanine aminotransferase (ALT), a hepatic enzyme which is commonly used as a biomarker to detect the presence and extent of ALF. All these findings indicate that miR-122 and miR-192 may serve as powerful biomarkers in diagnostic and prognostic of ALF due to APAP [109,110].

miR-122 has also been shown to be upregulated in patients with ALF due to malignancy and hepatitis [111]. It was reported that serum miR-122 levels were elevated approximately 100-fold in both ALF and chronic-HCV sera as compared to health individuals; however, there is no significant difference in serum miR-122 between chronic HCV patents with and without ALF [112]. Another study found that miR-122 levels in serum were strongly associated with spontaneous recovery from liver injury and higher levels of miR-122 predicted higher likelihood of liver regeneration following ALF. This study also observed that miR-122 also has increased expression in the livers of ALF patients [113]. This finding is clearly controversial to the role of miR-122 in inhibiting proliferation [22]. One possible explanation is that ALF leads to increased miR-122 rather than increased miR-122 promotes ALF. Another common cause of ALF is viral hepatitis. This often occurs due to inflammation and apoptosis. miRNA profiling revealed that many miRNAs including miR-155, miR-146a, miR-125a, miR-15b, and miR-16 are upregulated [114,115] and miR-1187 is downregulated in a TNF-dependent ALF mouse model. Reduced miR-1187 impaired hepatocyte proliferation by releasing its inhibition to caspase-8, a protein important in the apoptotic cascade [115]. In contrary, miR-15b and miR-16 downregulate the antiapoptotic protein Bcl-2 in TNF-mediated ALF, which potentially impaired the capacity of hepatocyte to proliferate in a TNF-mediated ALF mouse model [114]. Another study found that global miRNA deficiency in hepatocytes leads to increased apoptosis in an FAS-induced AFL mouse model [116]. Furthermore, miRNA profiling of mouse liver identified 11 miRNAs, which were upregulated in response to FAS-induced liver failure. One of these miRNAs is miR-221. Overexpression of miR-221 was able to protect hepatocytes from apoptosis and delay ALF in FAS-induced liver failure mice. miR-221 positively functions in ALF by inhibiting a proapoptotic protein, PUMA [116]. Interestingly, another study showed that miR-221 serum levels are increased in patients with spontaneous recovery from ALF, but are decreased in the livers of these same spontaneous recovery patients [113]. In addition, miR-221 targets that encode cell cycle inhibitor were downregulated in the livers of these spontaneous recovery patients. miR-221 may be a potential therapeutic target for ALF and a prognostic tool for predicting ALF outcomes in patients.

Despite the evidence that miRNAs are involved in the pathogenesis of ALF, the results from the different studies are controversial. Due to the complexity of ALF, a more detailed investigation is needed to address the mechanism by which miRNAs modulate the pathogenesis of ALFs caused by the different risk factors.

2.6 Liver Regeneration

The incidence of end stage liver disease has been rapidly increasing due to a variety of factors including hepatitis, NASH and the metabolic syndrome. Liver transplantation is still the only effective treatment for end-stage liver disease [117]. But donor organs are scarce, and some people succumb to liver failure before a donor is found. The study of liver regeneration in rodents has had a critical effect on medical practice, and allowed for the development of new liver-transplant technologies. The liver has the unique capacity to regenerate in response to injury [118], and liver regeneration plays a key role in reparation following viral infection, inflammation, and presence of toxins. The failure of regeneration is associated with ALF and can lead to death without liver transplantation [119]. Regeneration is also important in success following living-donor liver transplantation, wherein a part of a donor's liver is transplanted into a patient, and with successful regeneration, both donor and patient's livers grow to normal size [120]. Dysregulation of liver regeneration can also play causal roles in uncontrolled cell growth characteristic of HCC [119]. Liver regeneration occurs first by normal adult hepatocytes, and then by liver progenitor cell populations when liver injury is severe [121]. Liver regeneration has been studied extensively. However, up to date, its underlying mechanism is largely unknown, which impairs our ability to develop useful therapeutic strategies that can be used for the treatment of end-stage liver diseases.

The discovery of miRNAs has uncovered a new field of research on stem cell differentiation and organ regeneration. miRNAs have been implicated as key regulators in cell proliferation and cell cycle progression during liver regeneration. Many studies showed that miRNAs plays important role in the different stages of liver regeneration. One study showed that knockout of the essential miRNA biogenesis enzyme DGCR8 in hepatocytes of mice with partial hepatectomy (PH)–impaired cell cycle progression [122]. Liver regeneration still occurred in these mice due to the proliferation of liver progenitor cells that retained expression of DGCR8. This same study also found that miR-21 was upregulated in mice following PH and that miR-21 could play a key role in proliferation by antagonizing a cell cycle inhibitor gene. miR-378 downregulation was also observed in these mice, and research suggested that miR-378 suppression contributes to regeneration by releasing inhibition of *Odc1*, a gene-encoding enzyme necessary for DNA synthesis [122]. Another study found that miRNAs undergo dynamic changes during different stages of liver regeneration and miRNAs as a group plays an important role in the initial phase of liver regeneration [123].

Specific miRNAs have been identified in several studies as contributing to liver regeneration. miR-33 has been implicated in cell proliferation and research indicates that it represses genes involved in cell cycle progression, specifically *CDK6* and *CCND1*, resulting in cell cycle arrest. Liver regeneration following PH in mice was improved by inhibition of miR-33, suggesting that miR-33 inhibitor could be a valuable promoter of liver regeneration [124]. In contrast to miR-33, miR-150 has been found to be downregulated in the regenerating livers of mice following PH. It negatively regulates VEGF, a promoter of liver regeneration and miR-150 is downregulated by HIF1α in response to hypoxia commonly associated with hepatectomy and the initiation of liver regeneration [125]. miR-150 was also found to be downregulated in human patients who had undergone auxiliary liver transplantation and experienced successful liver regeneration. miR-150 upregulation was correlated with failed liver regeneration in the same study [119] further indicating the critical role of miR-150 in modulating liver regeneration.

miR-21 has been reported to have the most consistent altered expression in liver regeneration. miR-21 upregulation seems to contribute to hepatocyte proliferation during early stages of liver regeneration [122,123,126–130]. Chaveles et al. found miR-21 to be significantly upregulated in the regenerating livers of mice at the initial phase of liver regeneration following PH, and its expression changed more than any other miRNAs [131]. Results from a study by Marquez et al. showed similar results. They found that miR-21 upregulation was associated with NF-κB inhibition [126]. miR-21 upregulation in liver regeneration has also been associated with increases in cyclin D1, an essential cell cycle protein [128].

Regulation of miRNAs has been implicated in termination of liver regeneration. In one study, miR-34a was found to be highly upregulated following PH in rats in late-phase regenerating livers. The research suggested that miR-34a has a strong antiproliferative effect and causes significant G2/M arrest, thus upregulation of miR-34a may be important in the terminal phase of liver regeneration [132]. Another study found that downregulation of miR-23b is also involved in termination of liver regeneration in rats following PH. miR-23b is implicated in cell proliferation, migration, and differentiation [133]. miR-221 overexpression has also been shown to accelerate hepatocyte proliferation following PH in mice by inducing hepatocytes to rapidly enter S phase [134]. Current evidence show that miRNAs plays important role in the initiation and termination of liver regeneration. With the development of miRNA profiling techniques, it is estimated that more miRNAs controlling liver regeneration will be identified.

One study looked at the interaction between miRNAs and their mRNA targets at the RNA-induced silencing complex (RISC) in mice following PH. They hypothesized that this would be a more informative method of determining the functional role of miRNAs and the target genes they interacted with. They found that in many cases, changes in expression of miRNAs did not correspond with recruitment to the RISC. This complicates interpretations of results looking at miRNA expression changes and suggests that it may not be appropriate to infer that altered miRNA expression necessarily results in correlated target gene expression changes [127]. Examining miRNA:mRNA interactions at the RISC complex and miRNA RISC load may provide more powerful insights into the important miRNAs in liver regeneration.

2.7 Liver Fibrosis and Cirrhosis

Liver fibrosis is the excessive accumulation of extracellular matrix proteins (ECM) including collagen that occurs in most types of chronic liver diseases. The accumulation of ECM proteins distorts the hepatic architecture by forming a fibrous scar [135]. Cirrhosis is advanced liver fibrosis that has resulted in widespread distortion of normal hepatic architecture, which is characterized by regenerative nodules surrounded by dense fibrotic tissue. Late manifestations include portal hypertension, ascites, and, when decompensation occurs, liver failure. Diagnosis for both fibrosis and cirrhosis often requires liver biopsy.

Chronic liver hepatitis can often lead to fibrosis and cirrhosis. Fibrosis and cirrhosis commonly represent end stage disease progression of inflammatory liver conditions like viral hepatitis and NAFLD. In the Western countries, NASH is becoming a major reason of fibrosis development. Fibrosis develops due to repeated liver injury and subsequent cell death, and eventually replacement of hepatocytes with extracellular matrix. Fibrosis can progress to cirrhosis with continued inflammation and injury. Currently, cirrhosis is usually considered irreversible and there are no effective treatments for cirrhosis and fibrosis and prognosis of these patients is poor [136]. miRNAs have been implicated in development of these diseases and represent potential targets for diagnosis and therapeutics.

miR-122 dysregulation is implicated in most liver diseases. Cirrhosis and fibrosis are no exception. miR-122 has been shown to have an antiinflammatory role in hepatocytes, and knockout of miR-122 results in hepatic inflammation and fibrosis. This fibrogenesis seems to occur with miR-122 knockout due to the upregulation of the protein Ccl2, which recruits inflammatory immune cells that secrete cytokines like IL-6 and TNF to potentiate inflammation [22]. miR-29 has been also implicated in several studies as a key modulator in the development of fibrosis. Both fibrosis mouse models and biopsies from patients with advanced liver fibrosis show downregulation of miR-29 [137]. HSCs are responsible for producing the extracellular matrix characteristic of fibrogenesis. miR-29 is downregulated in these stellate cells, due to modulation by TGF-β and the inflammatory signals, lipopolysaccharide and NF-κB [137]. miR-29 overexpression in HSCs effectively inhibits collagen synthesis and can reverse the fibrotic phenotype caused by TGF-β overexpression [138]. Patients with fibrosis due to HCV-infection demonstrate similar downregulation of miR-29 in both HSCs and hepatocytes [139].

HSCs play a central role in hepatic fibrogenesis and several miRNAs besides miR-29 are dysregulated in HSCs. One good example is miR-31 that is upregulated in fibrotic HSCs by TGF-β and inhibition of miR-31 can prevent HSC activation [140]. In contrary, miR-133a has been shown to be downregulated specifically in HSCs in fibrosis due to expression alteration of TGF-β. Overexpression of miR-133a in HSCs can decrease collagen expression. Additionally, research suggests that miR-133a is increased in the serum of patients with progressive cirrhosis regardless of underlying etiology and that there is a correlation between level of miR-133a expression and stage of liver disease [141], suggesting its diagnostic and therapeutic potential for liver fibrosis.

Several studies have shown that miR-34a upregulation in liver tissues is associated with fibrosis [142,143]. miR-34a expression is increased in alcohol induced fibrosis, a mechanism that likely occurs due to ethanol-dependent hypomethylation of the miR-34a promoter [142]. miR-21 upregulation also seems to contribute to fibrosis and cirrhosis. The increase in miR-21 expression is seen in serum samples and liver tissues from patients with cirrhosis [144]. miR-21 overexpression has been shown to induce fibrosis in vitro via the PTEN/Akt signaling pathway [145].

Differential expression of miRNA combination is associated with the progression of fibrosis to cirrhosis. One study found that miR-199-5p/miR-199-3p and miR-221/-222 were upregulated in human fibrotic liver tissue, and their upregulation correlated with the level of disease progression [146]. miR-221/-222 were found to be upregulated specifically in activated HSCs [146]. Another study found that overexpression of miR-199a-5p, miR-199a*, miR-200a, and miR-200b are associated with the progression of liver fibrosis and expression of fibrotic genes in HSCs is increased with increased expression of these miRNAs [147].

With the development of deep sequencing approaches, it is anticipated that more differentially-expressed miRNAs will be identified during the pathogenesis of fibrosis and cirrhosis. Despite the potential of a single miRNA serving as a biomaker, a combination of several miRNAs might be a better choice.

As described earlier, misexpression or dysregulation of miRNA was discovered in liver tissues and serum of various types of human liver diseases such as NAFLD, HCC, fibrosis and cirrhosis, and hepatitis. Increasing evidence indicates that the introduction of specific miRNAs into diseased liver cells and livers induces favorable therapeutic responses [23,24,29]. The promise of miRNA therapy for liver diseases is perhaps greatest in liver diseases due to the apparent role of miRNAs in the pathogenesis of hepatic disorders. miRNA profiling has identified many miRNA signatures associated with diagnosis, staging, progression, prognosis, and response to treatment. With the development of detection techniques, more miRNAs that have dysregulated expression in liver diseases will be identified, which will make huge influences on the elucidation of pathogenesis of human liver diseases, and thereby provides new diagnostic biomarker and therapeutic approaches for human liver diseases.

3. MICRORNAs AS DIAGNOSTIC AND PROGNOSTIC MARKERS IN LIVER DISEASE

The management of liver disorders is challenging and there is a pressing need for informative and practical biomarkers that can facilitate early diagnostic, accurate prognostic, and treatment monitoring for liver disease. Dysregulation of many miRNAs in liver tissues and blood has been shown to be associated with liver diseases as described earlier. Furthermore, miRNAs are highly stable in formalin–fixed, paraffin–embedded tissue from clinical samples or in human blood due to the fact that they are short (~22 nt), protein bound, and often travel encapsulated in microvesicles. Based on these favorable characteristics of miRNAs, they might represent potential diagnostic markers for liver diseases.

3.1 Circulating MicroRNAs as Biomarkers in Liver Diseases

In addition to miRNAs that are dysregulated in livers of human individuals, circulating miRNAs have been found in blood samples [148]. miRNAs are stably present in human serum and detecting specific miRNAs in the blood has been shown to be easier than protein detection [148]. Protein-based biomarkers are a traditional tool for assessing diagnosis and prognosis; however, there are issues associated with this method. Proteins often have low abundance in the serum, are subject to posttranslational modifications, are difficult to capture within the serum, and become easily denatured. While there have been many innovations in tests involving protein biomarkers; specificity, sensitivity, and repetition remain to be improved for protein-based diagnostic and prognostic tests [149]. Thus, detection of specific miRNA biomarkers in serum may prove to be a more powerful and convenient diagnostic and prognostic tool.

Research has indicated that miRNAs are resistant to digestion by RNase A and remain stable in serum following numerous freeze–thaw cycles, boiling, high and low pH, and extended storage indicating that miRNAs are uniquely stable in harsh conditions [148,150]. Interestingly, synthesized miRNAs in a naked form are subject to RNAse digestion, but their endogenous form in the plasma is resistant to RNase degradation [148]. miRNA serum levels have been found to be generally consistent among healthy individuals, suggesting that levels are constant and tests for miRNAs as biomarkers would be reproducible and useful [150].

miRNA detection in serum can be accomplished by delivering complementary oligonucleotides to isolate the miRNA of interest, and then amplifying the miRNA by a standard PCR assay. Amplification by PCR is not necessary but improves detection and sensitivity [148]. Additionally, miRNAs can be detected and quantitated by qRT-PCR alone with miRNA specific primers. This method has been shown to have high specificity and sensitivity even with low concentrations of miRNAs [150,151].

It is important to recognize that not all miRNAs that are dysregulated in liver tissue exhibit correlated dysregulation in the serum. Further, patients with similar liver diseases of different etiologies can experience different types of changes in serum expression of miRNAs [109]. Quantifying whether liver disease-associated hepatocyte-specific changes in miRNAs also occur in serum is necessary prior to developing diagnostic and/or prognostic tests. miRNA serum levels are not always consistently correlated among all patients with a specific disease and can be confounded by the presence of other diseases. miRNA levels in serum need to be well investigated and confirmed among a diverse patient population to have utility in diagnostic and prognostic tests.

As detailed above, it might be a better choice to use a combination of several miRNAs as a biomarker of a specific liver disease, which will potentially increase the diagnostic accuracy.

Although miRNAs have characteristics that made them suitable biomarkers in clinical practice, such as high stability in tissues and fluids or good preservation in formalin-fixed, paraffin-embedded tissues, the detection of miRNA is challenging because of several intrinsic characteristics of these molecules: small size, sequence similarity among various members, lack of common features which could facilitate their selective amplification, low level, and tissue-specific or developmental stage-specific expression. Here, we summarized the main detection approaches for miRNAs and discussed the specific challenges in each miRNA detection method. The main detection approaches for miRNAs include miRNA microarray, quantitative real-time PCR, in situ hybridization, high throughput sequencing, and miRNA imaging. Each method has its strengths and weaknesses; thus, researchers and clinicians should choose appropriate methods based on their purpose and characteristics of a specific liver disease [152].

3.1.1 MicroRNA Reverse Transcription Polymerase Chain Reaction

Quantitative real-time reverse-transcription PCR (qRT-PCR) is the most common approach for gene expression quantification and one of the most common methods used to detect low levels of miRNA with high sensitivity. qRT-PCR is a sensitive and specific method for miRNA expression measurement that can be used for absolute quantification. After optimization by many laboratories [153–156], this method proved sensitive enough to detect as little as 220 miRNAs from the RNA of a single stem cell. In addition, the challenges related to the short length of mature miRNAs and hairpin structure of precursor miRNA have been overcome and the qRT-PCR represents the most convenient and reliable technique for miRNA expression analysis. However, this method fails to identify novel miRNAs. Another disadvantage of this approach is the medium throughput with respect to the number of samples per day. Fortunately, recently, Applied Biosystems developed a novel array-like qPCR card named Taqman Array MicroRNA Card. These arrays were designed to investigate all identified miRNAs in human, mouse, and rat. This method combined the characteristics of qRT-PCR sensitivity and high throughput of microarray.

However, this method still is unable to identify new miRNAs.

3.1.2 MicroRNA Microarray

miRNA microarray is the most common technique used for high-throughput miRNA profiling that is particularly effective to detect large numbers of miRNAs of more samples processed in a parallel in a single experiment. miRNA microarray technology is actually based on nucleic acid hybridization between target miRNAs and their corresponding complementary probes. Several microarray-based commercial platforms are now available for research use, including Illumina, Affymetrix, and Agilent. Recently, Agilent Technologies developed a single-color miRNA platform. This technology provides both sequence and size discrimination for mature miRNAs and generates results that are highly correlated with qRT-PCR results [157]. Despite its advantage as a high-throughput approach, microarray-based miRNA detection still faces several challenges. First, this method is unable to identify new miRNAs, and sometimes the result needs to be validated further using other methods. Some miRNAs have very similar sequences. The large dynamic range of miRNA expression and high miRNA sequence homology is another big challenge for this approach. Compared with qRT-PCR or RNA sequencing, this method has lower specificity. In addition, the lack of consensus on analytical methods and the absence of standard baseline data with known biological characteristics, make it difficult to evaluate platform performance and assess biological data quality. All these hurdles impair the capacity of miRNA microarray as a practical diagnostic approach for liver diseases.

3.1.3 RNA Sequencing

RNA sequencing is a new generation of high-throughput platform to identify new miRNAs and determine expression levels of both discovered and new miRNAs. This approach directly determines the nucleotide sequence of miRNAs, and involve RNA extraction, ligation of adapters to both ends of mature miRNAs. This approach has very high accuracy in distinguishing miRNAs that are very similar in sequence. Regarding this feature, deep sequencing is much more powerful than miRNA microarray. Deep sequencing method overcomes the disadvantages of microarrays and generates millions of small RNA sequence reads. This approach allows us to determine absolute abundance and to discover novel miRNAs. The low expression levels of novel miRNAs are effectively detected because of the high sensitivity of deep sequencing. Deep sequencing can be performed for miRNA detection in FFPE (formalin-fixed, paraffin-embedded) specimens, making large-scale clinical studies possible in the future. With the development of new sequencing technology and instrument, the cost of this technique is dropping dramatically, which makes this method available for routine laboratory work and potential clinical practice. However, this method needs substantial computational support for data analysis. Another disadvantage is that the deep sequencing needs large amount of RNA, which hamper its application for small samples.

Great progress in the detection approaches for circulating miRNAs have been made. However, challenges still exist in their clinical application and require further investigation.

A detection method that combines the advantages of PCR, miRNA microarray, deep sequencing is needed if we want to apply miRNA biomarkers to diagnosis.

4. MICRORNA-BASED THERAPEUTICS FOR LIVER DISEASE

Given that dysregulation of miRNAs is implicated in many liver diseases, knockdown or overexpression of these small molecules is attractive in the development of novel therapeutic approaches for hepatic disorders. More and more evidence showed that miRNA mimics and inhibitors are easy to be delivered to livers [23,34]. In general, there are two approaches to developing miRNA-based therapeutics: miRNA inhibitors and miRNA mimics (miRNA replacement therapy). Both miRNA mimics and inhibitors are easy to synthesize. miRNA inhibitor is chemically modified, single-stranded nucleic acids designed to specifically bind to and inhibit endogenous miRNA molecules. Chemical modifications of miRNA inhibitors can enhance the affinity for the target miRNA and trap the endogenous miRNA in a configuration that is unable to be processed, or alternatively, leads to degradation of the endogenous miRNA [158]. These inhibitors can be introduced into diseased liver tissues or cells, and enable detailed study of miRNA therapeutic effects [23]. miRNA mimics are small, chemically modified double-stranded RNAs that mimic endogenous miRNAs and enable miRNA functional analysis by upregulation of miRNA activity [158]. This approach is to generate non-natural double-stranded miRNA-like RNA fragments. Such an RNA fragment is designed to have its 5′-end bearing a partially complementary motif to the selected sequence in the 3′UTR unique to the target gene [158]. Once introduced into cells, this RNA fragment, mimicking an endogenous miRNA, can bind specifically to its target gene and produce post-transcriptional repression, more specifically translational inhibition, of the gene. If the reduction of one specific miRNA causes a liver disease, miRNA mimics can be introduced into diseased hepatic cells or tissues to recover to its normal level. In contrast, miRNA inhibitors will be used to knock down upregulated miRNA(s) that promotes liver diseases. The inhibitory approach is more commonly accepted and conceptually follows rules that also apply to small molecule inhibitors and short interfering

RNAs (siRNA). Although miRNA replacement therapy represents a new opportunity to explore the therapeutic potential of good miRNAs such as tumor suppressors (4, 7, 8, 11), this approach is relatively immature and is far from clinical application partly because low inhibitory effects of miRNA mimics on their targets in vivo.

With the development of delivery approaches, adeno-associated virus (AAV) or mini-circle-based miRNA replacement therapy have been developed recently for liver diseases. AAV virus-based approaches are widely used for gene therapy. Mini-circles are circular DNA elements that no longer contain antibiotic resistance markers or the bacterial origin of replication. These small vectors can provide for long-term transient expression of one or more transgenes without the risk of immunogenic responses. We have cloned miRNA precursors into the mini-circle or AAV vectors. The preliminary data has shown that these two approaches have high efficiency to deliver miRNAs into livers.

The application of miRNA therapy to liver diseases is relatively easier than other types of disorders. The liver filters a large fraction of blood per minute, and thus targeting the liver by miRNA therapies has been much easier than other organs in the body [159]. miRNA mimics are used to restore a loss of function. This approach aims to reintroduce miRNAs into diseased cells or tissues that are underexpressed under the health status. The reintroduction of these miRNAs leads to reactivation of pathways that are required for normal cellular and physiological function and block those that promote diseases.

Several miRNA therapeutics have already progressed into product and clinical development by Regulus Therapeutics, Santaris PharmaA/S, mirage Therapeutics, and miRNA Therapeutics. For example, a study indicates that a *let-*7 miRNA mimic can repress Kras, Myc, and HMGA-2, oncogenes mutated or upregulated in many cancer types, and therapeutic delivery of a *let-*7 miRNA mimic leads to robust inhibition of tumor growth in human non-small cell–lung cancer xenografts [160]. Additionally, miRNA antagonists have been developed to treat hepatitis C that specifically target hepatocytes [161]. miR-122 is a liver-specific miRNA that is hijacked by HCV to maintain viral abundance [162]. A miR-122 antagonist can successfully suppress HCV in serum in chimpanzees [161]. The inhibitor is currently in phase II clinical trials. The success of this liver-targeted miRNA antagonist suggests that miRNA therapeutics for many diseases could similarly move forward quickly into clinical trials given that experiments clearly elucidate their actions in human patients. Also, the binding of miRNAs to nonspecific targets has to be studied in more detail because miRNAs have multiple targets due to partial binding unlike siRNA.

4.1 MicroRNA Inhibitors as a Potential Therapeutic Approach

Some miRNAs silence mRNAs encoding "good" protein-coding genes that is required to keep our health status. For these miRNAs, their inhibitors have been designed for potential therapeutic agents. Various methods including directly delivering miRNA-ASO or expressing miRNA-ASO through viral vectors and minicircle vectors, have been developed for successful miRNA knockdown in cell culture system and in vivo. Recently, miR-ASO-based therapeutic approaches have been applied in humans, and ongoing clinical trials hold promise for treating liver disorders. Here we summarize the delivery, stability, and toxicity of miR-ASO, discuss important considerations when developing miRNA-ASO therapeutic agent, and review the status of clinical trials of miR-ASO as potential therapeutic approach.

4.2 Design of Chemically Modified Antisense Oligonucleotide of MicroRNA

Antagonizing endogenous miRNAs through miR-ASO needs to optimize miR-ASO for increased binding affinity and improved nuclease resistance and in vivo delivery. Several modifications have been carried out to confer nuclease resistance, facilitate cellular uptake, and increase the binding affinity of miR-ASO. To avoid miRNA-ASO degradation, ASO firstly was modified by adding a methyl group which binds with the 2′ oxygen of the ribose a 2′-O-methyl (2′-O-Me). The introduction of ASO with this 2′-O-Me into human breast cancer cells specifically inhibited endogenous miR-21 [163]. The 2′-O-Me modification as well as the 2′-O-methoxyethyl (2′-MOE) and 2′-fluoro (2′-F) chemistries are modified at the 2′ position of the sugar moiety, whereas locked nucleic acid (LNA) comprises a class of bicyclic RNA analogs in which the furanose ring in the sugar-phosphate backbone is chemically locked in an RNA mimicking N-type (C3′-endo) conformation by the introduction of a 2′-O,4′-C methylene bridge [164–168]. All these modifications confer nuclease resistance and increase the binding affinity of ASOs to their cognate miRNAs [164]. Among these, LNA possesses the highest affinity toward complementary RNA [168–170]. To silence miRNAs in vivo, Krutzfeldt et al. designed chemically modified, cholesterol-conjugated, single-stranded RNA analogs (termed antagomirs), which is able to efficiently inhibit the activity of the endogenous miRNAs in many organs including liver [66,171]. The conjugation of miR-ASO with lipophilic molecules such as high-density lipoprogein has also been used to deliver miR-ASO to specific organs [172,173]. miRNA sponges are a new form of miR-ASOs that are transiently expressed in cultured mammalian cells and organs [174–176]. miRNA sponges are transcripts expressed from strong promoters, containing multiple, tandem binding sites to an miRNA of interest. When vectors encoding these sponges are transiently transfected into cultured cells, sponges derepress miRNA targets at least as strongly as chemically modified miR-ASO [176]. Sponges also offer advantages over chemically modified ASOs. First, miRNA sponge transcripts generated by appropriate viral transduction in target cells provide an alternative way to overcome the limitation of low transfection efficiency of chemically modified miRNA-ASO. Second, this technology is especially useful in knocking down the expression of a family of miRNAs that shares an identical seed sequence, providing a powerful means for simultaneous inhibition of multiple miRNAs of interest simultaneously.

In addition to increasing stability by chemical modification, the phosphorothioate backbone modifications have been introduced into miR-ASO to increase plasma protein binding, thereby reducing clearance of the miR-ASOs by glomerular filtration and urinary excretion, which facilitates tissue delivery of miR-ASOs in vivo [177,178], as well as promote hepatic uptake of the antimiR while lowering delivery to other types of tissues [177,178].

4.3 MicroRNA Mimics as a Potential Therapeutic Approach

Some miRNAs has decreased expression in liver diseases as summarized earlier. Where loss of a specific miRNA inhibitory effect contributes to activation of some gene encoding pathogenic proteins such as oncogenes. In these situations, a rational treatment is miRNA replacement, which involves the re-introduction of an miRNA mimic to restore a loss of function. The development of miRNA replacement therapy represents a new strategy for the treatment of liver disease. However, targeted in vivo delivery of miRNA mimics faces many challenges including limited stability in blood, rapid blood clearance, off-target effects, and poor cellular uptake [179]. Chemical modifications have been developed for increasing the stability of miRNA mimics including 2′-O-methylation. However, scientists from Dharmacon found that the stabilized miRNA mimics through chemical modification are less functional than the natural miRNA mimics. In some cases, a reverse in dose response was observed for the stabilized mimics containing chemical modification; that is, a decrease in potency was seen at the higher concentration of stabilized mimic. Very recently, researchers from Exiqon designed miRNA mimics that can simulate naturally-occurring mature miRNAs. This type of miRNA mimics have a unique and novel innovative design. They are based on three RNA strands rather than the two RNA strands that characterize traditional miRNA mimics. The miRNA (guide) strand is an unmodified RNA strand with a sequence corresponding exactly to the annotation in miRBase. However, the passenger strand is divided in two LNA-enhanced RNA strands. The great advantage is that the segmented nature of the passenger strand ensures that only the miRNA strand is loaded into the RISC with no resulting miRNA activity from the two complementary passenger strands. Despite the progress to increase miRNA mimic stability by chemical modification, its application in vivo still is facing challenges such as low half-life and reduced functionality. A different technique that bypasses the disadvantages of miRNA mimics is vector-based miRNA expression. Many expression vectors of miRNAs have been developed. The main differences in these expression vectors are to use different promoters including polymerase III H1 RNA promoter [180], polymerase II promoter [181], or tissue-specific promoters. Currently, many expression vectors have been commercialized by companies including Origene, GeneCopoeia, and System Biosciences. Alternatively, miR-NAs can be delivered as a precursor miRNA encoded via viral vectors. Minicircles are episomal DNA vectors that are produced as circular expression cassettes devoid of any bacterial plasmid DNA backbone [182]. Their smaller molecular size enables more efficient delivery and offers sustained expression over a period of weeks as compared to standard plasmid vectors that may function for only a few days. In my laboratory, we used minicircle system and *TTR* (Transthyretin) promoter to highly and specifically over-express miRNAs in livers of NAFLD and HCC mice (data not published).

All these findings provided direct evidence that miRNA replacement therapy can be obtained by delivering miRNA mimics or miRNA expression vector into cell lines and livers of human and mouse. However, methods to deliver miRNA mimics or vector-based miRNA to the primary malignant liver tumor in animal models need to be further developed.

4.4 In Vivo Delivery of MicroRNA Inhibitors and MicroRNA Mimics

High level of delivery efficiency is very critical for miR-ASO and miRNA mimics as potential therapeutic approach. Currently, delivery of miR-ASO to liver is relatively easy due to its high stability and small size. However, there are few reports on the use of miRNAs mimics for in vivo therapy. Systemic delivery of miRNA replacement therapy and loss-of-function therapy has been developed in preclinical models including liposome, viral vectors and nanoparticles, and minicircle vectors. However, despite the early promise and exciting potential, critical hurdles often involving delivery of miRNA-targeting agents remain to be overcome before transition to clinical applications. Limitations that may be overcome by delivery include, but are not limited to, poor in vivo stability, inappropriate biodistribution, disruption, and saturation of endogenous RNA machinery. Both viral vectors and nonviral delivery systems can be developed to overcome these challenges. Viral vectors have high efficiency of delivery but their toxicity and immunogenicity limit their clinical application. In recent years, some nonviral vector has made great progress in delivering miRNAs. Herein, we review the recent advances in the lipid-based, viral and nonviral miRNA delivery systems and provide a perspective on the future of miRNA-based therapeutics for liver diseases.

4.5 Lipid-Based Delivery System

Liposomes are probably the most extensively studied materials for drug delivery and the most commonly used transfection reagents in vitro. However, safe and efficacious delivery in vivo is rarely achieved due to toxicity, nonspecific uptake, and unwanted immune response [183,184]. Liposome formulations have been used for delivering plasmids, miRNAs, and siRNAs [185]. Several liposomes have proven safe and efficient for delivering small molecule drugs in patients. For instance, Ortho Biotech has received FDA approval for the treatment of HIV-related Kaposi's sarcoma, breast cancer, ovarian cancer, and other solid tumors [186]. Liposomal amphotericin B (Ambisome; Gilead) is an approved antibiotic for the treatment of serious fungal infections [68]. For nucleic acid delivery, many of the lipid-based delivery vehicles self-assemble with siRNAs through electrostatic interactions with charged amines, generating multimellar lipoplexes with positively charged lipids and negatively charged siRNAs [187]. Recently, the miR-34a mimic encapsulated using an innovative liposomal formulation called SMARTICLES entered phase I clinical trial in patients with HCC [179,188]. This is notable in being the first miRNA mimic to be trialed in human patients, and also because it targets the liver, the only organ that receives pharmacologically relevant concentration of drug following systemic delivery of lipid-complexed siRNA or miRNA [188]. Two strategies have been used for in vivo delivery of antimiRs: (1) cholesterol conjugation and (2) modification of the phosphate backbone. These modifications significantly increase the stability of antimiRs, but also improve delivery efficiency of antimiRs. The first miRNA targeted drug for HCV treatment show efficacy in clinical trials [189]. Currently, various lipid compositions, including Lipofectamine 2000 (Life Technologies), Oligofectamine (Life Technologies), and TransIT-2020 (Mirus Bio), have been used routinely in the laboratory to transfect nucleic acids into cells. However, their uses in vivo are limited owing to toxicological concerns and poor stability [70].

4.6 Viral Delivery Systems

Expression vectors based on virus have been used widely for gene therapy. Modified adenovirus, AAV, and lentiviruses have been employed to successfully deliver siRNA into cells and stably integrate siRNA into targeted genome [188]. Due to the similarity between miRNAs and siRNAs, these virus-based approaches are feasible for miRNA delivery. Effective delivery of a specific miRNA into livers via lentiviruses and AAV has been established. Recently, miR-122 was highly and specifically expressed in liver using AAV and liver-specific promoter. Liver-specific miR-122 target sequences incorporated in AAV vectors efficiently inhibits transgene expression in the liver [190,191]. Lentivirus-based miRNA delivery led to inhibition of HCC in mice [192,193]. Using AAV, Moshiri et al. overexpressed miR-221 sponge in human liver cancer cells, which dramatically knocked down endogenous miR-221 [194]. Another good example for AAV-based miRNA replacement therapy is miR-26a. The reintroduction of miR-26a in a liver cancer mouse model using an AAV vector resulted in the inhibition of cancer cell proliferation, induction of tumor-specific apoptosis and protection from disease progression [194a]. AAVs have been one of the most promising methods for high effectiveness in in vivo gene therapy. However, they have some significant disadvantages. Some target cells have low ratios of appropriate primary and/or secondary adenoviral receptors, and this requires a high dose of vector application to cause target-cell cytotoxicity. The most serious problem in the use of AAV vectors is their tendency to cause strong immune and inflammatory responses at high doses, which significantly limits its usage for clinical practice. Lentivirus vectors present with some favorable features like the ability to transduce also nondividing cells and a potentially safer insertion profile. However, genetic modification with viral vectors in general and stable integration of the therapeutic gene into the host cell genome bear concerns with respect to different levels of personal or environmental safety. Among them, insertional mutagenesis by enhancer mediated dysregulation of neighboring genes or aberrant splicing is still the biggest concern [157]. Currently, several commercialized expression vectors for miRNAs based on adeno- and lentivirus have been developed, such as Vector Biolabs and System Biosciences. However, these vectors can only be used for research and their application for clinical practice still faces many challenges as described earlier.

4.7 Nanoparticles

Nanocapsules are vesicular systems in which a drug is confined to a cavity surrounded by a polymer membrane, whereas nanospheres are matrix systems in which the drug is physically and uniformly dispersed [195]. Nanoparticles are solid, colloidal particles consisting of macromolecular substances that vary in size from 10 to 1000 nm [195]. Delivery by nanoparticles has the advantages of being more cost effective, less immunogenic, less toxic, and oncogenic [196]. The nanoparticles can be targeted to specific organs or target cells by functionalizing them with target-specific ligands and by varying the size and composition of the nanoparticles [197]. It was reported that systemic delivery of a chemically stabilized antimiR-122 complexed with interfering nanoparticles (iNOPs) effectively silences the liver-expressed miR-122 in mice, suggesting that iNOPs can successfully deliver antimiR to specifically target and silence miRNA in clinically acceptable and therapeutically affordable doses [198]. Restoration of lost tumor suppressor miRNAs or other gate-keepers of human diseases using synthetic dsRNAs (with a delivery agent) has been successful in several models. miR-122, a tumor suppressor of liver cancer, was delivered into mice using nanoparticle and inhibited growth of engrafted tumor [199]. Compared with antimiR, the limitation of miRNA mimics as a potential therapeutic agent is that miRNA mimics have only a transient effect, are unstable, and require repeated treatment. Emerging evidence shows that nanoparticle delivery is not only as effective as, but also less toxic than liposome and viral delivery systems. However, the size, charge, and surface properties of nanocarriers have to be optimized to ensure effective delivery of the miRNA in clinical practice. The immune responses related to the nanocarriers and the double-stranded miRNA mimics delivery remains to be addressed.

4.8 Polyethylenimine-Based Delivery System

Polyethylenimines (PEIs) are positively charged, linear, or branched polymers that are able to form nanoscale complexes with small RNAs, leading to RNA protection, cellular delivery, and intracellular release [200]. As one of the most widely used and studied polymers for gene delivery, PEI has attracted attention from both basic and clinical researchers for miRNA mimic and inhibitors delivery recently. In 2015, Chien et al. for the first time used PEI to deliver miR-122 into liver, which accelerated differentiation of iPSCs into mature hepatocyte-like cells and improved stem cell therapy in a hepatic failure mouse model [201].

4.9 Poly (lactide-co-glycolide) Particles

Poly (lactide-co-glycolides) (PLGAs) are a family of water-insoluble polymers that have been widely used in biomedical applications over the last half a century. PLGA is biocompatible and biodegradable, exhibits a wide range of erosion times, has tunable mechanical properties, and most importantly, is an FDA-approved polymer [202]. In particular, PLGA has been extensively studied for the development of devices for controlled delivery of small molecule drugs, proteins, and other macromolecules in commercial use and in research [202]. The majority of the previously described polymers focus on the ability to condense miRNA and nonspecifically enter the cell while PLGA focus on these factors but also allow the release of miRNA over time [184]. It has the advantage of lower toxicity compared to cationic lipids and cationic polymers. PLGA nanocapsules have been used to encapsulate DNA, antisense oligodeoxynucleotides, and siRNA to improve the stability and to provide sustained release [203]. In mice, PLGA microspheres yielded sustained release of siRNA, antisense oligonucleotides, ribozymes, and DNAzymes for at least 7 days at the subcutaneous injection site [204]. Regarding the miRNA delivery using PLGA, PLGA particles overcome several limitations facing current miRNA therapy because they can protect nucleic acids from degradation, achieve high loading capacities, and afford multiple surface modifications so as to generate potentially favorable pharmacodynamics [184]. Successful delivery of miRNAs has been achieved and some ongoing clinical trials showed great potential [205].

4.10 Safety and Toxicity of MicroRNA Treatment

Accumulating evidence indicates that miRNAs are unique and novel therapeutic approaches for human disorders. The development of miRNA mimics or inhibitors for the treatment of human diseases is still in the early stage. Initial animal studies reveal that local or systemic administration of miRNA mimics or inhibitors does not exhibit short-term toxicity. Currently, there are several clinical trials examining miRNA treatment for the treatment of liver diseases. miR-34a mimic has entered into clinical trials phase II for the treatment of HCC. In vivo evidence for toxicity induced by miRNA mimics is still lacking. Regarding antimiR therapy, miR-122 inhibitor is being evaluated for the treatment of HCV [206]. In addition, no adverse effects associated with miR-122 inhibitors have been reported [206]. miR-122 inhibitor is currently being evaluated in phase II clinical trials and may be the first miRNA-based therapeutic that is brought to market. miRNAs are naturally occurring small molecules, suggesting their low adverse effects as potential therapeutic agents. As expected, both therapeutic agents based on miR-34a and miR-122 have very low adverse effect as reported. The evidence from my laboratory also showed that miR-206 mimic treatment has negligible effect on hepatic toxicity (data not shown). However, application of miRNA therapy to human faces many challenges. For example, miRNAs are able to target many genes simultaneously and has effects on expression of other miRNAs. Therefore, more in vivo studies remain to be done to evaluate the dose and long-term toxicity and adverse effects before applying miRNA therapy to humans.

4.11 Conclusion and Perspective

The safety, effectiveness, and ease of production and manufacturing are important considerations for selecting the appropriate delivery vehicle for miRNA drugs. Evaluation of the efficacy of miRNA therapeutics requires considerations of off-target effects and innate immune response. miRNAs are naturally occurring, small, noncoding RNAs that modulate several cellular and physiological processes by targeting many genes simultaneously. Their characteristic of naturally occurring small molecules suggests its low adverse effects, which is confirmed by two miRNA therapeutic agents under clinical trials and our recent findings. In addition, miRNAs are able to multiple genes functioning in different pathways, suggesting that miRNAs are also potential therapeutic approaches for some highly associated disorders such as hyperlipidemia with diabetes and NAFLD with HCC. The roles for individual miRNAs in liver disease have been well established during the past 5 years. Once the role of a specific miRNA in disease pathogenesis is established, selecting specific miRNA inhibitor or mimic and delivery strategies promises to be straightforward. Although effective and safe delivery of miRNA drugs remain difficult for many cell types such as brain and muscle, its delivery to liver is relatively easy since liver filters a lot of blood every minute. It is our hope that the miRNA therapeutics field will soon converge on a small number of "platform" technologies that allow a rapid and safe development path from academic discovery to effective drug for the treatment of liver disease.

LIST OF ACRONYMS AND ABBREVIATIONS

ALF Acute liver failure
HBV Hepatitis B Virus
HCC Hepatocellular carcinoma
HCV Hepatitis C virus
miRNA MicroRNA
NAFLD Nonalcoholic fatty liver disease
NASH Nonalcoholic steatohepatitis
SREBP Sterol regulatory element-binding protein
VAT Visceral adipose tissue

REFERENCES

[1] Filipowicz W, Bhattacharyya S, Sonenberg N. Mechanisms of post-transcriptional regulation by microRNAs: are the answers in sight? Nat Rev Genet 2008;9(2): 102–14.

[2] Bartel D. MicroRNAs: genomics, biogenesis, mechanism, and function. Cell 2004; 116(2):281–97.

[3] Esquela-Kerscher A, Slack F. Oncomirs—microRNAs with a role in cancer. Nat Rev Cancer 2006;6(4):259–69.

[4] Park S, Peter M. MicroRNAs and death receptors. Cytokine Growth Factor Rev 2008;19(3–4):303–11.

[5] Wienholds E, Plasterk R. MicroRNA function in animal development. FEBS Lett 2005;579(26):5911–22.

[6] Song G, Zhang Y, Wang L. MicroRNA-206 targets notch3, activates apoptosis, and inhibits tumor cell migration and focus formation. J Biol Chem 2009;284(46): 31921–7.

[7] Farrell GC, et al. Nash is an inflammatory disorder: pathogenic, prognostic and therapeutic implications. Gut Liver 2012;6(2):149–71.

[8] Postic C. Pathogenesis of fatty liver disease. Endocr Abstr 2012;29:S55.

[9] Cohen JC, Horton JD, Hobbs HH. Human fatty liver disease: old questions and new insights. Science 2011;332(6037):1519–23.

[10] Rosmorduc O. Relationship between hepatocellular carcinoma, metabolic syndrome and non-alcoholic fatty liver disease: which clinical arguments? Ann Endocrinol 2013;74(2).

[11] Gaggini M, et al. Non-alcoholic fatty liver disease (NAFLD) and its connection with insulin resistance, dyslipidemia, atherosclerosis and coronary heart disease. Nutrients 2013;5(5):1544–60.

[12] Levene AP, Goldin RD. The epidemiology, pathogenesis and histopathology of fatty liver disease. Histopathology 2012;61(2):141–52.

[13] Yilmaz Y. Review article: is non-alcoholic fatty liver disease a spectrum, or are steatosis and non-alcoholic steatohepatitis distinct conditions? Aliment Pharmacol Ther 2012;36(9):815–23.

[14] Cortez-Pinto H, de Moura MC, Day CP. Non-alcoholic steatohepatitis: from cell biology to clinical practice. J Hepatol 2006;44(1):197–208.

[15] Cheung O, et al. Nonalcoholic steatohepatitis is associated with altered hepatic MicroRNA expression. Hepatology 2008;48(6):1810–20.

[16] Dolganiuc A, et al. MicroRNA expression profile in Lieber-DeCarli diet-induced alcoholic and methionine choline deficient diet-induced nonalcoholic steatohepatitis models in mice. Alcohol Clin Exp Res 2009;33(10):1704–10.

[17] Li S, et al. Differential expression of microRNAs in mouse liver under aberrant energy metabolic status. J Lipid Res 2009;50(9):1756–65.

[18] Pogribny IP, et al. Difference in expression of hepatic microRNAs miR-29c, miR-34a, miR-155, and miR-200b is associated with strain-specific susceptibility to dietary nonalcoholic steatohepatitis in mice. Lab Invest 2010;90(10):1437–46.

[19] Wang B, et al. Role of microRNA-155 at early stages of hepatocarcinogenesis induced by choline-deficient and amino acid–defined diet in C57BL/6 mice. Hepatology 2009;50(4):1152–61.

[20] Hoekstra M, et al. Nonalcoholic fatty liver disease is associated with an altered hepatocyte microRNA profile in LDL receptor knockout mice. J Nutr Biochem 2012; 23(6):622–8.

[21] Alisi A, et al. Mirnome analysis reveals novel molecular determinants in the pathogenesis of diet-induced nonalcoholic fatty liver disease. Lab Invest 2011;91(2): 283–93.

[22] Hsu S-H, et al. Essential metabolic, anti-inflammatory, and anti-tumorigenic functions of miR-122 in liver. J Clin Invest 2012;122(8):2871–83.

[23] Esau C, et al. miR-122 regulation of lipid metabolism revealed by in vivo antisense targeting. Cell Metab 2006;3(2):87–98.

[24] Tsai W-C, et al. MicroRNA-122 plays a critical role in liver homeostasis and hepatocarcinogenesis. J Clin Invest 2012;122(8):2884–97.

[25] Edwards PA, et al. Regulation of gene expression by SREBP and SCAP. Biochim Biophys Acta BBA Mol Cell Biol Lipids 2000;1529(1–3):103–13.

[26] Horton JD, Goldstein JL, Brown MS. SREBPs: activators of the complete program of cholesterol and fatty acid synthesis in the liver. J Clin Invest 2002;109(9):1125–31.

[27] Ness GC, Chambers CM. Feedback and hormonal regulation of hepatic 3-hydroxy-3-methylglutaryl coenzyme a reductase: the concept of cholesterol buffering capacity. Exp Biol Med 2000;224(1):8–19.

[28] Jensen-Urstad APL, Semenkovich CF. Fatty acid synthase and liver triglyceride metabolism: housekeeper or messenger? Biochim Biophys Acta BBA Mol Cell Biol Lipids 2012;1821(5):747–53.

[29] Takaki Y, et al. Silencing of microRNA-122 is an early event during hepatocarcinogenesis from non-alcoholic steatohepatitis. Cancer Sci 2014;105(10):1254–60.

[30] Wang X-C, et al. MicroRNA-185 regulates expression of lipid metabolism genes and improves insulin sensitivity in mice with non-alcoholic fatty liver disease. World J Gastroenterol WJG 2014;20(47):17914–23.

[31] Bugianesi E, et al. Insulin resistance in non-diabetic patients with non-alcoholic fatty liver disease: sites and mechanisms. Diabetologia 2005;48(4):634–42.

[32] Sanyal AJ, et al. Nonalcoholic steatohepatitis: association of insulin resistance and mitochondrial abnormalities. Gastroenterology 2001;120(5):1183–92.

[33] Mattis AN, et al. A screen in mice uncovers repression of lipoprotein lipase by microRNA-29a as a mechanism for lipid distribution away from the liver. Hepatology 2015;61(1):141–52.

[34] Ng R, et al. Inhibition of microRNA-24 expression in liver prevents hepatic lipid accumulation and hyperlipidemia. Hepatology 2014;60(2):554–64.

[35] Gong Y, et al. Sterol-regulated ubiquitination and degradation of Insig-1 creates a convergent mechanism for feedback control of cholesterol synthesis and uptake. Cell Metab 2006;3(1):15–24.

[36] Yang T, et al. Crucial step in cholesterol homeostasis: sterols promote binding of SCAP to INSIG-1, a membrane protein that facilitates retention of SREBPs in ER. Cell 2002;110(4):489–500.

[37] Brown MS, Goldstein JL. A proteolytic pathway that controls the cholesterol content of membranes, cells, and blood. Proc Natl Acad Sci USA 1999;96(20):11041–8.

[38] Engelking LJ, et al. Overexpression of Insig-1 in the livers of transgenic mice inhibits SREBP processing and reduces insulin-stimulated lipogenesis. J Clin Invest 2004;113(8):1168–75.

[39] Engelking LJ, et al. Schoenheimer effect explained – feedback regulation of cholesterol synthesis in mice mediated by Insig proteins. J Clin Invest 2005;115(9):2489–98.

[40] Zheng L, et al. Effect of miRNA-10b in regulating cellular steatosis level by targeting PPAR-α expression, a novel mechanism for the pathogenesis of NAFLD. J Gastroenterol Hepatol 2010;25(1):156–63.

[41] Castro RE, et al. miR-34a/SIRT1/p53 is suppressed by ursodeoxycholic acid in the rat liver and activated by disease severity in human non-alcoholic fatty liver disease. J Hepatol 2013;58(1):119–25.

[42] Min H-K, et al. Increased hepatic synthesis and dysregulation of cholesterol metabolism is associated with the severity of nonalcoholic fatty liver disease. Cell Metab 2012;15(5):665–74.

[43] Lee J, et al. A pathway involving farnesoid X receptor and small heterodimer partner positively regulates hepatic Sirtuin 1 levels via MicroRNA-34a inhibition. J Biol Chem 2010;285(17):12604–11.

[44] Yamakuchi M, Ferlito M, Lowenstein CJ. miR-34a repression of SIRT1 regulates apoptosis. Proc Natl Acad Sci USA 2008;105(36):13421–6.

[45] Chang T-C, et al. Transactivation of miR-34a by p53 broadly influences gene expression and promotes apoptosis. Mol Cell 2007;26(5):745–52.

[46] Panasiuk A, et al. Expression of p53, Bax and Bcl-2 proteins in hepatocytes in non-alcoholic fatty liver disease. World J Gastroenterol WJG 2006;12(38):6198–202.

[47] Farrell GC, et al. Apoptosis in experimental NASH is associated with p53 activation and TRAIL receptor expression. J Gastroenterol Hepatol 2009;24(3):443–52.

[48] Derdak Z, et al. Inhibition of p53 attenuates steatosis and liver injury in a mouse model of non-alcoholic fatty liver disease. J Hepatol 2013;58(4):785–91.

[49] Cazanave SC, et al. A role for miR-296 in the regulation of lipoapoptosis by targeting PUMA. J Lipid Res 2011;52(8):1517–25.

[50] Miller AM, et al. MiR-155 has a protective role in the development of non-alcoholic hepatosteatosis in mice. PLoS ONE 2013;8(8):e72324.

[51] Sharma H, et al. Expression of genes for microRNA-processing enzymes is altered in advanced non-alcoholic fatty liver disease. J Gastroenterol Hepatol 2013;28(8):1410–5.

[52] Wree A, et al. Obesity affects the liver – the link between adipocytes and hepatocytes. Digestion 2011;83(1–2):124–33.

[53] Yamada H, et al. Associations between circulating microRNAs (miR-21, miR-34a, miR-122 and miR-451) and non-alcoholic fatty liver. Clin Chim Acta 2013;424:99–103.

[54] Tryndyak VP, et al. Plasma microRNAs are sensitive indicators of inter-strain differences in the severity of liver injury induced in mice by a choline-and folate-deficient diet. Toxicol Appl Pharmacol 2012;262(1):52–9.

[55] Gerhard GS, DiStefano JK. Micro RNAs in the development of non-alcoholic fatty liver disease. World J Hepatol 2015;7(2):226.

[56] Ceccarelli S, et al. Dual role of microRNAs in NAFLD. Int J Mol Sci 2013;14(4): 8437–55.

[57] Volinia S, et al. A microRNA expression signature of human solid tumors defines cancer gene targets. Proc Natl Acad Sci USA 2006;103(7):2257–61.

[58] Tomimaru Y, et al. MicroRNA-21 induces resistance to the anti-tumour effect of interferon-[alpha]/5-fluorouracil in hepatocellular carcinoma cells. Br J Cancer 2010;103(10):1617–26.

[59] Karakatsanis A, et al. Expression of microRNAs, miR-21, miR-31, miR-122, miR-145, miR-146a, miR-200c, miR-221, miR-222, and miR-223 in patients with hepatocellular carcinoma or intrahepatic cholangiocarcinoma and its prognostic significance. Mol Carcinog 2013;52(4):297–303.

[60] Gramantieri L, et al. MicroRNA involvement in hepatocellular carcinoma. J Cell Mol Med 2008;12(6a):2189–204.

[61] Hu S, et al. MicroRNA-21 promotes cell proliferation in human hepatocellular carcinoma partly by targeting HEPN1. Tumor Biol 2015:1–6.

[62] Bao L, et al. MicroRNA-21 suppresses PTEN and hSulf-1 expression and promotes hepatocellular carcinoma progression through AKT/ERK pathways. Cancer Lett 2013;337(2):226–36.

[63] Rong M, et al. Expression and clinicopathological significance of miR-146a in hepatocellular carcinoma tissues. Upsala J Med Sci 2013;119(1):19–24.

[64] Zhang Z, et al. microRNA-146a inhibits cancer metastasis by downregulating VEGF through dual pathways in hepatocellular carcinoma. Mol Cancer 2015; 14(1):5.

[65] Murakami Y, et al. Comprehensive analysis of microRNA expression patterns in hepatocellular carcinoma and non-tumorous tissues. Oncogene 2005;25(17): 2537–45.

[66] Pineau P, et al. miR-221 overexpression contributes to liver tumorigenesis. Proc Natl Acad Sci USA 2010;107(1):264–9.

[67] Fornari F, et al. MiR-221 controls CDKN1C/p57 and CDKN1B/p27 expression in human hepatocellular carcinoma. Oncogene 2008;27(43):5651–61.

[68] Gramantieri L, et al. MicroRNA-221 targets Bmf in hepatocellular carcinoma and correlates with tumor multifocality. Clin Cancer Res 2009;15(16):5073–81.

[69] Bae HJ, et al. MicroRNA-221 governs tumor suppressor HDAC6 to potentiate malignant progression of liver cancer. J Hepatol 2015;63(2).

[70] Morishita A, Masaki T. miRNA in hepatocellular carcinoma. Hepatol Res 2015; 45(2):128–41.

[71] Jopling C. Liver-specific microRNA-122: biogenesis and function. RNA Biol 2012;9(2):137–42.

[72] Wang B, et al. Reciprocal regulation of miR-122 and c-Myc in hepatocellular cancer: role of E2F1 and TFDP2. Hepatol Baltim MD 2014;59(2):555–66.

[73] Coulouarn C, et al. Loss of miR-122 expression in liver cancer correlates with suppression of the hepatic phenotype and gain of metastatic properties. Oncogene 2009;28(40):3526–36.

[74] Zoni E, et al. Epithelial plasticity in cancer: unmasking a MicroRNA network for TGF-β-, notch-, and wnt-mediated EMT. J Oncol 2015;2015.

[75] Wang S-C, et al. MicroRNA-122 triggers mesenchymal-epithelial transition and suppresses hepatocellular carcinoma cell motility and invasion by targeting RhoA. PLoS ONE 2014;9(7):e101330.

[76] Ning B-F, et al. Hepatocyte nuclear factor 4α suppresses the development of hepatocellular carcinoma. Cancer Res 2010;70(19):7640–51.

[77] Liu S, et al. MicroRNA-135a contributes to the development of portal vein tumor thrombus by promoting metastasis in hepatocellular carcinoma. J Hepatol 2012; 56(2):389–96.

[78] Ding J, et al. Gain of miR-151 on chromosome 8q24.3 facilitates tumour cell migration and spreading through downregulating RhoGDIA. Nat Cell Biol 2010; 12(4):390–9.

[79] Tao J, et al. Distinct anti-oncogenic effect of various microRNAs in different mouse models of liver cancer. Oncotartget 2015;6(9):6977.

[80] Liu AM, et al. Global regulation on microRNA in hepatitis B virus-associated hepatocellular carcinoma. OMICS 2011;15(3):187–91.

[81] Petrini E, et al. MicroRNAs in HBV-related Hepatocellular Carcinoma: functions and potential clinical applications. Panminerva Med 2015;57(4).

[82] Wen Y, et al. Plasma miRNAs as early biomarkers for detecting hepatocellular carcinoma. Int J Cancer 2015;137(7).

[83] Kumar M, et al. Endogenous antiviral microRNAs determine permissiveness for hepatitis B virus replication in cultured human fetal and adult hepatocytes. J Med Virol 2015;87(7).

[84] Meng F-L, Wang W, Jia W-D. Diagnostic and prognostic significance of serum miR-24-3p in HBV-related hepatocellular carcinoma. Med Oncol 2014;31(9):1–6.

[85] Li J-F, et al. Upregulation of MicroRNA-146a by hepatitis B virus X protein contributes to hepatitis development by downregulating complement factor H. mBio 2015;6(2).

[86] Moolla N, Kew M, Arbuthnot P. Regulatory elements of hepatitis B virus transcription. J Viral Hepat 2002;9(5):323–31.

[87] Sarkar N, et al. Expression of microRNA-155 correlates positively with the expression of Toll-like receptor 7 and modulates hepatitis B virus via C/EBP-β in hepatocytes. J Viral Hepat 2015;22(10).

[88] Huang J-Y, et al. MicroRNA-130a can inhibit hepatitis B virus replication via targeting PGC1α and PPARγ. RNA 2015;21(3).

[89] Zhao F, et al. MicroRNA-26b inhibits hepatitis B virus transcription and replication by targeting the host factor CHORDC1 protein. J Biol Chem 2014;289(50):35029–41.

[90] van der Ree MH, et al. MicroRNAs: role and therapeutic targets in viral hepatitis. Antivir Ther 2013;19(6):533–41.

[91] Guo H, et al. MicroRNAs-372/373 promote the expression of hepatitis B virus through the targeting of nuclear factor I/B. Hepatology 2011;54(3):808–19.

[92] Zhang X, et al. Modulation of hepatitis B virus replication and hepatocyte differentiation by MicroRNA-1. Hepatology 2011;53(5):1476–85.

[93] Bandopadhyay M, et al. Tumor suppressor micro RNA miR-145 and onco micro RNAs miR-21 and miR-222 expressions are differentially modulated by hepatitis B virus X protein in malignant hepatocytes. BMC Cancer 2014;14(1):721.

[94] Li CH, et al. Hepatitis B virus X protein promotes hepatocellular carcinoma transformation through interleukin-6 activation of microRNA-21 expression. Eur J Cancer 2014;50(15):2560–9.

[95] Damania P, et al. Hepatitis B virus induces cell proliferation via HBx-induced microRNA-21 in hepatocellular carcinoma by targeting programmed cell death protein4 (PDCD4) and phosphatase and tensin homologue (PTEN). PLoS ONE 2014;9(3):e91745.

[96] Dai X, et al. Modulation of HBV replication by microRNA-15b through targeting hepatocyte nuclear factor 1α. Nucleic Acids Res 2014:gku260.

[97] Wu CS, et al. Downregulation of microRNA-15b by hepatitis B virus X enhances hepatocellular carcinoma proliferation via fucosyltransferase 2-induced Globo H expression. Int J Cancer 2014;134(7):1638–47.

[98] Chung GE, et al. High expression of microRNA-15b predicts a low risk of tumor recurrence following curative resection of hepatocellular carcinoma. Oncol Rep 2010;23(1):113–9.

[99] Song G, et al. Studying the association of microRNA-210 level with chronic hepatitis B progression. J Viral Hepat 2014;21(4):272–80.

[100] Giray BG, et al. Profiles of serum microRNAs; miR-125b-5p and miR223-3p serve as novel biomarkers for HBV-positive hepatocellular carcinoma. Mol Biol Rep 2014;41(7):4513–9.

[101] Shepard CW, Finelli L, Alter MJ. Global epidemiology of hepatitis C virus infection. Lancet Infect Dis 2005;5(9):558–67.

[102] Stoll-Keller F, et al. Development of hepatitis C virus vaccines: challenges and progress. Expert Rev Vaccines 2009;8(3).

[103] Thibault PA, et al. Regulation of hepatitis C virus genome replication by Xrn1, and microRNA-122 binding to individual sites in the 5′UTR. J Virol 2015;89(12):6294–311.

[104] Hou W, et al. MicroRNA-196 represses Bach1 protein and hepatitis C virus gene expression in human hepatoma cells expressing hepatitis C viral proteins. Hepatology 2010;51(5):1494–504.

[105] Murakami Y, et al. Regulation of the hepatitis C virus genome replication by miR-199a. J Hepatol 2009;50(3):453–60.

[106] Kałużna EM. MicroRNA-155 and microRNA-196b: promising biomarkers in hepatitis C virus infection? Rev Med Virol 2014;24(3):169–85.

[107] Zhang Y, et al. Hepatitis C virus-induced up-regulation of microRNA-155 promotes hepatocarcinogenesis by activating Wnt signaling. Hepatology 2012;56(5):1631–40.

[108] Bihrer V, et al. Serum microRNA-21 as marker for necroinflammation in hepatitis C patients with and without hepatocellular carcinoma. PLoS ONE 2011;6(10):e26971.

[109] Wang K, et al. Circulating microRNAs, potential biomarkers for drug-induced liver injury. Proc Natl Acad Sci USA 2009;106(11):4402–7.

[110] Antoine DJ, et al. Mechanistic biomarkers provide early and sensitive detection of acetaminophen-induced acute liver injury at first presentation to hospital. Hepatology 2013;58(2):777–87.

[111] Starkey Lewis PJ, et al. Circulating microRNAs as potential markers of human drug-induced liver injury. Hepatology 2011;54(5):1767–76.

[112] Dubin PH, et al. Micro-RNA-122 levels in acute liver failure and chronic hepatitis C. J Med Virol 2014;86(9):1507–14.

[113] John K, et al. MicroRNAs play a role in spontaneous recovery from acute liver failure. Hepatology 2014;60(4):1346–55.

[114] An F, et al. miR-15b and miR-16 regulate TNF mediated hepatocyte apoptosis via BCL2 in acute liver failure. Apoptosis 2012;17(7):702–16.

[115] Yu DS, et al. The regulatory role of microRNA-1187 in TNF-alpha-mediated hepatocyte apoptosis in acute liver failure. Int J Mol Med 2012;29(4):663–8.

[116] Sharma AD, et al. MicroRNA-221 regulates FAS-induced fulminant liver failure. Hepatology 2011;53(5):1651–61.

[117] Kamath PS, et al. A model to predict survival in patients with end-stage liver disease. Hepatology 2001;33(2):464–70.

[118] Michalopoulos GK. Liver regeneration. J Cell Physiol 2007;213(2):286–300.

[119] Salehi S, et al. Human liver regeneration is characterized by the coordinated expression of distinct MicroRNA governing cell cycle fate. Am J Transplant 2013;13(5):1282–95.

[120] Kwon YJ, Lee KG, Choi D. Clinical implications of advances in liver regeneration. Clin Mol Hepatol 2015;21(1):7–13.

[121] Riehle KJ, et al. New concepts in liver regeneration. J Gastroenterol Hepatol 2011;26:203–12.

[122] Song G, et al. MicroRNAs control hepatocyte proliferation during liver regeneration. Hepatology 2010;51(5):1735–43.

[123] Shu J, et al. Genomewide microRNA down-regulation as a negative feedback mechanism in the early phases of liver regeneration. Hepatology 2011;54(2):609–19.

[124] Cirera-Salinas D, et al. Mir-33 regulates cell proliferation and cell cycle progression. Cell Cycle 2012;11(5):922–33.

[125] Yu ZY, et al. Expression of microRNA-150 targeting vascular endothelial growth factor-A is downregulated under hypoxia during liver regeneration. Mol Med Rep 2013;8(1):287–93.

[126] Marquez RT, et al. MicroRNA-21 is upregulated during the proliferative phase of liver regeneration, targets Pellino-1, and inhibits NF-κB signaling. Am J Physiol Gastrointest Liver Physiol 2010;298:G535–41.

[127] Schug J, et al. Dynamic recruitment of microRNAs to their mRNA targets in the regenerating liver. BMC Genomics 2013;14:264.

[128] Ng R, et al. A microRNA-21 surge facilitates rapid cyclin D1 translation and cell cycle progression in mouse liver regeneration. J Clin Invest 2012;122(3):1097–108.

[129] Dippold RP, et al. Chronic ethanol feeding enhances miR-21 induction during liver regeneration while inhibiting proliferation in rats. Am J Physiol Gastrointest Liver Physiol 2012;303:G733–43.

[130] Castro RE, et al. Identification of microRNAs during rat liver regeneration after partial hepatectomy and modulation by ursodeoxycholic acid. Am J Physiol Gastrointest Liver Physiol 2010;299:G887–97.

[131] Chaveles I, et al. MicroRNA profiling in murine liver after partial hepatectomy. Int J Mol Med 2012;29(5):747–55.

[132] Chen H, et al. Mir-34a is upregulated during liver regeneration in rats and is associated with the suppression of hepatocyte proliferation. PLoS ONE 2011;6(5):e20238.

[133] Yuan B, et al. Down-regulation of miR-23b may contribute to activation of the TGF-β1/Smad3 signalling pathway during the termination stage of liver regeneration. FEBS Lett 2011;585(6):927–34.

[134] Yuan Q, et al. MicroRNA-221 overexpression accelerates hepatocyte proliferation during liver regeneration. Hepatology 2013;57(1):299–310.

[135] Bataller R, Brenner DA. Liver fibrosis. J Clin Invest 2005;115(2):209.

[136] Roy S, et al. The role of miRNAs in the regulation of inflammatory processes during hepatofibrogenesis. Hepatobiliary Surg Nutr 2015;4(1):24–33.

[137] Roderburg C, et al. Micro-RNA profiling reveals a role for miR-29 in human and murine liver fibrosis. Hepatology 2011;53(1):209–18.

[138] Kwiecinski M, et al. Hepatocyte growth factor (HGF) inhibits collagen I and IV synthesis in hepatic stellate cells by miRNA-29 induction. PLoS ONE 2011;6(9):e24568.

[139] Bandyopadhyay S, et al. Hepatitis C virus infection and hepatic stellate cell activation downregulate miR-29: miR-29 overexpression reduces hepatitis C viral abundance in culture. J Infect Dis 2011;203(12):1753–62.

[140] Hu J, et al. The role of miR-31/FIH1 pathway in TGFbeta-induced liver fibrosis. Clin Sci Lond 2015;129(4).

[141] Roderburg C, et al. miR-133a mediates TGF-β-dependent derepression of collagen synthesis in hepatic stellate cells during liver fibrosis. J Hepatol 2013;58(4):736–42.

[142] Meng F, et al. Epigenetic regulation of miR-34a expression in alcoholic liver injury. Am J Pathol 2012;181(3):804–17.

[143] Li W-Q, et al. The rno-miR-34 family is upregulated and targets ACSL1 in dimethylnitrosamine-induced hepatic fibrosis in rats. FEBS J 2011;278(9):1522–32.

[144] Zhao J, et al. MiR-21 simultaneously regulates ERK1 signaling in HSC activation and hepatocyte EMT in hepatic fibrosis. PLoS ONE 2014;9(10):e108005.

[145] Wei J, et al. MicroRNA-21 activates hepatic stellate cells via PTEN/Akt signaling. Biomed Pharmacother 2013;67(5):387–92.

[146] Ogawa T, et al. MicroRNA-221/222 upregulation indicates the activation of stellate cells and the progression of liver fibrosis. Gut 2012;61(11):1600–9.

[147] Murakami Y, et al. The progression of liver fibrosis is related with overexpression of the miR-199 and 200 families. PLoS ONE 2011;6(1):e16081.

[148] Mitchell PS, et al. Circulating microRNAs as stable blood-based markers for cancer detection. Proc Natl Acad Sci USA 2008;105(30):10513–8.

[149] Ebert MPA, et al. Advances, challenges, and limitations in serum-proteome-based cancer diagnosis. J Proteome Res 2006;5(1):19–25.

[150] Chen X, et al. Characterization of microRNAs in serum: a novel class of biomarkers for diagnosis of cancer and other diseases. Cell Res 2008;18(10):997–1006.

[151] Gilad S, et al. Serum MicroRNAs are promising novel biomarkers. PLoS ONE 2008;3(9):e3148.

[152] Pritchard CC, Cheng HH, Tewari M. MicroRNA profiling: approaches and considerations. Nat Rev Genet 2012;13(5):358–69.

[153] Chen C, et al. Real-time quantification of microRNAs by stem–loop RT–PCR. Nucleic Acids Res 2005;33(20):e179.

[154] Jiang J, et al. Real-time expression profiling of microRNA precursors in human cancer cell lines. Nucleic Acids Res 2005;33(17):5394–403.

[155] Schmittgen TD, et al. A high-throughput method to monitor the expression of microRNA precursors. Nucleic Acids Res 2004;32(4):e43.

[156] Tang F, et al. 220-plex microRNA expression profile of a single cell. Nat Protoc 2006;1(3):1154–9.

[157] Ach RA, Wang H, Curry B. Measuring microRNAs: comparisons of microarray and quantitative PCR measurements, and of different total RNA prep methods. BMC Biotechnol 2008;8(1):69.

[158] Wang Z. The guideline of the design and validation of MiRNA mimics. In: MicroRNA and cancer. Springer; 2011. p. 211–23.

[159] Wang Z. The guideline of the design and validation of MiRNA mimics. In: Wu W, editor. MicroRNA and cancer. Totowa, NJ: Humana Press; 2011. p. 211–23.

[160] Trang P, et al. Regression of murine lung tumors by the let-7 microRNA. Oncogene 2010;29(11):1580–7.

[161] Lanford RE, et al. Therapeutic silencing of MicroRNA-122 in primates with chronic hepatitis C virus infection. Science 2010;327(5962):198–201.

[162] Jopling CL. Modulation of hepatitis C virus RNA abundance by a liver-specific MicroRNA. Science 2005;309(5740):1577–81.

[163] Meister G, et al. Sequence-specific inhibition of microRNA-and siRNA-induced RNA silencing. RNA 2004;10(3):544–50.

[164] Stenvang J, et al. Inhibition of microRNA function by antimiR oligonucleotides. Silence 2012;3(1):1.

[165] Davis S, et al. Improved targeting of miRNA with antisense oligonucleotides. Nucleic Acids Res 2006;34(8):2294–304.

[166] Davis S, et al. Potent inhibition of microRNA in vivo without degradation. Nucleic Acids Res 2009;37(1):70–7.

[167] Esau CC. Inhibition of microRNA with antisense oligonucleotides. Methods 2008;44(1):55–60.

[168] Petersen M, Wengel J. LNA: a versatile tool for therapeutics and genomics. Trends Biotechnol 2003;21(2):74–81.

[169] Koshkin AA, et al. LNA (Locked Nucleic Acids): synthesis of the adenine, cytosine, guanine, 5-methylcytosine, thymine and uracil bicyclonucleoside monomers, oligomerisation, and unprecedented nucleic acid recognition. Tetrahedron 1998;54(14):3607–30.

[170] Braasch DA, Corey DR. Locked nucleic acid (LNA): fine-tuning the recognition of DNA and RNA. Chem Biol 2001;8(1):1–7.

[171] T-shirt GFB. Silencing of microRNAs in vivo with 'antagomirs'. Nature 2005; 438:685–9.

[172] Liu X, et al. HDL drug carriers for targeted therapy. Clin Chim Acta 2013;415: 94–100.

[173] Wolfrum C, et al. Mechanisms and optimization of in vivo delivery of lipophilic siRNAs. Nat Biotechnol 2007;25(10):1149–57.

[174] Valastyan S, et al. A pleiotropically acting microRNA, miR-31, inhibits breast cancer metastasis. Cell 2009;137(6):1032–46.

[175] Hansen TB, et al. Natural RNA circles function as efficient microRNA sponges. Nature 2013;495(7441):384–8.

[176] Ebert MS, Neilson JR, Sharp PA. MicroRNA sponges: competitive inhibitors of small RNAs in mammalian cells. Nat Methods 2007;4(9):721–6.

[177] van Rooij E, Kauppinen S. Development of microRNA therapeutics is coming of age. EMBO Mol Med 2014;6(7):851–64.

[178] van Rooij E, Olson EN. MicroRNA therapeutics for cardiovascular disease: opportunities and obstacles. Nat Rev Drug Discov 2012;11(11):860–72.

[179] Bader AG. miR-34–a microRNA replacement therapy is headed to the clinic. Front Genet 2012;3:120.

[180] Takamizawa J, et al. Reduced expression of the let-7 microRNAs in human lung cancers in association with shortened postoperative survival. Cancer Res 2004; 64(11):3753–6.

[181] Johnson CD, et al. The let-7 microRNA represses cell proliferation pathways in human cells. Cancer Res 2007;67(16):7713–22.

[182] Chen Z-Y, He C-Y, Kay MA. Improved production and purification of minicircle DNA vector free of plasmid bacterial sequences and capable of persistent transgene expression in vivo. Hum Gene Ther 2005;16(1):126–31.

[183] Lv H, et al. Toxicity of cationic lipids and cationic polymers in gene delivery. J Controlled Release 2006;114(1):100–9.

[184] Zhang Y, Wang Z, Gemeinhart RA. Progress in microRNA delivery. J Controlled Release 2013;172(3):962–74.

[185] Zhang Q, et al. Simultaneous delivery of therapeutic antagomirs with paclitaxel for the management of metastatic tumors by a pH-responsive anti-microbial peptide-mediated liposomal delivery system. J Controlled Release 2015;197:208–18.

[186] Wang R, Billone PS, Mullett WM. Nanomedicine in action: an overview of cancer nanomedicine on the market and in clinical trials. J Nanomater 2013;2013:1.

[187] Zhou J, et al. Nanoparticle-based delivery of RNAi therapeutics: progress and challenges. Pharmaceuticals 2013;6(1):85–107.

[188] Wang V, Wu W. MicroRNA-based therapeutics for cancer. BioDrugs 2009;23(1): 15–23.

[189] Janssen HL, et al. Treatment of HCV infection by targeting microRNA. N Engl J Med 2013;368(18):1685–94.

[190] Qiao C, et al. Liver-specific microRNA-122 target sequences incorporated in AAV vectors efficiently inhibits transgene expression in the liver. Gene Ther 2011;18(4): 403–10.

[191] Knabel MK, et al. Systemic delivery of scAAV8-encoded MiR-29a ameliorates hepatic fibrosis in carbon tetrachloride-treated mice. PLoS ONE 2015;18(4).

[192] Zheng F, et al. Systemic delivery of MicroRNA-101 potently inhibits hepatocellular carcinoma in vivo by repressing multiple targets. PLoS Genet 2015;11(2).

[193] Tu X, et al. MicroRNA-101 suppresses liver fibrosis by targeting the TGFβ signalling pathway. J Pathol 2014;234(1):46–59.

[194] Moshiri F, et al. Inhibiting the oncogenic mir-221 by microRNA sponge: toward microRNA-based therapeutics for hepatocellular carcinoma. Gastroenterol Hepatol Bed Bench 2014;7(1):43.

[194a] Kota J, Chivukula RR, O'Donnell KA, Wentzel EA, Montgomery CL, Hwang HW, Chang TC, Vivekanandan P, Torbenson M, Clark KR. Herapeutic delivery of miR-26a inhibits cancer cell proliferation and induces tumor-specific apoptosis. Cell 137 2009;1005.

[195] Singh R, Lillard JW. Nanoparticle-based targeted drug delivery. Exp Mol Pathol 2009;86(3):215–23.

[196] Wu Y, et al. Therapeutic delivery of microRNA-29b by cationic lipoplexes for lung cancer. Mol Ther Nucleic Acids 2013;2(4):e84.

[197] Sahoo SK, Labhasetwar V. Nanotech approaches to drug delivery and imaging. Drug Discov Today 2003;8(24):1112–20.

[198] Su J, et al. Silencing microRNA by interfering nanoparticles in mice. Nucleic Acids Res 2011;39(6):e38.

[199] Hsu S-H, et al. Cationic lipid nanoparticles for therapeutic delivery of siRNA and miRNA to murine liver tumor. Nanomed Nanotechnol Biol Med 2013;9(8):1169–80.

[200] Höbel S, Aigner A. Polyethylenimines for siRNA and miRNA delivery in vivo. Wiley Interdiscip Rev Nanomed Nanobiotechnol 2013;5(5):484–501.

[201] Chien Y, et al. Synergistic effects of carboxymethyl-hexanoyl chitosan, cationic poly-urethane-short branch PEI in miR122 gene delivery: accelerated differentiation of iPSCs into mature hepatocyte-like cells and improved stem cell therapy in a hepatic failure model. Acta Biomater 2015;13:228–44.

[202] Makadia HK, Siegel SJ. Poly lactic-co-glycolic acid (PLGA) as biodegradable con-trolled drug delivery carrier. Polymers 2011;3(3):1377–97.

[203] Wang J, et al. Delivery of siRNA therapeutics: barriers and carriers. AAPS J 2010; 12(4):492–503.

[204] Khan A, et al. Sustained polymeric delivery of gene silencing antisense ODNs, siRNA, DNAzymes and ribozymes: in vitro and in vivo studies. J Drug Target 2004;12(6):393–404.

[205] Babar IA, et al. Nanoparticle-based therapy in an in vivo microRNA-155 (miR-155)-dependent mouse model of lymphoma. Proc Natl Acad Sci USA 2012;109(26): E1695–704.

[206] Bader AG, Lammers P. The therapeutic potential of microRNAs. Innovations Pharm Technol 2011:52–5.

CHAPTER 6

MicroRNAs in Inflammatory Lung Disease

C. Bime, C.I. Gurguis, L. Hecker, A.A. Desai, T. Wang, J.G.N. Garcia

Key Concepts

This chapter updates current concepts and knowledge regarding the role of microR-NAs in development of inflammatory lung disease and the contribution to the severity of these disorders. This includes up to date information of microRNA expression and regulation in the major types of inflammatory lung diseases [asthma, ARDS, CF, chronic obstructive pulmonary disease (COPD), and sarcoidosis] with a summary of the clinical implications of microRNA studies in these diseases.

1. INTRODUCTION

Inflammatory lung disease is a big category of lung disease defined with clear lung inflammation features, including high neutrophil count. Typical inflammatory lung disease includes both chronic (eg, asthma) and acute (eg, acute respiratory distress syndrome or ARDS). Both acute and chronic inflammatory lung diseases are associated with airway inflammation and are influenced by a combination of environmental, genetic, and epigenetic components. Epigenetic regulation of gene expression, especially by microRNA (miRNA), significantly contributes to inflammatory lung disease development and severity.

MicroRNAs are small, endogenous noncoding RNAs with 20–25 nucleotides that negatively regulate gene expression at a posttranscriptional level by inducing mRNA degradation or translation inhibition. With the advancement of "omics" technology, miRNA expression and function in inflammatory lung diseases has been well studied, leading to a clear conclusion that miRNAs are differentially expressed under inflammatory lung disease conditions and contribute to genomic dysregulation in these diseases. Here we summarized the most up-to-date findings of miRNA in the inflammatory lung diseases including asthma, ARDS, COPD, cystic fibrosis (CF), and sarcoidosis, and clinical implications of miRNA in these diseases.

Translating MicroRNAs to the Clinic
ISBN 978-0-12-800553-8

2. MICRORNAs IN ASTHMA

Asthma, a common chronic inflammatory disorder of the airways that affects more than 250 million people worldwide, is characterized by episodic and reversible airflow obstruction, airway hyperresponsiveness (AHR), and underlying inflammation of the airways [1]. Even though asthma is defined in routine clinical care and research by reversible airflow obstruction and or AHR, the complex and heterogeneous nature of asthma is increasingly recognized. By using unbiased cluster analyses and linking specific clusters to distinct pathophysiological mechanisms, researchers now subdivide the asthma syndrome into distinct endotypes. In general, the pathogenesis of asthma is attributed to genetic risk factors as well as other environmental factors [1]. However, each endotype is driven by a distinct set of triggers, cell types, and cytokines. Allergic asthma, associated with atopic status, is characterized by a Th2-driven inflammation, which in the presence of allergen, leads to expression of high-affinity IgE receptors on inflammatory cells and mucosal infiltration with eosinophils, CD4$^+$ cells, mast cells. Intrinsic (nonatopic) asthma is also associated with an increase in Th-2 cells, activation of mast cells, and eosinophilic infiltration but allergens do not seem to drive the process and the role of local or systemic IgE is unclear. In noneosinophilic asthma, airway inflammation is characterized by a predominance of neutrophils. Aspirin-intolerant asthma is a severe form of asthma, characterized by an intense eosinophilic inflammation of the nasal and bronchial tissues. IgE does not appear to play a role and overproduction of cystenyl leukotrienes has been implicated as a major pathogenic feature. Another subgroup of asthmatics is characterized by extensive remodeling of the airways with minimal inflammation, hypertrophy of airway smooth muscles, thickening of the small airways, alveolar detachment, goblet cell hyperplasia, mucous production, and angiogenesis. Since asthma has a strong environmental component, miRNAs likely play a critical role in the pathological processes that determine the different asthma endotypes. Understanding the role of different miRNAs in determining disease susceptibility and response to therapy will improve our understanding of the different endotypes. This also has the potential to identify novel therapeutic targets. Here, we describe the miRNAs that have been identified to be differentially expressed in asthma and the genes targeted by these miRNAs. The translational implications of these findings are also discussed (Table 6.1).

2.1 Profiling of Differentially Expressed MicroRNAs in Asthma, Genes Targeted, and Translational Approaches

MiR-21. Interleukin-13 (IL-13), a Th2 cell-derived effector cytokine, plays a key role in the pathogenesis of allergic asthma. Lu and colleagues defined

miRNA	Sample type	Validation method	Predicted targeted genes	Translational implication	References
Upregulated miRNAs					
miR-21	Whole lung tissue, macrophages, dendritic cells (mice)	• Microarray • pRT-PCR • ISH	IL-12p35	• Represses IL-12 thus inhibiting T_H1 differentiation in favor of T_H2 polarization	[2,3]
miR-106a	• Macrophage—RAW267.7 (mice) • Lung cells and tissue (mice)	• qRT-PCR • Northern blotting	IL-10—a proinflammatory cytokine	• Represses expression of IL-10 • Anti-mmu-miR106a ameliorates airway inflammation, reduces T_H2 response and other asthma features	[14]
miR-126	Lung cells and lower airway tissue (mice)	• Microarray • qRT-PCR	• Oct binding factor 1 (OBF.1) • OBF.1 negatively regulates TLR4 and T_H2 expression	• Upregulates OBF.1 • Inhibition of miRNA-126 function suppresses airway hyperresponsiveness	[16,20]
miR-145	Lower airway tissue (mice)	• qRT-PCR	T_H2 cytokines IL-13, IL-5, IFN-γ	• House dust mite exposure increases expression of miR-145 in airways • Inhibition of miR-145, leads to reduction of eosinophilic inflammation, mucus production, T_H2 cytokine production	[4]
miR-146a, -146b, -150, -181a	Splenic CD4+ T lymphocytes (mice)	• qRT-PCR		• Increased expression in splenic CD4+ T cells of ova-sensitized mice compared to control • Dexamethasone reduces expression of miR-146a	[26]

Continued

Table 6.1 MicroRNAs in asthma—cont'd

miRNA	Sample type	Validation method	Predicted targeted genes	Translational implication	References
Upregulated miRNAs					
miR-155	• Cell lines—RAW264.7, THP-1 Macrophages	• Microarray • qRT-PCR • Transfections	Glucocorticoids (GCs)	Downregulation or inhibition of miR-155 enhances the antinflammatory effect of GCs	[27]
miR-221	• Mast cell line (mice) • Airway smooth muscle (human)	• qRT-PCR • Transfection assay	Multiple gene targets • cytoskeleton • p21^{WAF1}, p27^{kip1}	• Regulates cell cycle, cytoskeleton, degranulation, cytokine production, and cell adherence • Regulates enhanced proliferation and IL-6 release in severe asthma	[30,31]
Downregulated miRNAs					
Let-7 family (miR-98, let-7d, -7f, -7g, -7i)	• A549 cells • Cultured T-cells (Cells)	• qRT-PCR • Northern blots	3′-UTR of IL-13	Exogenous let-7 reduces IL-13 levels	[5]
miR-20b	Alveolar macrophages (mice)	• qRT-PCR • Transfection	VEGF	In alveolar macrophages downregulation of miR-20b leads to upregulation ofVEGF	[11]

miR–133a	• Bronchial smooth muscle (human) • Bronchial tissues (mice)	• qRT-PCR	RhoA	Expression level of RhoA is negatively regulated by endogenous miR–133a	[22]
miR–146a, –164b, –28–5p	• Circulating T cells (human)	• Microarray • qRT-PCR	Genes involved in activation of CD4(+) and CD8(+) cells	• Several mRNAs upregulated in CD8(+) cells in patients with severe asthma • Likely mediated by differential expression of miR–28–5p	[29]
Differentially expressed					
Let-7 (a-e); miR–200 (200b, 141) families	Exosomes from human BAL fluid (human)	• Microarray • qRT-PCR	244 target genes. (IL–6, IL–8, IL–10, IL–13)	Significant alteration of MAPK, JAK, and STAT pathways	[6]

a signature of 21 differentially expressed miRNAs in whole lung tissue using an IL-13-induced murine model of allergic asthma with miR-21 as the most upregulated miRNA [2,3]. They further identified IL-12p35 as a target gene for miR-21 with induction of miR-21 expression reducing IL-12p35 expression. IL-12 is a key cytokine involved in adaptive immune responses that involve Th1 cell polarization and IL-12 levels are decreased in experimental asthma models. The ability of miR-21 to downregulate IL-12 indicates an important potential therapeutic target for future investigations. Inhibition of miR-21 could drive Th1 polarization whereas administering miR-21 may prime Th2- and IL-13 responses. Attempts have been made to use antimiR-21 to modulate allergic asthma responses. Collison and colleagues tested, unsuccessfully, if intranasal antimiR-21 administration to sensitized mice 1 day prior to aeroallergen challenge with house dust mite (HDM) would reduce eosinophil recruitment or Th2 cytokine production [4]. The lack of effect could be because the antimiR-21 was administered after the Th1 versus Th2 polarization had already occurred. Other limitations to using antagomirs include lack of specificity and difference in effects depending on mode or route of administration. MiR-21 plays a key role in the pathogenesis of allergic asthma since an increase in its expression of miR-21 in lung tissue likely contributes to the action of IL-13. It is a potential target for more translational research into the pathogenesis of allergic asthma and possible therapeutic targets.

Let-7 family (miR-98, let-7d, -7f, -7g, -7i). Computational analyses identified 11 miRNAs with potential binding to the IL-13 3'UTR [5] with five miRNAs belonging to the let-7 family (miR-98, let-7d, -7f, -7g, and -7i). Experimental validation was performed in A549 cells and T cells isolated from human peripheral blood mononuclear cells (PBMCs), demonstrating that let-7 miRNAs target IL-13 3'UTR, resulting in inhibition of IL-13. An exogenous let-7 mimic, delivered via the intranasal route to the lungs of murine model of allergic airway inflammation significantly reduced levels of IL-13 in the lung tissue and bronchoalveolar lavage (BAL) fluid that were associated with a reduction in AHR and airway infiltration by inflammatory cells [5]. Consistent with the important role of the let-7 family of miRNAs in asthma, isolated exosomes from BAL fluid obtained from 10 normal controls and 10 asthma patients revealed significant differences in baseline levels of 24 exosomal miRNAs between healthy controls and asthmatics including let-7 family members and the miRNA-200 family (200b, 141) [6]. Collectively, these findings demonstrate the important role of let-7 miRNA family in the posttranslational regulation of IL-13 and highlight the potential therapeutic role of miRNAs in asthma and potentially other inflammatory lung diseases.

MiR-20b. Vascular endothelial growth factor (VEGF) is a potent proangiogenic factor that stimulates endothelial cell (EC) migration and proliferation [7], and plays an important role in the pathophysiology of asthma [8]. Compared to controls, asthma patients have increased VEGF levels in induced sputum, BAL fluid, and bronchial biopsies [9]. In murine models, overexpression of VEGF induces an asthma-like phenotype [9,10]. Alveolar macrophages (AMs) appear to be an important source of VEGF and AM-derived VEGF was required for allergic airway inflammation in asthma [11]. In these studies quantitative RT-PCR showed expression of miR-20b was fourfold lower in AMs of asthmatic mice compared to control mice and high miR-20b expression was inversely correlated with the expression of VEGF. An inhibitor to miR-20b in AM of control mice increased VEGF protein levels. Transfection of AM from asthmatic mice with miR-20b resulted in a significantly decreased VEGF levels. These findings suggest a potential therapeutic role for targeting miR-20b and VEGF in asthma [11].

MiR-106a. Interleukin-10 (IL-10) is an antiinflammatory cytokine and its secretion is defective in many inflammatory diseases including asthma [12]. In a murine model, miR-106a inhibits IL-10 expression in lymphoid and myeloid cells [13] and inhibition of mmu-miR-106a with anti-mmu-miR-106a increased IL-10 levels and alleviated features of allergic airway inflammation [14].

MiR-126. Toll-like receptors (TLRs) are important for the innate immune response but also play an important role in sensing molecular patterns and in programming adaptive T-cell responses [15]. It has been suggested that signals derived from early activation of the innate immune system play an important role in the generation of aberrant Th2 responses [16]. It is also known that activation of TLRs by microbial and viral bioproducts can control the expression of miRNAs [17,18]. In a well-characterized model of HDM-induced allergic asthma, a selective set of miRNAs, including miR-126, were rapidly unregulated in the airway wall after allergen exposure [19]. Increased miRNA expression and the development of HDM-induced Th2-mediated allergic inflammation were dependent on the TLR4/myD88 pathway. The specific role of miR-126 was assessed by exposing airways to a specific antisense inhibitor, antimiR-126, resulting in abolished expression of miR-126 in response to HDM and complete suppression of AHR, attenuation of mucus hypersecretion, and inhibition of eosinophil accumulation in the airways and lung tissue. In a well-established model of chronic asthma, compared to scrambled control, antagomirs to miR-126 (antimiR-126) significantly reduced recruitment of intraepithelial eosinophils [20].

MiR-133a. RhoA is a GTP-binding protein that plays a key role in the contraction of smooth muscles [21] and Rho upregulation is associated with increased contractility of bronchial smooth muscle (BSM) in animal models of allergic asthma [21]. MiR-133a negatively regulates RhoA expression in cardiomyocytes and miR-133a inhibition caused upregulation of RhoA in cultured hBSMCs. Transfection of hBSMCs with premiR-133a caused a downregulation of RhoA, and treatment of hBSMCs with IL-13 caused upregulation of RhoA [22]. These findings indicate that the expression of RhoA in hBSMCs is negatively regulated by miR-133a and downregulation of miR-133a causes an upregulation of RhoA [23]. Mir-133a is therefore a potential therapeutic target for airway hypersensitivity in allergic asthma [24,25].

MiR-145. HDM challenge increases the expression of miR-145, miR-21, and let-7b in murine airways [4] with inhibition of miR-145 suppressing eosinophilic inflammation induced by HDM [4] in a manner equivalent to treatment with dexamethasone. Inhibition of miR-145 prior to HDM challenge significantly reduced AHR, more so than the effect of dexamethasone [4]. These results again highlight the promise of specific miRNA and antagomir as novel targeted antiinflammatory therapies in asthma.

Other important miRNAs in asthma (miR-146a, -146b, -150, -155, -164b, -181a, -221, -28-5p). Th2 polarization is important in allergic asthma and miR-146a, -146b, -150, -155, and -181a participate in Th2 inflammation of asthma as detected in splenic CD4$^+$ T lymphocytes in murine models of acute asthma [26]. MiR-155 also appears important in regulating the immune response as glucocorticoids inhibit the expression of miR-155 in a GC-receptor and NF-κB-dependent manner [27,28] suggesting that miR-155 may be a novel target through which glucocorticoids exert their antiinflammatory effect on the LPS-induced macrophage inflammatory response [27]. Transcriptomic analysis of circulating T cells showed significant changes in circulating CD8$^+$ T cells in severe asthma that correlated with downregulation of miR-164a/b and miR28-5p [29]. Though intriguing, the translational implications of these findings remain unclear [29]. The role of miR-221 in regulating airway smooth muscle proliferation and activity in severe asthma has been explored [30,31]. MiR-221 regulates IL-6 release and enhanced proliferation in severe asthma [30], and may play a role in the regulation of cell cycle and cytoskeleton in stimulated mast cells [31]. Taken together, targeting miR-221 expression might lead to improved airflow obstruction and better asthma control [30,32].

3. ACUTE RESPIRATORY DISTRESS SYNDROME

The ARDS is a clinical syndrome characterized by acute onset (within 1 week of a known clinical insult or new or worsening respiratory symptoms) of refractory hypoxemia with bilateral infiltrates on chest imaging that is not fully explained by cardiac failure or fluid overload [33]. ARDS affects more than 200,000 patients annually and has a mortality that approaches 40% in the United States [34]. ARDS is a heterogeneous syndrome in terms of the disorders associated with its development (sepsis, pneumonia, aspiration of gastric contents, blood transfusion, major trauma, etc.), the severity, and progression of disease [33,35]. The classic pathological finding in ARDS is a combination of widespread epithelial injury with denuding of the alveolar epithelium, endothelial injury leading to interstitial and alveolar edema, accumulation of neutrophils, macrophages, red blood cells, surfactant dysfunction, and deposition of hyaline membranes in the alveoli [35]. As the disease progresses, there is infiltration of fibroblasts and collagen deposition [35]. The physiologic consequences of these changes include impaired gas exchange, decreased lung compliance, and increased work of breathing. There are several pathways involved in the development of ARDS but a key feature is inflammation due to activation of neutrophils. Several proinflammatory mediators such as tumor necrosis factor (TNF), IL-1, and IL-6 are involved in ARDS pathogenesis [36]. There is an imbalance between proinflammatory and antiinflammatory cytokines, oxidants and antioxidants, procoagulants and anticoagulants, proteases and protease inhibitors, and neutrophil recruitment and clearance. This similar pathogenesis process is observed in sepsis, which is one of the major causes of ARDS [37]. There is a complex interplay between the innate and adaptive immune responses in ARDS. The role of miRNAs in regulating the molecular mechanisms that underlie the pathogenesis of ARDS is now being appreciated. An understanding of the role of miRNAs in regulating lung inflammation presents an attractive approach to control inflammation that occurs in lung injury. The susceptibility to and mortality from ARDS are influenced by genetic and environmental factors [35]. This underscores the importance of studying the role of miRNA–gene interactions in determining risk and the potential for new therapeutic options.

3.1 Profiling of Differentially Expressed MicroRNAs and Genes Targeted in ARDS

Several miRNAs have been implicated in immune responses and ARDS. For example, miR-146a/b has an important role in the regulation of TLR and cytokine signaling, a key feature of the innate immune response [18] and in endotoxin-induced tolerance [38]. Expression of miR-9 is increased by TLR4-activated NF-κB and operates a feedback control to fine-tune the expression of key elements of the innate immune system in ARDS [39]. MicroRNA-148 and MiR-152 regulate IL-12, IL-6, and TNF-α production [40]. MiR-21 is an important negative regulator of TLR4 signaling by targeting of programmed cell death protein 4 (PDCD4), a proinflammatory protein that promotes that activation of NF-κB and suppresses IL-10. Mice deficient in PDCD4 are protected from LPS-induced cell death [41] (Table 6.2).

Simultaneous miRNA and mRNA microarray analyses in a rat model of ARDS induced by saline lavage and mechanical ventilation, identified miRNA–mRNA interactions [42]. Early B-cell factor 1 (Ebf10) inversely correlated with miR-26a, and superoxide dismutase 2 (SOD2) inversely correlated with let-7a, b, c, and f. Many miRNAs target more than one mRNA and several mRNAs were targeted by multiple miRNAs confirming the complexity of the miRNA–mRNA interaction in the regulation of ARDS [43]. More than 50% of the upregulated genes in ARDS were involved in cellular homeostasis, more than 33% of the genes that are upregulated were involved in regulation of apoptosis [42].

A cardinal feature of ARDS and sepsis is increased vascular permeability in the lung [44] due to loss of the integrity of the EC barrier. We previously demonstrated that the nonmuscle myosin light chain kinase isoform (nmMLCK, *MYLK*) is critical to pulmonary EC barrier regulation [45]. Using Targetscan and *in Silico* bioinformatics tools, we identified several miRNAs as potential regulators of the *MYLK* gene expression: miR-374a, miR-374b, miR-529c-3p, and miR-1290 [44]. These findings indicate that the collective action of these miRNAs on nmMLCk represent a potential novel therapeutic target for reducing lung vascular permeability in ARDS [44]. Another outstanding candidate gene in ARDS and ventilator-induced lung injury (VILI), which can cause or worsen ARDS [46], is pre-B-cell colony-enhancing factor (NAMPT/PBEF aka *NAMPT*), a proinflammatory cytokine, which is implicated in the pathogenesis of ARDS and VILI [47]. We showed posttranslational regulation of PBEF/NAMPT is epigenetically regulated by miR-374a and miR-568 [48]. Other studies

Table 6.2 MicroRNAs in acute respiratory distress syndrome (ARDS)

miRNA	Sample type	Validation method	Predicted targeted genes	Translational implications	References
Upregulated miRNAs					
Let-7; miR-21; -146, -155	Lung tissue (mice)	Taqman Loe density arrays and qRT-PCR	IL-6, SMAD, SOCS	Many miRNAs target more than one mRNA and several mRNAs were targeted by multiple miRNAs	[49]
miR-32; -466-5p; -466-3p	Alveolar epithelia cells (rat)	qRT-PCR		Increased permeability induced by cyclic stretch was associated with up- or downregulation of many miRNAs. Inhibition of miR-466d-5p and miR-466f-3p attenuated this permeability	[50]
miR-146a	Lung macro-phage cell line		TNF-α, IL-6, IL-1β, IRAK-1, TRAF-6	Upregulation of miR-146a suppresses inflammatory mediators in LPS induced-ALI mode and therefore miR-146a may be therapeutically targeted as a mean to repress inflammatory response following ALI	[51,52]

Continued

Table 6.2 MicroRNAs in acute respiratory distress syndrome (ARDS)—cont'd

miRNA	Sample type	Validation method	Predicted targeted genes	Translational implications	References
Upregulated miRNAs					
miR–181b	Cell line	qRT-PCR	NF–κB signaling	Treatment of mice with proinflammatory stimuli reduced miR–181b expression and miR–181b inhibition exacerbated endotoxin-induced NF–κB activity, leukocyte influx, and lung injury	[53]
Down-regulated miRNAs					
miR–127	Mouse macrophage cell line	Luciferase assay and microarray	IgG FcγRI (CD64)	MiR–127 directly targeted 3'-UTR of CD64, resulting in the downregulation of CD64	[54]
Differentially expressed					
Let–7a, b, c, f miR–26a	Rat model of ARDS	Microarray and qRT-PCR	Gabrb1, Sod2, Eif2ak1, Fbln5	Highlights the complexity of the miRNA–mRNA interaction in the regulation of ARDS	[42]

demonstrated that several miRNAs were differentially expressed between high tidal volume ventilation (HTV)-treated mice and controls and that the miRNA expression was altered according to the length of HTV treatment [49]. Many of these differentially expressed miRNAs are associated with regulation of inflammation.

In summary, miRNAs are an attractive biomarker in ARDS because their expression patterns are dynamic and reflect the changing intracellular and extracellular environments. They are stable over time and can be detected in a variety of sources. As with nmMLCK and NAMPT/NAMPT/PBEF, specific miRNAs such as miR-374a, represent promising novel therapeutic strategies to reduce lung vascular permeability in ARDS.

4. CHRONIC OBSTRUCTIVE PULMONARY DISEASE

COPD is a complex chronic inflammatory lung disease characterized by poor airflow and breathing-related problems [55]. COPD, an umbrella term which includes chronic bronchitis, emphysema, and overlaps with asthma, is a leading cause of death worldwide [55]. Although smoking is a primary cause of COPD, the etiology of the disease is diverse. Additionally, the economic costs of COPD represent a substantial economic burden [56]. Therefore, developing and improving approaches to treatment of COPD represent a major problem in healthcare today. Understanding the molecular basis of the disease and dysregulation of molecular networks involved in COPD is critical to this goal.

Recently, miRNAs have received a great deal of attention for their role in the molecular pathology and disease progression of COPD [57]. Many miRNAs are dysregulated in COPD, and have been frequently implicated in inflammation [58–64]. For example, 11 separate miRNAs have TLR4 as a predicted gene target, suggesting many opportunities for dysregulation of inflammatory responses as TLR4 is implicated in lung inflammation via NF-κB signaling [65]. Similarly, let-7, miR-328, and miR-21 were implicated in regulating many interleukins (IL-1b, IL-5, IL-8, IL-13, and IL-17) [64], while differentially expressed miR-145-5p leads to eosinophil recruitment via Th2 response [63].

miRNAs have also been implicated in alterations to muscle dysfunction [59,66,67], oxidative stress [68], and antiapoptotic or proliferative effects [61,68]. Hsa-miR-133 [66,67] has been implicated in muscle's ability to adapt to overload on the diaphragm and hsa-miR-206 in induction of myoblast differentiation. These results suggest a role of miRNAs' role in mediating compensatory responses to disease burden. MiR-638 is highly pleiotropic and implicated in a number of oxidative response pathways [68] by targeting *CARHSP1*. In addition, miR-638 regulates a number of genes (*ADAM15, ARHGDIA, COMMD1, DHCR7, HDAC5, MAD2L2, PFKL, YPEL3*) that function in regulating cell proliferation [68]. These examples illustrate the fact that miRNAs are involved in a large number of functions related to disease progression of COPD.

miRNAs may also serve as potential biomarkers for COPD progression as dysregulation of miRNAs successfully predicts staging of COPD based on the Global Initiative for Chronic Obstructive Lung Disease (GOLD) classification system. GOLD staging (I–IV) describes progression of COPD on the basis of forced expiratory volume in 1 s (FEV1) and complications (stage IV is marked by chronic respiratory failure or right heart failure). Patterns of dysregulation are also useful for distinguishing COPD patients from healthy controls, COPD smokers from COPD nonsmokers, and COPD smokers from non-COPD smokers [63,69]. Together, these patterns of dysregulation in miRNAs may provide useful prognostic tools for clinicians [70,71].

A wide variety miRNAs are upregulated in COPD, suggesting that they are promising, actionable therapeutic targets for the disease. For example, miR-7 is upregulated in airway smooth muscle cells leading to downregulation of Epac1 and enhanced inflammatory response. Thus, targeting miR-7 to alleviate inflammation in COPD is a potential therapeutic strategy [59]. To date, however, there are no clinical trials or drugs under investigation through the FDA that would target miRNAs in COPD. More work is needed to identify and investigate potential miRNAs as therapeutic targets in COPD.

Most miRNAs, however, are highly pleiotropic and are predicted to target a wide variety of genes and functions. Differences in expression of miRNAs have been observed across sample types (eg, in miR-342-3p in plasma [71], BAL fluid [72], and induced sputum supernant [73]), and dysregulation of miRNAs is not necessarily consistent during progression of COPD [60,66,67]. These complexities offer both an opportunity to more fully understand the basic biology of miRNAs and a challenge to developing effective therapies for the disease. For example, miR-1 was found to be upregulated in human plasma in COPD [60], but downregulated in lung diaphragm muscles [66,67]. These differential effects would make targeting miR-1 with therapeutics more difficult, despite clear functions in the disease process.

Network approaches are likely to provide crucial insights into the extent of effects of dysregulated miRNAs [58,68]. Future longitudinal studies will be useful for providing a complete understanding of how miRNAs are linked to disease progression and how patterns of miRNA expression and regulation of genes change over time. Further, studies should examine dysregulation of miRNAs in more cell types (eg, endothelium in addition to parenchymal cells) as many studies have focused on lung fibroblasts (Table 6.3).

5. CYSTIC FIBROSIS

CF is the most common genetic disorder among Caucasians affecting 1 in every 3000 newborns [82,83]. CF is defined by autosomal recessive mutations of the CF transmembrane conductance regulator (CFTR) gene, which encodes a chloride channel transcript expressed in several tissues, most prominently in specialized epithelia of the airways and in select exocrine tissues [82,83]. Although defective CFTR affects many organs (such as lungs, pancreas, liver, and the reproductive system), progressive airway tract obstruction, ultimately resulting in respiratory failure, is the leading cause of death, contributing up to 85% of the mortality [82–84]. CFTR conducts Cl^-, HCO_3^-, and other anions regulating the composition and volume of airway surface liquid and essential for maintaining sweat, digestive juices, and mucus ion balance (Table 6.4).

While more than 1500 pathogenetic variants have been described in CF, F508del is the most prevalent [85]. CFTR mutations can manifest in multiple organs including lung inflammation with vulnerability to infections, pancreatic insufficiency, intestinal obstruction, and male infertility [85]. Thick copious airway mucus is the most common clinical sequelae with recurrent inflammation, chronic microbial infections, followed by chronic bacterial colonization with highly antibiotic-resistant *Pseudomonas aeruginosa* communities and ultimately, by injury and deterioration of lung functions. Despite harboring a Mendelian mutation, CF is characterized by significant phenotypic heterogeneity including differences in life span.

A number of factors appear to impact this clinical variability such as genotype, environmental factors, epigenetics, and modifier genes; recently, much attention has been focused on epigenetic signaling with miRNA regulation [86]. Recently work unique miRNA profiles were identified in the CF lung and analysis of target genes revealed important roles for miRNAs in the regulation of CFTR expression and function and inflammation [83].

Table 6.3 MicroRNAs in chronic obstructive pulmonary disease

miRNA	Sample type	Validation method	Predicted genes targeted	Translational implications	References
Upregulated miRNAs					
miR–7	Airway smooth muscle cells (human)	RT-PCR	Epac1	Downregulation of Epac1, leading to inflammation	[59]
Hsa-miR–1208, hsa-miR–943	Lung fibroblasts (human)	Hybridization and microarray	TLR4, TGFB2	Dysregulation of proinflammatory responses	[58]
Composite of 8 miRNAs	Lung fibroblasts (human)	Hybridization and microarray	TLR4	Dysregulation of proinflammatory responses	[58]
Hsa-miR–199b–5p, hsa-miR–328	Lung fibroblasts (human)	Hybridization and microarray	TGFB2	Dysregulation of proinflammatory responses	[58]
Hsa-miR–885–3p	Lung fibroblasts (human)	Hybridization and microarray	TLR4, HHIP	Dysregulation of proinflammatory responses	[58]
Hsa-miR–92b	Lung fibroblasts (human)	Hybridization and microarray	TGFB2, HHIP	Dysregulation of proinflammatory responses	[58]
Hsa-miR–1184, hsa-miR–224	Lung fibroblasts (human)	Hybridization and microarray	RASSF2, IL1R1, PTGS1, TGFBR3	Dysregulation of proinflammatory responses	[58]
Hsa-miR–27a-star	Lung fibroblasts (human)	Hybridization and microarray	TRPA1, RASSF2, PTGS1, TGFBR3, PLAU, FGL2	Dysregulation of proinflammatory responses	[58]
miR–34c	Bronchial epithelial cells (human) and fetal lung fibroblasts	qRT-PCR and microarray	SERPINE1, MAP4K4, ZNF3, ALDOA, HNF4A	Upregulation of mi-34c downregulates SERPINE1	[74]

(human)		microarray	DIA, COMMD1, DHCR7, HDAC5, MAD2L2, PFKL, YPEL3, ATG9A, GANAB, DHCR7, ERAL1, SLC25A1, STARD3, TOMM40, APBB1, KRT7, MAD2L2, CARHSP1, LTBP4	autophagy and protein degradation, mitochondrial function, DNA damage response, oxidative stress response, and ECM remodeling	
miR-1	Plasma (human)	qRT-PCR	N/A	Modestly, negatively correlated with FEV_1	[60]
miR-499	Plasma (human)	qRT-PCR	NF-κB p50, TNFα, IL2, IL5	Dysregulation of proinflammatory responses, circulating cytokines	[60]
miR-133, miR-206	Plasma (human), serum (rat)	qRT-PCR	TNFα, IL2, IL5	Dysregulation of proinflammatory responses, circulating cytokines	[60]
miR-101	Lung tissue (human)	qRT-PCR	CFTR	Upregulation of miR-101 suppresses CFTR protein	[75]
miR-7	Plasma (human)	qRT-PCR	N/A	Potential diagnostic biomarkers for COPD	[76]
miR-34a	Lung tissue (human)	qRT-PCR	pAKT	Upregulation of miR-34a leads to lowered expression of pAKT	[70]

Continued

Table 6.3 MicroRNAs in chronic obstructive pulmonary disease—cont'd

miRNA	Sample type	Validation method	Predicted genes targeted	Translational implications	References
Upregulated miRNAs					
miR–199a–5p	Lung tissue (human)	qRT-PCR	HIF1α, pAKT	Upregulation of miR–199a–5p leads to lowered expression of HIF1α, and lowered expression of pAKT	[70]
Composite of 10 miRNAs	Lung tissue (human)	qRT-PCR and microarray	SMAD7	Potential biomarkers; dysregulation of miR–15b leads to dysregulation of TGF-β via its effects on SMAD7	[68]
miR–183	Lung fibroblasts (human)	qRT-PCR	IDH2	Increased expression of miR–183 decreases expression of IDH2	[77]
miR–21	Serum (human)	Microarray	RECK	Increased levels of miR–21 can enhance MMP activity or inhibit RECK expression, promoting MMP-9 expression	[61]
Composite of 10 miRNAs	BAL fluid (human)	Reverse transcription and microassay	N/A	Potential diagnostic biomarkers for COPD; distinguish adenocarcinoma from COPD	[72]

Downregulated miRNAs

Composite of 9 miRNAs	Plasma (human)	qRT-PCR and microarray	N/A	Potential diagnostic biomarkers for COPD	[78]
Hsa-miR–1	Diaphragm muscle (human)	Reverse transcription and qRT-PCR	IGF1, SRF, HDAC4, IGF1	Promotes cell differentiation and innervation, negatively associated with phosphorylation of the kinase Akt	[66,67,79]
Hsa-miR–133a	Diaphragm muscle (human)	Reverse transcription and qRT-PCR	SRF	May promote adaptation to overload in diaphragm muscles	[66,67,74]
Hsa-miR–206	Diaphragm muscle (human)	Reverse transcription and qRT-PCR	N/A	Promotes cell differentiation and innervation, induces myoblast differentiation	[66,67]
Composite of 7 miRNAs	Plasma (human)	qRT-PCR and microarray	N/A	Distinguish COPD from heart failure patients and from other breathless patients	[71]
miR–206	Serum (rat) and pulmonary artery smooth muscle cells (rat)	qRT-PCR	HIF1α, Fhl-1	Decrease in miR–206 leading to increase in HIF1α and Fhl-1 proteins, cell proliferation also enhanced	[80]

Continued

Table 6.3 MicroRNAs in chronic obstructive pulmonary disease—cont'd

miRNA	Sample type	Validation method	Predicted genes targeted	Translational implications	References
Downregulated miRNAs					
miR–20a, miR–28-3p, miR–34c-5p, miR–100	Plasma (human)	qRT-PCR	N/A	Potential diagnostic biomarkers for COPD	[76]
Composite of 10 miRNAs	Lung tissue (human)	qRT-PCR and microarray	N/A	N/A	[68]
miR–181a	Serum (human)	Microarray	N/A	Decreased expression in miR–181a promotes antiapoptotic activity, proinflammatory responses	[61]
Composite of 9 miRNAs	BAL fluid (human)	Reverse transcription and microassay	N/A	Potential diagnostic biomarkers for COPD; distinguish adenocarcinoma from COPD	[72]
miR–146a	Lung fibroblasts (human)	RT-PCR	PGE2, COX-2	Decreased expression in miR146a prolongs COX-2 mRNA half-life; dysregulation of proinflammatory responses	[62]

Composite of 7 miRNAs	Induced sputum supernatant (human)	qRT-PCR	N/A	Possible diagnostic biomarkers for COPD; these were also downregulated in cigarette-exposed mice	[73]
Hsa-let-7c	Induced sputum supernatant (human)	qRT-PCR	TNFR-II	let7c is inversely correlated with sTNFR-II	[73]
Differentially expressed					
miR-1229-3p	Plasma (human)	Microarray	FGFBP3	Proinflammatory responses	[63]
miR-145-5p	Plasma (human)	Microarray	IFI30, FGFBP3	Th2 response and eosinophil recruitment, reduction of proinflammatory cytokines	[63]
miR-338-3p	Plasma (human)	Microarray	LTK, IFITM1, TNFRSF1b, IGF2R	Proinflammatory responses; also known to regulate AATK mRNA, apoptosis, differentiation, and tissue degeneration	[63]
miR-3620-3p	Plasma (human)	Microarray	TNFRSF1b, IGF2R, FGFBP3, GGT6	Proinflammatory responses	[63]

Continued

Table 6.3 MicroRNAs in chronic obstructive pulmonary disease—cont'd

miRNA	Sample type	Validation method	Predicted genes targeted	Translational implications	References
Differentially expressed					
miR–4707–3p	Plasma (human)	Microarray	TNFRSF1b, IGF2R, GGT6	Proinflammatory responses	[63]
miR–4485	Plasma (human)	Microarray	N/A	N/A	[63]
miR–636	Plasma (human)	Microarray	IGF2R	Proinflammatory responses	[63]
Let–7a	Exhaled breath condensates (human)	qRT-PCR	IL-13, TGF-b receptor, TLR4	Proinflammatory responses	[64]
miR–328	Exhaled breath condensates (human)	qRT-PCR	IL-13, IL-5, IL-1b, IL-8, TLR2	Proinflammatory responses	[64]
miR–21	Exhaled breath condensates (human)	qRT-PCR	IL-13 receptor a subunit, IL-17, STAT3, IL-1b	Proinflammatory responses	[64]
Composite of 18 miRNAs	Lung fibroblasts (human)	Microarray	N/A	Study demonstrated certain dysfunctions in miRNAs from COPD lung fibroblasts are invariant to inflammatory responses mediated by IL-1β and TNF-α	[81]

Table 6.4 MicroRNAs in cystic fibrosis

miRNA	Sample type	Validation method	Validated targeted genes	Translational implications	References
Upregulated					
miR–155	• ΔF508–CFTR and wild type–CFTR lung epithelial cell lines • CF neutrophils	• miRNA expression array, qRT-PCR	SHIP1	• CF lung epithelium, hyperexpress miR–155 • Elevated miR–155 stabilizes IL–8 in CF lung epithelium through loss of SHIP1, activated PI3K/Akt signaling, and downstream activation of MAPK signaling	[97]
24 total miRNAs	• Ileum tissue (Gitract) from WT versus BALBc/J Cftr^m1UNC (BALB CF) mice	• miRNA array, qRT-PCR	Hundreds of putative targets; none functionally validated	• 24 miRNAs significantly differentially expressed in tissue from CF mice compared to wild type • Overlap of genes with decreased expression in the CF intestine to that of genes putatively targeted by the 24 miRNAs revealed 155 genes (20.4%) to overlap with predicted target	[98]
miR–155, –215	CF lung epithelium	• qRT-PCR	IL–8	Reduced miR–155 suppresses IL–8	[99]

Continued

Table 6.4 MicroRNAs in cystic fibrosis—cont'd

miRNA	Sample type	Validation method	Validated targeted genes	Translational implications	References
Upregulated					
miR–145, miR–223, and miR–494	• CF bronchial brushings cells • 16HBE14o⁻ and ΔF508 homozygous CFBE41o⁻ cell lines • HEK293 cells	• qRT-PCR • luciferase assay	CFTR	• Overexpression of premiR–145, –223, and –494 decreases CFTR mRNA and protein expression in 16HBE14o⁻ cells • Knockdown of miR–145, miR–223, and miR–494 enhances ΔF508 CFTR expression • Functional inhibition of CFTR increases expression of miR–145, –223, and –494 • Inhibition or overexpression of miR–145, miR–223, and miR–494 alters intracellular chloride levels in CF and non-CF cells • Increased expression of miR–145, miR–223, miR–494 attributed to *Pseudomonas* colonization within the CF lung	[100]
miR–155	CF lung epithelial cells	qRT-PCR	Not relevant	• KSRP and TTP have an antagonistic role in miR–155 biogenesis • TTP overexpression in CF lung epithelium suppresses miR–155 • TTP induces miR–1 expression, a regulator of miR–155 biogenesis in CF lung epithelial cells	[101]

Continued

miRNA	Sample	Method	Target	Findings	Ref
hsa-let-7e/ hsa-miR-125a	expressing cells	RT-PCR	Not relevant	• Three SNPs identified in the hsa-miR-99b/hsa-let-7e/hsa-miR-125a cluster hsa-miR-99b and -125a were significantly increased in F508-CFTR expressing cells	[93]
miR-199a-5p	Human and murine CF macrophages and murine CF lungs	RT-PCR	Cav1	• Reduced CAV1 with increased TLR4 signaling caused by high miR-199a-5p levels, which are PI3K/AKT-dependent • Downregulation of miR-199a-5p or increased AKT signaling restores CAV1 expression • Celecoxib reestablishes the AKT/miR-199a-5p/CAV1 axis in CF macrophages, and ameliorates lung hyperinflammation in Cftr-deficient mice	[90]
miR-122, -25, -21, -210, -148a, -19a	• Circulating serum miRNA levels in patients with CF liver disease	RT-PCR	Not relevant	• miR-122 elevated in CF liver disease; miR-25 and miR-21 elevated in patients without CF liver disease • 6 miRNAs (-122, -21, -25, -210, -148a, -19a) distinguished F0 from F3–F4 fibrosis	[94]
miR-101, miR-145, miR-384	• Primary human nasal epithelial cells from healthy individuals and CF patients	RT-PCR	CFTR	• FOXA, C/EBP and miRs (-101, -145, -384) regulate switch from strong fetal to low CFTR expression after birth • miR-101 directly acts on its cognate site in CFTR-3′UTR • miRNA-binding blocker oligonucleotides rescued CFTR channel activity/expression	[89]

Table 6.4 MicroRNAs in cystic fibrosis—cont'd

miRNA	Sample type	Validation method	Validated targeted genes	Translational implications	References
Upregulated					
miR-509-3p	• A549 human lung cancer cells	• RT-pCR • Luciferase assay	CFTR	• 7-base long peptide nucleic acid-RNA hybrid complementary to the seed region of miR-509-3p reverts the expression • Potential of small PNAs as effective antimiRNA agents	[102]
miR-509-3p and miR-494	• CF airway epithelia	RT-PCR	CFTR	• Two miRNAs act cooperatively in regulating CFTR expression • Inflammatory mediators (TNF, IL-1B, staph infection) regulate these miRNAs • Expression of both miRNAs was responsive to NF-κB signaling	[88]
miR-145 and miR-494	Nasal epithelial tissues from CF patients	qRT-PCR Luciferase assay	CFTR–miR494 SMAD3-mir-145	• miR-145 synthetic mimics suppressed SMAD3 3′-UTR expression • miR-494 levels showed a trend of correlation with reduced CFTR expression	[85]

Downregulated

miR–126	• Bronchial epithelial cells brushed from patients with CF • Airway epithelial cell line from CF	TaqMan low-density arrays and qRT-PCR	TOM1 (target of Myb1, a validated miR–126 target)	• miR–126 expression is regulated by ER stress (induced by thapsigargin) • Knockdown of TOM1 increases IL–8 protein production in response to LPS, IL–1β, or lipopeptide in CFBE41o⁻ cells • Overexpression of Tom1 inhibits LPS- or IL–1β-induced NF-κB reporter gene expression in CFBE41o⁻ cells	[103]
miR–126	• CF airway epithelial cells,	• RT-PCR	TOM1	• Polymer polyethyleneimine (PEI) more effective than chitosan in knockdown of TOM1 after complexing with miR–126 as nanoparticles at low nitrogen/phosphate (N/P) ratios	[104]
miR–31	• CF airway epithelial cells	qRT-PCR	IRF-1	• CF airway epithelial cells express and secrete significantly more Cathepsin S (CTSS) than non-CF control cells in the absence of proinflammatory stimulation • IRF-1 correlated with increased levels of its target CTSS • miR–31 mimic decreased IRF-1 protein levels with concomitant knockdown of CTSS expression and secretion	[91]

Continued

Table 6.4 MicroRNAs in cystic fibrosis—cont'd

miRNA	Sample type	Validation method	Validated targeted genes	Translational implications	References
Unknown					
Hsa-miR-101, -145, -331-3p, -376b, -377, -384, -494, -600, -607, -939, -1246, -1290, -1827	• Caco-2 colon carcinoma cells and PANC-1 pancreatic adenocarcinoma cells • 16HBE14o⁻ bronchial epithelial cells • Primary human airway epithelial cells • HT29 (human colorectal carcinoma cell line)	qRT-PCR	SLC12A2 (NKCC1) and CFTR	• hsa-miR-145 and hsa-miR-494, which regulate CFTR expression by directly targeting discrete sites in the CFTR 3′ UTR • At least 12 miRNAs are capable of repressing endogenous CFTR mRNA expression in the Caco-2 cell line • At least three are expressed in primary human airway epithelial cells • sa-miR-384, miR-494 and miR-1246, inhibit Na⁺ − K⁺ − Cl⁻ cotransporter SLC12A2	[105]
miR-101 and miR-494	HEK293 cell line	Luciferase assay	CFTR	• Both miR-101 and miR-494 synthetic mimics significantly inhibited the expression of a reporter construct containing the 3′-UTR of CFTR in luciferase assays	[106]

miR–125b	Preimplantation embryos (2 cell stage) from CFTR-knockout mice	qRT-PCR	p53 and p21	• HCO_3^- regulates miR–125b expression through CFTR–mediated influx during preimplantation embryo development • Knockdown of miR–125b mimics the effect of HCO_3^- removal and CFTR inhibition, while injection of miR–125b precursor reverses it • Downregulation of miR–125b upregulates p53 cascade in both human and mouse embryos • The activation of miR–125b is shown to be mediated by sAC/PKA-dependent nuclear shuttling of NF-κB	[107]
miR–193b	PC–3 and LNcap cell lines	Not tested directly	Urokinase plasminogen activator (uPA)	• Overexpression of CFTR suppresses uPA by upregulating the recently described tumor suppressor miR–193b (miR–193b), and overexpression of premiR–193b significantly reverses CFTR knockdown–enhanced malignant phenotype and abrogates elevated uPA activity in prostate cancer cell line	[108]

Continued

Table 6.4 MicroRNAs in cystic fibrosis—cont'd

miRNA	Sample type	Validation method	Validated targeted genes	Translational implications	References
Unknown					
miR–138	• Primary cultures of human airway epithelia • Calu-3 cell line • HeLa cells • Primary human *CFTR* null airway epithelia	Luciferase assay qRT-pCR	*SIN3A* (*SIN3 homolog A*)	• mir–138 regulates CFTR expression and function by relieving SIN3A–mediated repression • miR–138 mimic increased *CFTR* mRNA and also enhanced CFTR abundance and transepithelial Cl⁻ permeability independent of elevated mRNA levels • miR–138 altered expression of genes encoding proteins associated with CFTR and may influence its biosynthesis • Manipulating the miR–138 regulatory network improved biosynthesis of CFTR–ΔF508 and restored Cl⁻ transport to cystic fibrosis airway epithelia	[87]
miR–101 and miR–144	• Human airway epithelial cells • mice lungs	qRT-pCR Luciferase assay	*CFTR*	• Cigarette smoke and cadmium upregulate the expression of two miRNAs (miR–101 and miR–144) • Mice exposed to cigarette smoke upregulate miR–101 and suppress CFTR protein in their lungs • miR–101 is highly expressed in lung samples from patients with severe chronic obstructive pulmonary disease	[109]
15 composite gene miRNA partners	Human airway epithelium	–?	Multiple	• 9 out of 15 composite feed-forward loops confirmed in patient samples and CF epithelial cells lines	[110]

While miRNAs have been predicted to have thousands of gene targets, both novel and established gene targets (including CFTR directly) and signaling cascades have been observed in CF [83].

A recent study profiled miRNA expression in well-differentiated primary cultures of human CF and non-CF airway epithelia, and discovered that miR-509-3p and miR-494 levels were increased in CF epithelia [87,88]. Human non-CF airway epithelia, transfected with the mimics of miR-509-3p or miR-494, showed decreased CFTR expression, whereas their respective antimiRs exerted the opposite effect. Upon infecting non-CF airway epithelial cells with *Staphylococcus aureus*, or upon stimulating them with the proinflammatory cytokines TNF-α or IL-1β, increased expression of both miRNAs and a concurrent decrease in CFTR expression and function were revealed, suggesting that inflammatory mediators may regulate these miRNAs. Transfecting epithelia with antimiRs for miR-509-3p and miR-494, or inhibiting NF-κB signaling before stimulating cells with TNFα or IL-1β, suppressed these responses, suggesting that the expression of both miRNAs was responsive to NF-κB signaling. Thus, miR-509-3p and miR-494 are dynamic regulators of CFTR abundance and function in normal, non-CF airway epithelia [88].

Similarly, both miR-145 and miR-494 are significantly upregulated in nasal epithelial tissues from CF patients compared with healthy controls [85]. Only miR-494 levels showed a trend of correlation with reduced CFTR mRNA expression and positive sweat test values, supporting the negative regulatory role of this miRNA on CFTR synthesis. Using in silico analyses and luciferase reporter assays, SMAD family member 3 (SMAD3), a key element of the TGF-β1 inflammatory pathway, was identified as a target of miR-145 with 40% suppression of SMAD3 3′-UTR reporter by miR-145 synthetic mimics [85].

miRNAs are posttranscriptional regulators that are expressed in a tissue-specific or developmental stage-specific manner, thereby greatly contributing to cell/tissue-specific protein expression profiles, including during lung organogenesis [89]. In fact, temporal expression in CF has also been reported for miRNAs. For example, while miR-101 and miR-145, negatively regulate CFTR transcription in adult lung cells, they have no effect in fetal lung cells. In addition, miR-101 decreases luciferase activity in an embryonic kidney cell line, whereas it does not affect CFTR mRNA stability in pancreatic cell lines, suggesting a potential developmental- and tissue-specific role [89].

Macrophages play a key role in CF inflammation. Preclinical models utilizing LPS administration in CF mice have provided evidence of this observation with reduced levels of the scaffold protein caveolin 1 (CAV1) with subsequent uncontrolled TLR4 signaling [90]. A recent study found that reduced CAV1 and, consequently, increased TLR4 signaling, in human and murine CF macrophages and murine CF lungs, is caused by high miR-199a-5p levels, that are PI3K/AKT-dependent. Downregulation of miR-199a-5p or increased AKT signaling led to increased CAV1 levels with reduced inflammation in CF macrophages. Celecoxib, an FDA-approved drug reduced lung inflammation in CFTR-deficient mice via restoration of the AKT/miR-199a-5p/CAV1 axis in CF macrophages [90].

Free and active proteases also contribute to the destruction of lung tissue and other soluble proteins in the respiratory tract. An imbalance of protease–antiprotease activity has been observed in pediatric CF with persistent abnormal levels sustained with age. Similar to the role of neutrophil elastase in progressive lung inflammation and destruction, cathepsins including cathepsin S (CTSS) have been implicated in CF [91]. In addition to antigen processing, CTSS promotes remodeling of the extracellular matrix via its potent elastinolytic activity. CTSS also inactivates antimicrobials in the CF airways including surfactant protein, lactoferrin, and members of the **b**-defensin family [91]. Importantly, a recent study has demonstrated the role of miR-31 in altering CTSS levels via regulation of IRF-1. IRF-1 is a transcription factor that plays an important role in the regulation of cellular responses in host defense, such as the innate and adaptive immune responses, antigen presentation, and cellular apoptosis [91]. In fact, evidence of increased IRF-1 expression has been reported in CF blood neutrophils. With basic mimic studies in CF bronchial epithelial cells, IRF-1 and CTSS expression and secretion levels were both reduced. Altered miR-31 expression has been reported in diseases such as cancer and psoriasis, and a number of target genes have been identified including RhoA, PP2A regulatory subunit B a isoform, and serine/threonine kinase 40. MiR-31 is also one of the most highly expressed miRNAs present in human non-CF airway epithelial cells and data demonstrate that miR-31 expression is significantly decreased in CF airway epithelial cell lines and primary cells [91].

Similar to other human genes, miRNA expression can be altered by several mechanisms, such as chromosomal abnormalities, mutations, defects in their biogenesis machinery, epigenetic silencing (DNA methylation, histone modification), or by transcription factors. MiR-31 expression is regulated by several distinct signaling networks in an intricate lineage- and cell type–dependent manner. MiR-31 expression levels can be enhanced by TNF, bone morphogenetic protein-2, and transforming growth factor-**b**1 and by the transcription factor C/EBP-**b** [91]. Other work has found that miR-31 is subject to epigenetic and nonepigenetic silencing in different human breast cancer cell lines. In addition, reduced miR-31 levels may arise because of the defective posttranscriptional processing of the miR-31 RNA precursor rather than transcriptional repression of the miR-31 gene itself. With these common emerging themes across disease states and expression regulation, mir-31 becomes an attractive and novel CF target in the proteolytic pathway [91].

Given the central role of *P. aeruginosa* in CF, another study evaluated the impact of an infected state in several CF relevant cell lines. Using luciferase assays, they reported that miR-93 [84], which is highly expressed in basal conditions, decreases during infection in parallel with increased expression of the IL-8 gene. Sequence analysis showed that the 3'-UTR region of IL-8 mRNA is a potential target of miR-93 and that the consensus sequence is highly conserved throughout molecular evolution. Upmodulation of IL-8 after *P. aeruginosa* infection was counteracted in cells by premiR-93 transfection. In addition, IL-8 was upregulated in uninfected cells treated with antagomiR-93 [84].

The presence of single nucleotide polymorphisms (SNPs) in the 3' untranslated region (3'UTR) may also be responsible for a variable phenotype from disease severity to organ specificity. SNPs may influence several different regulatory epigenetic mechanisms involving miRNAs. They may directly modify the expression of miRNAs, whose target sequence is present in the 3'UTR of a target CF gene. Alternatively, as supported by a previous report, SNPs may be associated with a decrease of CFTR protein expression that, in turn, increases the affinity for a specific miRNA in the 3'UTR of the CFTR gene [92]. These data suggest that the molecular analysis of the 3'UTR region should be performed in all cases in which clinical symptoms are present but no mutation has been identified [92].

Another study evaluated a cohort of F508del patients for potential SNPs in the hsa-miR-99b/hsa-let-7e/hsa-miR-125a cluster (19q13.2–19q13.4 region) [93]. Three SNPs were identified in the cluster. Based on in vitro results, they found that hsa-miR-99b and hsa-miR-125a were significantly increased in F508del-CFTR expressing cells. While derived from the same precursor, their expression levels differed suggesting differential maturation of these miRNAs in CF. In silico analysis revealed that two out of the three SNPs could modulate miRNA maturation affecting hsa-miR-99b/hsa-let-7e/hsa-miR-125a expression levels [93].

More recent work has reported on the utility of miRNA therapeutics, with a focus on clinical translation and application. For example, one report designed miRNA-binding blocker oligonucleotides (MBBOs) that target the miRNA-binding sites in the CFTR 3′UTR instead of the miRNA itself [89]. Specifically, the report demonstrated that lung development-specific transcription factors (FOXA, C/EBP) and miRNAs (miR-101, miR-145, miR-384) regulate the switch from strong fetal to very low CFTR expression after birth. Using miRNA profiling and gene reporter assays, miR-101 and miR-145 were specifically upregulated in adult lung and that miR-101 directly acts on its cognate site in the CFTR-3′UTR in combination with an overlapping AU-rich element [89]. MBBO-based inhibition led to the correction of CFTR channel activity through CFTR mRNA stabilization and subsequent increase in CFTR protein levels in nasal epithelial cells from patients homozygous for F508del mutation. Given the association of reductions in miR-101 and miR-145 with cancer progression and lung cancer, the specificity of MBBOs to block the cognate CFTR mRNA motif may have potential therapeutic benefits with stabilization of CFTR transcription and providing adequate functional protein to improve patients' disease phenotype without disturbing other signaling cascades [89].

Additionally, given their pleiotropic impact on many genes/pathways, miRNA literature also supports their utility as a biomarker for severity of and organ involvement in CF. One study profiled 84 circulating serum miRNA levels in patients with CF liver disease (CFLD, $n = 52$), patients with CF without liver disease (CFnoLD, $n = 30$), and non-CF pediatric controls ($n = 20$) [94]. Seven candidate miRNAs identified were validated by reverse transcription-quantitative polymerase chain reaction, normalizing data to geNorm-determined stable reference genes, miR-19b and miR-93. MiR-122 was significantly elevated in patients with CFLD versus patients with CFnoLD and controls; in contrast, miR-25 ($P = .0011$) and miR-21 were elevated in patients with CFnoLD versus patients with CFLD and controls. CFLD was discriminated by both miR-122 [area under the curve (AUC) 0.71] and miR-25 (AUC 0.65). Logistic regression combining three miRNAs (-122, -25, -21) was greatly predictive of detecting CFLD (AUC 0.78). A combination of six miRNAs (-122, -21, -25, -210, -148a, -19a) distinguished F0 from F3–F4 fibrosis (AUC 0.73), and miR-210 combined with miR-22 distinguished F0 fibrosis from any fibrosis, that is, F1–F4 (AUC 0.72). These biomarker studies provide the first glimpse of stratifying subphenotypes of CF via circulating serum miRNA levels [94].

In summary, evidence for the role of miRNA function in CF is growing rapidly highlighting miRNAs both as biomarkers of disease and as pathophysiolgically relevant. With the advent of targeted or "personalized" molecular therapies such as Ivacaftor [95] and Lumacaftor [96] for variant-specific CF, the potential for miRNA as a novel therapeutic in CF is only now being realized.

6. SARCOIDOSIS

Sarcoidosis, a systemic inflammatory syndrome of unknown cause, is characterized by granulomatous inflammation in multiple organs but with >90% occurrence in the lung (pulmonary sarcoidosis) [111,112]. Though sarcoidosis may be asymptomatic and/or chronic, some patients may develop complications with gradual damage to lung, which is designated as complicated sarcoidosis. In serious cases, sarcoidosis leads to scar tissue in lung, which may significantly impair pulmonary function. In addition, significant racial and gender differences in disease development and prognosis have also been reported (eg, highest annual incidence in African-American females) [113]. Several protein-coding genes (eg, *MMP12* and *ADAMDEC1* [114]) and inflammatory pathways (eg, T-cell signaling pathway [115] and interferon signaling pathway [116]) have been associated with sarcoidosis pathology. However, how these genes/pathways interact with miRNA has been largely unexplored.

In recent years, several groups attempted to use genomic approaches to identify miRNAs implicated in sarcoidosis. Using customized miRNA microarray, Crouser et al. identified 60 miRNAs differentially expressed in lung tissues between controls and patients with active pulmonary sarcoidosis [117]. This miRNA array analysis also identified 214 deregulated miRNAs in PBMCs from sarcoidosis patients [117]. However, no significant overlap was observed between lung tissue and PBMC deregulated miRNAs [117]. An intensive study indicated that a subgroup of the deregulated miRNAs in lung tissues potentially target transforming growth factor (TGF-β) signaling pathways, while the deregulated miRNAs identified in PBMCs are predicted to regulate the TGF-β-regulated "wingless and integrase-1" (WNT) pathway [117]. Both TGF-β and WNT pathways have been previously incriminated in the pathogenesis of sarcoidosis [117]. In another study, miRNA expression in whole blood from sarcoidosis and tuberculosis (TB) patients was examined by Agilent Human miRNA Microarray. In total, 189 miRNAs were identified as differentially expressed between sarcoidosis patients and healthy controls, while 191 miRNAs were differentially expressed between TB patients and healthy controls [118]. The miRNA expression signatures in sarcoidosis show highly similar patterns in TB, with a significant overlap of 145 miRNAs [118]. The most strongly upregulated miRNA in both diseases is miR-144, an miRNA potentially regulating cellular response to oxidative stress [118], whose passenger strand, miR-144*, was found to modulate T-cell cytokine production [118,119]. Four miRNAs (ie, miR-182, miR-355, miR-15b*, miR-340) were differentially expressed between sarcoidosis and TB [118].

All the above evidences suggest that miRNAs may serve as biomarkers with diagnostic applications in sarcoidosis. Accordingly, we evaluated the discriminative performance of the miRNA expression in a published data set from Saarland University (GEO accession number: GSE26409), in which genome-wide miRNA expression data from peripheral blood are available for 55 healthy controls and 45 sarcoidosis patients. Principal component analysis (PCA) indicated that genome-wide miRNA expression significantly distinguish sarcoidosis patients from healthy controls (Fig. 6.1A). Next, a classification score was assigned to each subjects based on the first principal component of PCA [120]. We found that the classification score can distinguish sarcoidosis patients from healthy controls with a fairly good accuracy: the area under the receiver operating characteristic (ROC) curve (AUC) was 0.822 (Fig. 6.1B). These results suggest that profiling

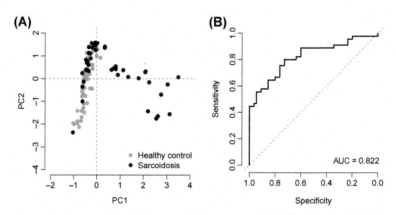

Figure 6.1 *Classification performance of genome-wide miRNA expression in sarcoidosis and controls.* (A) Principal component analysis (PCA) on miRNA expression. *X*-axis: the first principal component; *Y*-axis: the second principal component. (B) The receiver operating characteristic curve of the classification score. A classification score was assigned to each subjects based on the value of the first principal component of PCA.

whole genome miRNA expression could provide an opportunity to explore miRNA-based biomarkers associated with sarcoidosis and advance precision medicine into clinical practice.

7. SUMMARY AND CONCLUSION

Extensive studies have been carried out to characterize miRNA expression and function in inflammatory lung diseases. miRNAs are valued as biomarkers, diagnostic tools, as well as therapeutic targets of inflammatory lung diseases. Although progress has been made in this field (eg, differential expression profiling of miRNA in inflammatory lung diseases), translational research and develop of miRNA-based diagnostic/therapeutic tools are still in immense need.

ACKNOWLEDGMENT

This study is supported by National Institutes of Health P01HL126609 and R01HL125615 (JGNG), R01HL91899 (JGNG and TW).

REFERENCES

[1] Martinez FD, Vercelli D. Asthma. Lancet 2013;382(9901):1360–72.
[2] Lu TX, Munitz A, Rothenberg ME. MicroRNA-21 is up-regulated in allergic airway inflammation and regulates IL-12p35 expression. J Immunol 2009;182(8):4994–5002.
[3] Lu TX, Rothenberg ME. Diagnostic, functional, and therapeutic roles of microRNA in allergic diseases. J Allergy Clin Immunol 2013;132(1):3–13. quiz 14.
[4] Collison A, Mattes J, Plank M, Foster PS. Inhibition of house dust mite-induced allergic airways disease by antagonism of microRNA-145 is comparable to glucocorticoid treatment. J Allergy Clin Immunol 2011;128(1):160–7. e164.
[5] Kumar M, Ahmad T, Sharma A, et al. Let-7 microRNA-mediated regulation of IL-13 and allergic airway inflammation. J Allergy Clin Immunol 2011;128(5):1077–85. e1–10.
[6] Levanen B, Bhakta NR, Torregrosa Paredes P, et al. Altered microRNA profiles in bronchoalveolar lavage fluid exosomes in asthmatic patients. J Allergy Clin Immunol 2013;131(3):894–903.
[7] Maniscalco WM, Watkins RH, D'Angio CT, Ryan RM. Hyperoxic injury decreases alveolar epithelial cell expression of vascular endothelial growth factor (VEGF) in neonatal rabbit lung. Am J Respir Cell Mol Biol 1997;16(5):557–67.
[8] Kristan SS, Malovrh MM, Silar M, et al. Airway angiogenesis in patients with rhinitis and controlled asthma. Clin Exp Allergy 2009;39(3):354–60.
[9] Lee CG, Link H, Baluk P, et al. Vascular endothelial growth factor (VEGF) induces remodeling and enhances TH2-mediated sensitization and inflammation in the lung. Nat Med 2004;10(10):1095–103.
[10] Kanazawa H. VEGF, angiopoietin-1 and -2 in bronchial asthma: new molecular targets in airway angiogenesis and microvascular remodeling. Recent Pat Inflamm Allergy Drug Discov 2007;1(1):1–8.

[11] Song C, Ma H,Yao C,Tao X, Gan H.Alveolar macrophage-derived vascular endothelial growth factor contributes to allergic airway inflammation in a mouse asthma model. Scand J Immunol 2012;75(6):599–605.

[12] Asadullah K, Sterry W,Volk HD. Interleukin-10 therapy—review of a new approach. Pharmacol Rev 2003;55(2):241–69.

[13] Sharma A, Kumar M, Aich J, et al. Posttranscriptional regulation of interleukin-10 expression by hsa-miR-106a. Proc Natl Acad Sci 2009;106(14):5761–6.

[14] Sharma A, Kumar M, Ahmad T, et al. Antagonism of mmu-mir-106a attenuates asthma features in allergic murine model. J Appl Physiol (1985) 2012;113(3):459–64.

[15] Medzhitov R. Toll-like receptors and innate immunity. Nat Rev Immunol 2001;1(2):135–45.

[16] Mattes J, Collison A, Plank M, Phipps S, Foster PS. Antagonism of microRNA-126 suppresses the effector function of TH2 cells and the development of allergic airways disease. Proc Natl Acad Sci USA 2009;106(44):18704–9.

[17] O'Connell RM, Taganov KD, Boldin MP, Cheng G, Baltimore D. MicroRNA-155 is induced during the macrophage inflammatory response. Proc Natl Acad Sci 2007;104(5):1604–9.

[18] Taganov KD, Boldin MP, Chang K-J, Baltimore D. NF-κB-dependent induction of microRNA miR-146, an inhibitor targeted to signaling proteins of innate immune responses. Proc Natl Acad Sci 2006;103(33):12481–6.

[19] Mattes J,Yang M, Foster PS. Regulation of microRNA by antagomirs: a new class of pharmacological antagonists for the specific regulation of gene function? Am J Respir Cell Mol Biol 2007;36(1):8–12.

[20] Collison A, Herbert C, Siegle JS, Mattes J, Foster PS, Kumar RK. Altered expression of microRNA in the airway wall in chronic asthma: miR-126 as a potential therapeutic target. BMC Pulm Med 2011;11:29.

[21] Chiba Y, Ueno A, Shinozaki K, et al. Involvement of RhoA-mediated Ca^{2+} sensitization in antigen-induced bronchial smooth muscle hyperresponsiveness in mice. Respir Res 2005;6:4.

[22] Chiba Y, Tanabe M, Goto K, Sakai H, Misawa M. Down-regulation of miR-133a contributes to up-regulation of Rhoa in bronchial smooth muscle cells. Am J Respir Crit Care Med 2009;180(8):713–9.

[23] Chiba Y, Misawa M. MicroRNAs and their therapeutic potential for human diseases: MiR-133a and bronchial smooth muscle hyperresponsiveness in asthma. J Pharmacol Sci 2010;114(3):264–8.

[24] Chiba Y, Misawa M. MicroRNA miR-133a and bronchial smooth muscle hyperresponsiveness in allergic bronchial asthma. J Pharmacol Sci 2010;112(Suppl. 1). 24P.

[25] Chiba Y, Matsusue K, Misawa M. RhoA, a possible target for treatment of airway hyperresponsiveness in bronchial asthma. J Pharmacol Sci 2010;114(3):239–47.

[26] Feng MJ, Shi F, Qiu C, Peng WK. MicroRNA-181a, -146a and -146b in spleen $CD4^+$ T lymphocytes play proinflammatory roles in a murine model of asthma. Int Immunopharmacol 2012;13(3):347–53.

[27] Zheng Y, Xiong S, Jiang P, et al. Glucocorticoids inhibit lipopolysaccharide-mediated inflammatory response by downregulating microRNA-155: a novel anti-inflammation mechanism. Free Radic Biol Med 2012;52(8):1307–17.

[28] Rodriguez A,Vigorito E, Clare S, et al. Requirement of bic/microRNA-155 for normal immune function. Science 2007;316(5824):608–11.

[29] Tsitsiou E,Williams AE, Moschos SA, et al.Transcriptome analysis shows activation of circulating $CD8^+$ T cells in patients with severe asthma. J Allergy Clin Immunol 2012;129(1):95–103.

[30] Perry MM, Baker JE, Gibeon DS, Adcock IM, Chung KF. Airway smooth muscle hyperproliferation is regulated by microRNA-221 in severe asthma.Am J Respir Cell Mol Biol 2014;50(1):7–17.

[31] Mayoral RJ, Deho L, Rusca N, et al. MiR-221 influences effector functions and actin cytoskeleton in mast cells. PLoS ONE 2011;6(10):e26133.

[32] Perry MM, Adcock IM, Chung KF. Role of microRNAs in allergic asthma: present and future. Curr Opin Allergy Clin Immunol 2015;15(2):156–62.

[33] The A.D.T.F.. Acute respiratory distress syndrome: the berlin definition. JAMA 2012;307(23):2526–33.

[34] Rubenfeld GD, Caldwell E, Peabody E, et al. Incidence and outcomes of acute lung injury. N Engl J Med 2005;353(16):1685–93.

[35] Matthay MA, Zemans RL. The acute respiratory distress syndrome: pathogenesis and treatment. Annu Rev Pathol 2011;6:147–63.

[36] Crosby LM, Waters CM. Epithelial repair mechanisms in the lung. Am J Physiol Lung Cell Mol Physiol 2010;298.

[37] Gao L, Barnes KC. Recent advances in genetic predisposition to clinical acute lung injury. Am J Physiol Lung Cell Mol Physiol 2009;296.

[38] Nahid MA, Pauley KM, Satoh M, Chan EKL. miR-146a is critical for endotoxin-induced tolerance: implication in innate immunity. J Biol Chem 2009;284(50): 34590–9.

[39] Bazzoni F, Rossato M, Fabbri M, et al. Induction and regulatory function of miR-9 in human monocytes and neutrophils exposed to proinflammatory signals. Proc Natl Acad Sci 2009;106(13):5282–7.

[40] Liu X, Zhan Z, Xu L, et al. MicroRNA-148/152 impair innate response and antigen presentation of TLR-triggered dendritic cells by targeting CaMKIIα. J Immunol 2010;185(12):7244–51.

[41] Sheedy FJ, Palsson-McDermott E, Hennessy EJ, et al. Negative regulation of TLR4 via targeting of the proinflammatory tumor suppressor PDCD4 by the microRNA miR-21. Nat Immunol 2010;11(2):141–7.

[42] Huang C, Xiao X, Chintagari N, Breshears M, Wang Y, Liu L. MicroRNA and mRNA expression profiling in rat acute respiratory distress syndrome. BMC Med Genomics 2014;7(1):46.

[43] Bhaskaran M, Xi D, Wang Y, et al. Identification of microRNAs changed in the neonatal lungs in response to hyperoxia exposure. Physiol Genomics 2012;44(20):970–80.

[44] Adyshev DM, Moldobaeva N, Mapes B, Elangovan V, Garcia JGN. MicroRNA regulation of nonmuscle myosin light chain kinase expression in human lung endothelium. Am J Respir Cell Mol Biol 2013;49(1):58–66.

[45] Mirzapoiazova T, Moitra J, Moreno-Vinasco L, et al. Non–muscle myosin light chain kinase isoform is a viable molecular target in acute inflammatory lung injury. Am J Respir Cell Mol Biol 2011;44(1):40–52.

[46] Slutsky AS, Ranieri VM. Ventilator-induced lung injury. N Engl J Med 2013;369(22):2126–36.

[47] Ye SQ, Simon BA, Maloney JP, et al. Pre-B-cell colony-enhancing factor as a potential novel biomarker in acute lung injury. Am J Respir Crit Care Med 2005;171(4):361–70.

[48] Adyshev DM, Elangovan VR, Moldobaeva N, Mapes B, Sun X, Garcia JGN. Mechanical stress induces pre-B-cell colony-enhancing factor/NAMPT expression via epigenetic regulation by miR-374a and miR-568 in human lung endothelium. Am J Respir cell Mol Biol 2013;50(2):409–18.

[49] Vaporidi K, Vergadi E, Kaniaris E, et al. Pulmonary microRNA profiling in a mouse model of ventilator-induced lung injury. Am J Physiol Lung Cell Mol Physiol 2012;303(3):L199–207.

[50] Yehya N, Yerrapureddy A, Tobias J, Margulies S. MicroRNA modulate alveolar epithelial response to cyclic stretch. BMC Genomics 2012;13(1):154.

[51] Zeng Z, Gong H, Li Y, et al. Upregulation of miR-146a contributes to the suppression of inflammatory responses in LPS-induced acute lung injury. Exp Lung Res 2013;39(7):275–82.

[52] Perry MM, Moschos SA, Williams AE, Shepherd NJ, Larner-Svensson HM, Lindsay MA. Rapid changes in microRNA-146a expression negatively regulate the interleukin-1β induced inflammatory response in human lung alveolar epithelial cells. J Immunol (Baltim Md, 1950) 2008;180(8):5689–98.

[53] Sun X, Icli B, Wara AK, et al. MicroRNA-181b regulates NF-κB-mediated vascular inflammation. J Clin Invest 2012;122(6):1973–90.

[54] Xie T, Liang J, Liu N, et al. MicroRNA-127 inhibits lung inflammation by targeting IgG Fcgamma receptor I. J Immunol 2012;188(5):2437–44.

[55] Rycroft CE, Heyes A, Lanza L, Becker K. Epidemiology of chronic obstructive pulmonary disease: a literature review. Int J Chron Obstruct Pulmon Dis 2012;7:457–94.

[56] Ford ES, Croft JB, Mannino DM, Wheaton AG, Zhang X, Giles WH. COPD surveillance–United States, 1999–2011. Chest 2013;144(1):284–305.

[57] Chavali S, Bruhn S, Tiemann K, et al. MicroRNAs act complementarily to regulate disease-related mRNA modules in human diseases. RNA 2013;19(11):1552–62.

[58] Basma H, Gunji Y, Iwasawa S, et al. Reprogramming of COPD lung fibroblasts through formation of induced pluripotent stem cells. Am J Physiol Lung Cell Mol Physiol 2014;306(6):L552–65.

[59] Oldenburger A, van Basten B, Kooistra W, et al. Interaction between Epac1 and miRNA-7 in airway smooth muscle cells. Naunyn Schmiedeb Arch Pharmacol 2014;387(8):795–7.

[60] Donaldson A, Natanek SA, Lewis A, et al. Increased skeletal muscle-specific microRNA in the blood of patients with COPD. Thorax 2013;68(12):1140–9.

[61] Xie L, Wu M, Lin H, et al. An increased ratio of serum miR-21 to miR-181a levels is associated with the early pathogenic process of chronic obstructive pulmonary disease in asymptomatic heavy smokers. Mol Biosyst 2014;10(5):1072–81.

[62] Sato T, Liu X, Nelson A, et al. Reduced miR-146a increases prostaglandin E(2)in chronic obstructive pulmonary disease fibroblasts. Am J Respir Crit Care Med 2010;182(8):1020–9.

[63] Wang M, Huang Y, Liang Z, et al. Plasma miRNAs might be promising biomarkers of chronic obstructive pulmonary disease. Clin Respir J 2014;10(1).

[64] Pinkerton M, Chinchilli V, Banta E, et al. Differential expression of microRNAs in exhaled breath condensates of patients with asthma, patients with chronic obstructive pulmonary disease, and healthy adults. J Allergy Clin Immunol 2013;132(1):217–9.

[65] Janardhan KS, McIsaac M, Fowlie J, et al. Toll like receptor-4 expression in lipopolysaccharide induced lung inflammation. Histol Histopathol 2006;21(7):687–96.

[66] Puig-Vilanova E, Aguilo R, Rodriguez-Fuster A, Martinez-Llorens J, Gea J, Barreiro E. Epigenetic mechanisms in respiratory muscle dysfunction of patients with chronic obstructive pulmonary disease. PLoS ONE 2014;9(11):e111514.

[67] Puig-Vilanova E, Ausin P, Martinez-Llorens J, Gea J, Barreiro E. Do epigenetic events take place in the vastus lateralis of patients with mild chronic obstructive pulmonary disease? PLoS ONE 2014;9(7):e102296.

[68] Christenson SA, Brandsma CA, Campbell JD, et al. miR-638 regulates gene expression networks associated with emphysematous lung destruction. Genome Med 2013;5(12):114.

[69] Ezzie ME, Crawford M, Cho JH, et al. Gene expression networks in COPD: microRNA and mRNA regulation. Thorax 2012;67(2):122–31.

[70] Mizuno S, Bogaard HJ, Gomez-Arroyo J, et al. MicroRNA-199a-5p is associated with hypoxia-inducible factor-1alpha expression in lungs from patients with COPD. Chest 2012;142(3):663–72.

[71] Ellis KL, Cameron VA, Troughton RW, Frampton CM, Ellmers LJ, Richards AM. Circulating microRNAs as candidate markers to distinguish heart failure in breathless patients. Eur J Heart Fail 2013;15(10):1138–47.

[72] Molina-Pinelo S, Pastor MD, Suarez R, et al. MicroRNA clusters: dysregulation in lung adenocarcinoma and COPD. Eur Respir J 2014;43(6):1740–9.

[73] Van Pottelberge GR, Mestdagh P, Bracke KR, et al. MicroRNA expression in induced sputum of smokers and patients with chronic obstructive pulmonary disease. Am J Respir Crit Care Med 2011;183(7):898–906.

[74] Savarimuthu Francis SM, Davidson MR, Tan ME, et al. MicroRNA-34c is associated with emphysema severity and modulates SERPINE1 expression. BMC Genomics 2014;15:88.

[75] Hassan T, Carroll TP, Buckley PG, et al. miR-199a-5p silencing regulates the unfolded protein response in chronic obstructive pulmonary disease and alpha1-antitrypsin deficiency. Am J Respir Crit Care Med 2014;189(3):263–73.

[76] Akbas F, Coskunpinar E, Aynaci E, Oltulu YM, Yildiz P. Analysis of serum micro-RNAs as potential biomarker in chronic obstructive pulmonary disease. Exp Lung Res 2012;38(6):286–94.

[77] Vohwinkel CU, Lecuona E, Sun H, et al. Elevated CO_2 levels cause mitochondrial dysfunction and impair cell proliferation. J Biol Chem 2011;286(43):37067–76.

[78] Soeda S, Ohyashiki JH, Ohtsuki K, Umezu T, Setoguchi Y, Ohyashiki K. Clinical relevance of plasma miR-106b levels in patients with chronic obstructive pulmonary disease. Int J Mol Med 2013;31(3):533–9.

[79] Lewis A, Riddoch-Contreras J, Natanek SA, et al. Downregulation of the serum response factor/miR-1 axis in the quadriceps of patients with COPD. Thorax 2012;67(1):26–34.

[80] Yue J, Guan J, Wang X, et al. MicroRNA-206 is involved in hypoxia-induced pulmonary hypertension through targeting of the HIF-1alpha/Fhl-1 pathway. Lab Invest 2013;93(7):748–59.

[81] Ikari J, Smith LM, Nelson AJ, et al. Effect of culture conditions on microRNA expression in primary adult control and COPD lung fibroblasts in vitro. Vitro Cell Dev Biol Anim 2015;51(4).

[82] Corvol H, Thompson KE, Tabary O, le Rouzic P, Guillot L. Translating the genetics of cystic fibrosis to personalized medicine. Transl Res J Lab Clin Med 2015;168.

[83] Sonneville F, Ruffin M, Guillot L, et al. New insights about miRNAs in cystic fibrosis. Am J Pathol 2015;185(4):897–908.

[84] Fabbri E, Borgatti M, Montagner G, et al. Expression of microRNA-93 and Interleukin-8 during *Pseudomonas aeruginosa*-mediated induction of proinflammatory responses. Am J Respir Cell Mol Biol 2014;50(6):1144–55.

[85] Megiorni F, Cialfi S, Cimino G, et al. Elevated levels of miR-145 correlate with SMAD3 down-regulation in cystic fibrosis patients. J Cyst Fibros 2013;12(6): 797–802.

[86] Mattick JS, Makunin IV. Non-coding RNA. Hum Mol Genet 2006;15(Spec No 1): R17–29.

[87] Ramachandran S, Karp PH, Jiang P, et al. A microRNA network regulates expression and biosynthesis of wild-type and DeltaF508 mutant cystic fibrosis transmembrane conductance regulator. Proc Natl Acad Sci USA 2012;109(33):13362–7.

[88] Ramachandran S, Karp PH, Osterhaus SR, et al. Post-transcriptional regulation of cystic fibrosis transmembrane conductance regulator expression and function by microRNAs. Am J Respir Cell Mol Biol 2013;49(4):544–51.

[89] Viart V, Bergougnoux A, Bonini J, et al. Transcription factors and miRNAs that regulate fetal to adult CFTR expression change are new targets for cystic fibrosis. Eur Respir J 2015;45(1):116–28.

[90] Zhang PX, Cheng J, Zou S, et al. Pharmacological modulation of the AKT/microRNA-199a-5p/CAV1 pathway ameliorates cystic fibrosis lung hyper-inflammation. Nat Commun 2015;6:6221.

[91] Weldon S, McNally P, McAuley DF, et al. miR-31 dysregulation in cystic fibrosis airways contributes to increased pulmonary cathepsin S production. Am J Respir Crit Care Med 2014;190(2):165–74.

[92] Amato F, Seia M, Giordano S, et al. Gene mutation in microRNA target sites of CFTR gene: a novel pathogenetic mechanism in cystic fibrosis? PLoS ONE 2013;8(3):e60448.

[93] Endale Ahanda ML, Bienvenu T, Sermet-Gaudelus I, et al. The hsa-miR-125a/hsa-let-7e/hsa-miR-99b cluster is potentially implicated in Cystic Fibrosis pathogenesis. J Cyst Fibros 2015;14(5).

[94] Cook NL, Pereira TN, Lewindon PJ, Shepherd RW, Ramm GA. Circulating microRNAs as noninvasive diagnostic biomarkers of liver disease in children with cystic fibrosis. J Pediatr Gastroenterol Nutr 2015;60(2):247–54.

[95] Accurso FJ, Rowe SM, Clancy JP, et al. Effect of VX-770 in persons with cystic fibrosis and the G551D-CFTR mutation. N Engl J Med 2010;363(21):1991–2003.

[96] Boyle AP, Hong EL, Hariharan M, et al. Annotation of functional variation in personal genomes using RegulomeDB. Genome Res 2012;22(9):1790–7.

[97] Bhattacharyya S, Balakathiresan NS, Dalgard C, et al. Elevated miR-155 promotes inflammation in cystic fibrosis by driving hyperexpression of interleukin-8. J Biol Chem 2011;286(13):11604–15.

[98] Bazett M, Paun A, Haston CK. MicroRNA profiling of cystic fibrosis intestinal disease in mice. Mol Genet Metab 2011;103(1):38–43.

[99] Tsuchiya M, Kumar P, Bhattacharyya S, et al. Differential regulation of inflammation by inflammatory mediators in cystic fibrosis lung epithelial cells. J Interferon Cytokine Res 2013;33(3):121–9.

[100] Oglesby IK, Chotirmall SH, McElvaney NG, Greene CM. Regulation of cystic fibrosis transmembrane conductance regulator by microRNA-145, -223, and -494 is altered in DeltaF508 cystic fibrosis airway epithelium. J Immunol 2013;190(7): 3354–62.

[101] Bhattacharyya S, Kumar P, Tsuchiya M, Bhattacharyya A, Biswas R. Regulation of miR-155 biogenesis in cystic fibrosis lung epithelial cells: antagonistic role of two mRNA-destabilizing proteins, KSRP and TTP. Biochem Biophys Res Commun 2013;433(4):484–8.

[102] Amato F, Tomaiuolo R, Nici F, et al. Exploitation of a very small peptide nucleic acid as a new inhibitor of miR-509-3p involved in the regulation of cystic fibrosis disease-gene expression. Biomed Res Int 2014;2014:610718.

[103] Oglesby IK, Bray IM, Chotirmall SH, et al. miR-126 is downregulated in cystic fibrosis airway epithelial cells and regulates TOM1 expression. J Immunol 2010;184(4): 1702–9.

[104] McKiernan PJ, Cunningham O, Greene CM, Cryan SA. Targeting miRNA-based medicines to cystic fibrosis airway epithelial cells using nanotechnology. Int J Nanomed 2013;8:3907–15.

[105] Gillen AE, Gosalia N, Leir SH, Harris A. MicroRNA regulation of expression of the cystic fibrosis transmembrane conductance regulator gene. Biochem J 2011;438(1): 25–32.

[106] Megiorni F, Cialfi S, Dominici C, Quattrucci S, Pizzuti A. Synergistic post-transcriptional regulation of the Cystic Fibrosis Transmembrane conductance Regulator (CFTR) by miR-101 and miR-494 specific binding. PLoS ONE 2011;6(10):e26601.

[107] Lu YC, Chen H, Fok KL, et al. CFTR mediates bicarbonate-dependent activation of miR-125b in preimplantation embryo development. Cell Res 2012;22(10):1453–66.

[108] Xie C, Jiang XH, Zhang JT, et al. CFTR suppresses tumor progression through miR-193b targeting urokinase plasminogen activator (uPA) in prostate cancer. Oncogene 2013;32(18):2282–91. 2291 e2281–87.

[109] Hassan F, Nuovo GJ, Crawford M, et al. MiR-101 and miR-144 regulate the expression of the CFTR chloride channel in the lung. PLoS ONE 2012;7(11):e50837.

[110] Liu Z, Borlak J, Tong W. Deciphering miRNA transcription factor feed-forward loops to identify drug repurposing candidates for cystic fibrosis. Genome Med 2014;6(12):94.

[111] Iannuzzi MC, Rybicki BA, Teirstein AS. Sarcoidosis. N Engl J Med 2007;357(21): 2153–65.

[112] Newman LS, Rose CS, Maier LA. Sarcoidosis. N Engl J Med 1997;336(17):1224–34.

[113] Rybicki BA, Major M, Popovich Jr J, Maliarik MJ, Iannuzzi MC. Racial differences in sarcoidosis incidence: a 5-year study in a health maintenance organization. Am J Epidemiol 1997;145(3):234–41.

[114] Crouser ED, Culver DA, Knox KS, et al. Gene expression profiling identifies MMP-12 and ADAMDEC1 as potential pathogenic mediators of pulmonary sarcoidosis. Am J Respir Crit Care Med 2009;179(10):929–38.

[115] Zhou T, Zhang W, Sweiss NJ, et al. Peripheral blood gene expression as a novel genomic biomarker in complicated sarcoidosis. PLoS ONE 2012;7(9):e44818.

[116] Koth LL, Solberg OD, Peng JC, Bhakta NR, Nguyen CP, Woodruff PG. Sarcoidosis blood transcriptome reflects lung inflammation and overlaps with tuberculosis. Am J Respir Crit Care Med 2011;184(10):1153–63.

[117] Crouser ED, Julian MW, Crawford M, et al. Differential expression of microRNA and predicted targets in pulmonary sarcoidosis. Biochem Biophys Res Commun 2012;417(2):886–91.

[118] Maertzdorf J, Weiner 3rd J, Mollenkopf HJ, et al. Common patterns and disease-related signatures in tuberculosis and sarcoidosis. Proc Natl Acad Sci USA 2012;109(20):7853–8.

[119] Liu Y, Wang X, Jiang J, Cao Z, Yang B, Cheng X. Modulation of T cell cytokine production by miR-144* with elevated expression in patients with pulmonary tuberculosis. Mol Immunol 2011;48(9–10):1084–90.

[120] Venet D, Dumont JE, Detours V. Most random gene expression signatures are significantly associated with breast cancer outcome. PLoS Comput Biol 2011;7(10):e1002240.

CHAPTER 7

MicroRNAs in Idiopathic Pulmonary Fibrosis: Partners in Health and Disease

K.V. Pandit, N. Kaminski

Key Concepts

- Idiopathic pulmonary fibrosis (IPF) has a distinct microRNA expression profile.
- MicroRNA expression changes in IPF lead to feed-forward loops and sustained fibrosis.
- MicroRNA expression and functional studies using mimics and inhibitors are routinely used to elucidate targets and roles of microRNAs.
- MicroRNAs are novel therapeutic targets that can be targeted in vivo.

1. MICRORNAs

MicroRNAs are small noncoding RNAs ranging from about 18 to 25 nucleotides in length that either degrade protein-coding transcripts or inhibit their translation into proteins. Their discovery changed the central dogma of biology that DNA is transcribed into RNA, which in turn is translated into protein. Some microRNAs are ubiquitous in different cell types or tissues, but others are tissue specific. Each microRNA targets numerous messenger RNAs (mRNAs) and each mRNA is targeted by many microRNAs. It has been estimated that a single microRNA may target up to hundreds of mRNAs, but to a varying level of potency. These varied combinations of microRNA/mRNA enable microRNAs to affect a wide variety of physiologic and pathologic processes affecting multiple organ systems at different levels. MicroRNAs are present not only in humans, but also in other animals, plants, and viruses. Many of them are evolutionarily conserved.

MicroRNAs are encoded either within introns, exons, or in intergenic regions. Some microRNAs are in close proximity to each other and are transcribed from a common promoter [1]. The canonical pathway of

Translating MicroRNAs to the Clinic
ISBN 978-0-12-800553-8

microRNA biogenesis is explained and illustrated by Ha and Kim [2]. Briefly, following transcription by RNA Polymerase II or III [3,4], the primary microRNA transcript (primicroRNA) containing a stem-loop structure is generated in the nucleus. The microRNA is located in the stem-loop part of the transcript. The primicroRNA is cleaved by an RNAse III endonuclease Drosha to form the precursor microRNA (premicroRNA), which is transported to the cytoplasm [5]. Here it is cleaved again into the mature microRNA duplex by another RNAse III endonuclease named Dicer. The duplex is loaded onto the Argonaute protein to form the RNA-induced silencing complex. The strand with the relatively unstable 5′ terminus is preferentially selected as the guide strand. However, the passenger strand (microRNA*) is also demonstrated to form functional microRNAs in numerous studies [6]. Better the complementarity between the microRNA and its target mRNA, the greater the probability of cleavage of the target transcript.

The miRBase database has a list of all microRNAs that are currently identified in all organisms including humans [7]. Specific criteria for the experimental verification of microRNAs and conventional naming of microRNAs have been defined [7,8]. There are numerous algorithms for microRNA target predictions such as TargetScan [9], Pictar [10], miRanda [11,12], and Pita [13] to name a few.

2. IDIOPATHIC PULMONARY FIBROSIS

The overall incidence and prevalence of IPF in the United States is 42.7 and 16.3 per 100,000 persons respectively [14]. The median survival of IPF patients ranges from 2.5 to 3.5 years [15]. The mean age of presentation at initial diagnosis is 63 years with familial cases presenting earlier at 59.4 years [16]. Common symptoms include chronic dry cough, dyspnea, bilateral inspiratory crackles, and clubbing. The international guidelines for diagnosis and management of IPF [15] require that IPF be diagnosed based on the presence of a usual interstitial pneumonia (UIP) pattern on a high resolution computed tomography (HRCT) without a lung biopsy or combinations of HRCT patterns and a UIP histology on biopsy as well as the absence of obvious etiological factors that may cause this pattern [17]. The clinical spectrum can vary considerably from slow progression over many years to acute exacerbations and rapid decline in lung function [18–20]. Cigarette smoking is strongly associated with IPF [21]. Environmental exposures such as exposure to metal and asbestos, occupations involving dust or fume exposure including diamond polishing, industrial car cleaning, welding, and technical dental work have significant association with IPF [22–25]. Chronic viral infections such as Epstein–Barr virus, hepatitis C, and cytomegalovirus have also been reported in IPF patients [26–28], but the association is unclear.

2.1 Pathobiology of Idiopathic Pulmonary Fibrosis

Initially, IPF was thought to result from initiation of an inflammatory cascade from an unknown insult [29]. The current hypothesis of pathogenesis of IPF is that repeated epithelial cell injury causes dysregulated epithelial–mesenchymal interaction and recruitment of fibroblasts to form myofibroblasts [30]. The epithelial cell injury may possibly be the consequence of unfolded protein response, reactive oxygen species, or viral infection. The myofibroblasts may originate from resident fibroblasts, circulating fibrocytes [31], or other cells. The contribution of epithelial–mesenchymal transition to the pathogenesis of IPF is debated due to conflicting results from lineage-tracing studies [32,33]. IPF is characterized by a UIP pattern on lung biopsy with predominantly paraseptal and subpleural honeycombing, patchy parenchymal involvement with normal areas adjacent to fibrotic regions. The classical histology finding in IPF is a fibroblast focus which is a subepithelial aggregation of proliferating fibroblasts and myofibroblasts [34]. There is absence of inflammation, organizing pneumonia, granuloma, and extensive hyaline membrane formation.

2.2 Staging of Idiopathic Pulmonary Fibrosis

Staging IPF allows the use of standardized nomenclature by clinicians, guide treatment options, and aid patient counseling. Traditionally, IPF was divided into mild, moderate, and severe stages or early and advanced stages. These stages were based on pulmonary function tests. However, these stages do not reflect biological or clinical phenotypes and do not have much therapeutic or prognostic relevance. To circumvent these drawbacks, Du Bois et al. developed a scoring system comprising four predictors: age, recent hospitalization, baseline forced vital capacity (FVC), and 24-week change in FVC [35]. The composite score obtained from these parameters estimates the 1-year risk of death. Another staging system developed by Ley et al. [36] was named the GAP model because it consisted of four variables: gender (G), age (A), and two lung physiology variables (P) (FVC and DLCO). The calculated GAP index was divided into three stages, I, II, and III with 1-year mortalities of 6%, 16%, and 39%, respectively.

2.3 Gene Expression Changes in Idiopathic Pulmonary Fibrosis

The transcriptome of IPF lungs was extensively studied by analyzing high-throughput microarray data [37,38]. Differentially expressed genes include genes associated with extracellular matrix formation and degradation, smooth muscle markers, growth factors, chemokines, and complement pathway. Pathways associated with lung development are significantly enriched. These include transforming growth factor-β (TGF-β) pathway, Wnt pathway, Sonic Hedgehog pathway, and PTEN pathway [39].

2.4 Biomarkers in Idiopathic Pulmonary Fibrosis

A significant positive correlation has been found with rapidly declining lung function and/or decreased survival with high serum or plasma levels of epithelial or macrophage-related proteins such as SP-A, SP-D, KL-6, CCL18, and MMP7 [40–43]. Serum SP-A, SP-D, and serum DNA are significantly increased in IPF compared to other interstitial lung diseases [44–46]. Moeller et al. reported significantly elevated circulating fibrocytes in IPF patients and even more during acute exacerbations [47]. Patients with >5% fibrocytes had a mean survival of 7.5 months compared to 27 months for patients with <5% fibrocytes. A concise list of potential biomarkers has been tabulated by Borensztajn et al. [48].

2.5 Clinical Management of Idiopathic Pulmonary Fibrosis

The standard treatment for IPF was essentially supportive unless the patients received a lung transplant. Recently, two drugs, pirfenidone and Nintedanib, were approved by the US Food and Drug Administration (FDA) for the treatment of IPF following demonstration of a lower rate of functional decline in clinical trials [49,50]. Pirfenidone possibly acts by modulating TGF-β and inhibiting fibroblasts [51]. Nintedanib is an intracellular inhibitor of tyrosine kinases which decreases the rate of decline in lung function over 12 months of treatment in patients with mild to moderate disease [50,52,53].

3. MICRORNA STUDIES IN IDIOPATHIC PULMONARY FIBROSIS

We have summarized a few of these studies in an earlier review [54]. Studies performed till date are detailed in Table 7.1.

Table 7.1 MicroRNA studies in idiopathic pulmonary fibrosis

MicroRNA	[a]Targets	Organism	[b]Cell types	References
Antifibrotic				
Let-7d	HMGA2	Human	A549	[71]
		Mouse	HBEC	
		N/A	NHLF	[72]
			phFLF	
miR-17–92	DNMT1	Human	pNHLF	[87]
		Mouse	pFHLF	
miR-26a	CTGF	Human	MRC-5	[84]
		Mouse		
		Mouse	A549	[83]
miR-29	ITGA11	Mouse	IMR-90	[73]
	ADAMTS9		FG293	
	ADAM12			
	NID1			
	COL1A1/2	Human	pNHLF	[74,115]
	COL3/4A1		pFHLF	
	FBN1			
	ELN1			
miR-200	N/A	Mouse	MRC-5	[80]
			pmAECs	
			prAECs	
			RLE-6TN	
miR-326	TGF-β	Human	A549, NHBE, MCF7	[86]
		Mouse	NHF, MiaPaCa2 Raji cells	
Profibrotic				
miR-21	SMAD7	Human	MRC-5	[55]
		Mouse	IMR-90	
			HEK-293	
miR-96	FOXO3a	Human	pNHLF	[62]
			pHFLF	
miR-145	KLF4	Mouse	MCR-5	[58]
			pmLF	
miR-154	p15	Human	NHLF	[56]
			hFLF	
			pNHLF	
			pHFLF	
miR-199a–5p	CAV1	Human	MRC-5	[61]
		Mouse	hFL1	
			A549	
			HEK-293	
miR-210	MNT	Human	pNHLF	[68]
			pHFLF	
miR-200	N/A	Human	MDCK cell	[82]

[a]Direct binding of microRNAs to 3′UTR of predicted targets was proven with luciferase assays except for p15. Reduced protein level of p15 were determined by Western blot.
[b]*HBEC*, human bronchial epithelial cells; *hFL1*, human fetal lung fibroblasts; *NHBE*, normal human bronchial epithelial cells; *NHF*, normal human fibroblasts; *NHLF*, normal human lung fibroblasts; *pFHLF*, primary fibrotic human lung fibroblasts; *phFLF*, primary human fetal lung fibroblasts; *pmLF*, primary mouse lung fibroblasts; *pNHLF*, primary normal human lung fibroblasts.

Adapted from Pandit KV, Milosevic J. MicroRNA regulatory networks in idiopathic pulmonary fibrosis. Biochem Cell Biol April 2015;93(2):129–37.

3.1 MicroRNAs Upregulated in Idiopathic Pulmonary Fibrosis

3.1.1 MiR-21

MiR-21 is increased in human IPF lungs and in the lungs of bleomycin-treated mice [55]. MiR-21 colocalized with α-SMA in myofibroblasts in bleomycin-treated mouse lungs. It targets the TGF-β inhibitor SMAD7 and also induced SMAD2 phosphorylation accelerating the fibrotic process. Bleomycin mice pretreated with miR-21 did not demonstrate α-SMA expression or myofibroblast accumulation. Administration of miR-21 antagomir, before or after bleomycin treatment, reduced the mRNA and protein of ECM components Col1a1, Col1a2, and fibronectin.

3.1.2 MiR-154

The miR-154 family located on chr14q32 is the most abundant family with increased microRNA expression in IPF [56]. Computational analysis discovered nine SMAD3 binding elements upstream of miR-154 family members. Two SMAD3-binding sites upstream of miR-154 were experimentally verified by chromatin immunoprecipitation. The microRNA profile of lung fibroblasts stimulated with TGF-β was compared to that of IPF lungs. Of the 14 microRNAs upregulated in both groups, 12 were located on chr14q32, 7 being members of miR-154 family suggesting that the upregulation of some of the chr14q32 microRNAs observed in IPF is partly due to TGF-β activation. Transfection of miR-154 in primary cultures of normal human and fetal lung fibroblasts significantly increased their migration and proliferation. Transfected miR-154 also significantly reduced expression of the cell cycle inhibitor, CDKN2B (p15), a predicted target of miR-154 by *TargetScan v6.2* [57]. Inhibition of the WNT/β-catenin pathway reversed proliferation induced by miR-154 suggesting that the effects of miR-154 on lung and fetal fibroblasts were partly mediated through p15 and WNT/β-catenin pathway.

3.1.3 MiR-145

Increased miR-145 in IPF activates latent TGF-β. MiR-145 directly targets Kruppel-like factor 4 (KLF4) [58], a negative regulator of α-SMA [59]. Fibroblasts expressing α-SMA have almost twice the contractility than that of α-SMA-negative fibroblasts in culture [60]. MiR-145 increased contractility of lung fibroblasts and increased focal and fibrillar adhesion. MiR-$145^{-/-}$ mice treated with bleomycin developed less fibrosis in comparison to wild-type mice treated with bleomycin.

3.1.4 MiR-199a-5p

MiR-199a-5p is upregulated in IPF lungs and in lungs of bleomycin-treated mice [61]. TGF-β induces miR-199a-5p which targets caveolin-1, a main coat protein of caveolae. The lipid-raft internalization of caveoli degrades TGF-β receptors consequently decreasing TGF-β signaling. Profibrotic miR-199a-5p acts through both a caveolin-dependent and caveolin-independent pathway. Comparison of gene expression profiles of cells transfected separately with miR-199a-5p and caveolin siRNA, revealed enrichment of fibrotic processes such as IL-1 signaling, acute phase response signaling, and p38 MAPK signaling pathways in the miR-199a-5p-transfected cells but not with caveolin-1 inhibition. The authors also compared the gene expression of lung fibroblasts following stimulation with TGF-β and overexpression of miR-199a-5p. The TGF-β response seen in lung fibroblasts was in most part through the action of miR-199a-5p as indicated by a significant overlap in these two profiles. Similar to miR-199a-5p, TGF-β also induces miR-21, which is overexpressed in IPF. Gene expression profiles of lung fibroblasts transfected separately with miR-199a-5p and miR-21 mimics showed a partial overlap indicating that both microRNAs cooperatively activate fibroblasts by altering genes in different pathways.

3.1.5 MiR-96

The expression of miR-96 is upregulated in IPF fibroblasts compared to control fibroblasts when cultured on collagen [62]. MiR-96 is expressed mainly in the fibroblastic foci of IPF lungs. MiR-96 directly targets forkhead box O3a (FoxO3a) [63]. FoxO3a activity is changed in IPF fibroblasts due to high Akt activity that contributes to the pathological phenotype of IPF fibroblasts [64]. MiR-96 inhibition in IPF fibroblasts increased FoxO3a target protein expression. This in turn led to a reduction in FoxO3a targets viz p27, p21, and Bim. FoxO3a inhibits cell proliferation and promotes cell death by increasing p27, p21, and Bim [65–67].

3.1.6 MiR-210

MiR-210 expression is increased in IPF and is localized to fibroblasts where it is coexpressed with hypoxia-inducible transcription factor HIF-2α [68]. Hypoxia activates HIF-2α which in turn stimulates the expression of miR-210. Previous studies have demonstrated that miR-210 regulates cell proliferation by targeting MNT, a c-myc inhibitor [69,70]. The authors demonstrate that knockdown of miR-210 increases MNT and inhibits proliferation of IPF fibroblasts in response to hypoxia [68]. Silencing HIF-2α inhibits hypoxia-mediated increase in miR-210, thus blocking proliferation of IPF fibroblasts.

3.2 MicroRNAs Downregulated in Idiopathic Pulmonary Fibrosis

3.2.1 Let-7d

Let-7d was the first microRNA to be studied in IPF [71]. The transcription factor SMAD3 negatively regulates let-7d. HMGA2, a direct target of the let-7 family is increased in IPF. Also, SMAD4 transcriptionally activates HMGA2. HMGA2 is a regulator of EMT and aids in transcription of SNAI1 and TWIST, transcription factors that promote EMT. Inhibition of let-7d in vivo reduced expression of CDH1 and TJP1 and increased expression of COL1A1, HMGA2, and α-SMA. However, these changes are not only mediated through HMGA2. Let-7d also targets other fibrosis-relevant genes such as RAS, IGF-1, IGF1R which are downregulated in IPF. Thus the fibrotic changes observed after let-7d inhibition, may be a result of release of let-7d microRNA inhibitory signals on a complex network of profibrotic molecules. We studied the effects of overexpressing let-7d in fibroblasts [72]. Transfection of let-7d in fibroblasts induced changes to their phenotype including, decreased expression of mesenchymal markers such as α-SMA, N-cadherin, and fibronectin and reduced proliferation and migration [72].

3.2.2 MiR-29

MiR-29 family of microRNAs is downregulated in IPF [71]. TGF-β suppresses miR-29 resulting in upregulation of profibrotic genes [73]. In an elegant study, Parker et al. showed that IPF-associated extracellular matrix proteins are direct targets of miR-29 [74]. Fibroblasts cultured on IPF-derived extracellular matrix had reduced miR-29 levels. Upon restoration of miR-29, ECM targets returned to baseline levels. Cushing et al. knocked down miR-29 in IMR-90 cells and performed high-throughput expression profiling. They detected upregulation of genes involved in organ fibrosis and ECM remodeling [75–77]. SMAD3$^{-/-}$ mice were protected from fibrosis and miR-29 loss [78]. TGF-β1/SMAD3/CTGF signaling in fibrotic lung was reduced in bleomycin-treated mice overexpressed with miR-29b. We injected synthetic miR-29 duplexes intravenously in bleomycin-treated mice and successfully decreased collagen expression in the lung and reversed bleomycin-induced lung fibrosis [79].

3.2.3 MiR-200

MiR-200 family is expressed predominantly in alveolar epithelial cells and is decreased in IPF and bleomycin-induced lung fibrosis in mice [80]. TGF-β reduces expression of miR-200 family in rat alveolar epithelial cells

(RLE-6TN). TGF-β-induced expression of mesenchymal cell markers and mesenchymal cell morphological characteristics was attenuated by overexpression of miR-200b and miR-200c [81]. Transfection of miR-200 mimic in rat AECs led to a significant decrease in mRNA levels of GATA3, ZEB1, and ZEB2, transcriptional factors involved in EMT regulation. These transcription factors are not direct targets of miR-200. Mice that received the miR-200 mimic with bleomycin treatment had less collagen deposition in their lungs compared to those that received a control mimic. Though Yang et al. observed a reduction in miR-200 in IPF, Cho et al. reported an increase [82], while we did not observe a change in miR-200 in IPF [71]. The cause of this discrepancy may be due to differences in patient samples (biopsy vs lung explant) or the choice of controls tissue. The choice of control may significantly influence the results; lung transplant rejected tissues may be affected by inflammation, prolonged ventilation or processing while normal histology edges of lungs resected for lung cancer may have cancer field effects.

3.2.4 MiR-26a

MiR-26a is decreased in IPF and bleomycin-treated mice [83,84]. TGF-β negatively regulates miR-26a via direct SMAD3 binding to the promoter of miR-26a [84]. MiR-26 targets CTGF [84] and COL1A2 [85]. CTGF is a cytokine which increases collagen formation and deposition. MiR-26 also targets SMAD4 thus inhibiting the nuclear localization of p-SMAD2/SMAD3. Loss of miR-26a expression also promotes EMT by directly targeting HMGA2 [83]. EMT observed in vitro in A549 lung cancer cells and in bleomycin-induced pulmonary fibrosis was rescued partially by exogenous miR-26a.

3.2.5 MiR-326

MiR-326 is the only microRNA studied so far that targets TGF-β and hence plays a role in lung fibrosis [86]. MiR-326 is decreased in lungs of patients and in bleomycin-treated mice. NIH/3T3 fibroblasts transfected with miR-326 reduced the expression of profibrotic genes COL1A2, COL3A1, SMAD3 and increased expression of antifibrotic genes SMAD7, BMP7, and IL10. These genes are not proven direct targets of miR-326. Inhalation of miR-326 mimic reduced TGF-β expression and attenuated the fibrotic process in mice. Stimulation of A549 cells and human bronchial epithelial cells with IL-13/TNF-α increased TGF-β production and reduced miR-326 levels. Upon transfecting miR-326 these cells were rescued from EMT, but upon knocking down miR-326, TGF-β expression increased which in turn promoted expression of EMT markers.

3.2.6 MiR-17–92 Cluster

Dakhlallah et al. studied the miR-17–92 cluster and demonstrated its epigenetic silencing by methylation in lung tissue and fibroblasts from IPF patients [87]. Transfection of this cluster in IPF fibroblasts reduced VEGF, CTGF, and COL3A1 all direct targets of this cluster. Treatment with a DNA methylation inhibitor, 5′-aza-2′-deoxycytidine, restored the expression of four of six members of the cluster in IPF fibroblasts. This cluster also targets DNA methyltransferase 1 (DNMT1) whose expression increases with severity of the disease, and inversely correlated to the expression of the microRNA cluster, thus forming a feed-forward loop. 5′-aza-2′-deoxycytidine-mitigated bleomycin-induced DNA methylation of the miR-17–92 cluster and reduced DNMT1 compared with mice treated with bleomycin alone. The gene collagens, VEGF, CTGF, and DNMT-1, were significantly reduced but not differentially methylated between IPF and controls suggesting that changes in their expression are a consequence of increased expression of the microRNA cluster and not by changes in their methylation status. Another study investigated the role of miR-92a in regulating WNT1-inducible signaling pathway protein 1 (WISP1), a profibrotic mediator in IPF [88,89]. MiR-92a reversed TGFβ-induced WISP1 expression in lung fibroblasts.

A common theme to many of the microRNAs implicated in IPF is their interaction with TGF-β1 regulatory networks. While most reports studied microRNAs in isolation, we believe that it is important to take into consideration the complex interactions of microRNAs in pulmonary fibrosis. Fig. 7.1 depicts the network generated after we integrate all the microRNA studies involving TGF-β, and it highlights the presence of distinct regulatory loops governed by microRNAs. While in normal cells changes in a protein or a microRNA following injury are counteracted by negative regulation, in the case of pulmonary fibrosis, changes in expression of key microRNAs lead to feed-forward loops and sustained fibrosis.

4. LABORATORY TECHNIQUES

We have divided this discussion about laboratory techniques into two parts: (1) the methods required for investigating fibrotic phenotypes and (2) microRNA manipulation techniques.

4.1 Lung Fibrosis Techniques

4.1.1 Bleomycin Administration

Animal models aid in studying the pathobiology of diseases. Exact modeling of IPF in an animal model is not yet possible since the etiology and natural

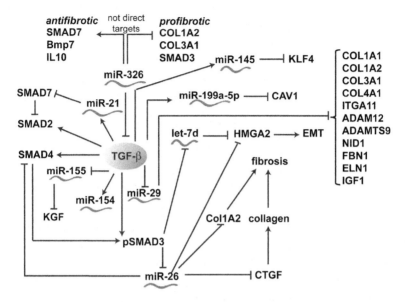

Figure 7.1 *Transforming growth factor-β (TGF-β) microRNA–mRNA signaling in the fibrotic process.* The microRNAs in the figure either target TGF-β or are regulated by TGF-β as indicated. These microRNAs in turn regulate fibrosis-relevant genes by binding to their 3′ untranslated region. The genes in the figure are direct targets of the microRNAs unless stated otherwise (except miR-326). *Adapted from Pandit KV, Milosevic J. MicroRNA regulatory networks in idiopathic pulmonary fibrosis. Biochem Cell Biol April 2015;93(2):129–37.*

history of IPF is yet unknown. Common methods of causing lung fibrosis in animals include radiation damage, bleomycin instillation, silica, asbestos, or gene transfer of cytokines. The most common mode is instillation of bleomycin in the lungs. Bleomycin is an antineoplastic drug having lung fibrosis as one of its side effects. Bleomycin hydrolase, the enzyme that inactivates bleomycin, is present in low quantities in lungs and hence lungs are more prone to bleomycin-induced fibrosis. Following intratracheal bleomycin instillation, there is an initial inflammatory phase not observed in IPF and then development of an active phase of fibrosis which most often resolves over time [90].

4.1.2 Sircol Assay

Fibrosis involves increased production of extracellular matrix including collagen in the lung. The sircol collagen assay is a dye-binding technique for the quantitation of acid and pepsin-soluble collagens.

4.1.3 Masson's Trichrome Staining

Masson's trichrome stains tissue sections with specific colors: keratin and muscle fibers are stained red, collagen is stained blue, cytoplasm is stained pink, and nuclei are stained dark brown to black.

4.1.4 Cell Proliferation and Migration

IPF is characterized by uncontrolled proliferation of fibroblasts. This proliferation can be quantified by colorimetric methods that measure cell metabolic activity. Enzymes produced by viable cells reduce the dyes to colored compounds. The absorbance of this colored solution is measured at specific wavelengths using a spectrophotometer.

4.2 MicroRNA Techniques

4.2.1 Global MicroRNA Expression

Microarray technology is employed to study the expression of multiple microRNAs. There are multiple microarray platforms but all utilize detection of microRNAs using hybridization-specific probes and detection by measuring a fluorescent signal [91]. The microarrays are scanned to measure fluorescent intensity. Statistical analysis provides a list of microRNAs that are differentially expressed between two or more groups of samples being compared. Other methods of detection by hybridization such as Nanostring [92] or high-throughput PCR such as Fluidigm technology [93] are also available.

Recently, global microRNA expression is being determined by sequencing. Unlike microarrays, sequencing does not require species-specific or transcript-specific probes. The main advantage of this technique is the ability to discover novel microRNAs or the presence of sequence variants. Coverage depth can be increased to detect rare transcripts, transcript variants, or low-expressed microRNAs.

4.2.2 Specific MicroRNA Expression

Northern blotting remains the gold standard for detection of microRNAs, especially novel microRNAs in tissues. It is the easiest and most reliable method to detect transcript size. However, the large amount of RNA required for northern blotting is a serious disadvantage, especially when sample quantity available is limited.

Quantitative real-time polymerase chain reaction is widely used by researchers for studying known microRNAs that are annotated in miRBase. The RNA template is transcribed into complementary DNA (cDNA) using retroviral reverse transcriptase. This cDNA is exponentially amplified by PCR. RT-PCR is extremely sensitive. Minute contamination with genomic DNA can produce false results, which can be prevented by pretreatment of RNA with DNAse.

Another method of microRNA detection is in situ hybridization. In situ hybridization uses a labeled complementary probe to stain and localize the microRNA in a tissue section to identify the type of cells the microRNA is expressed. The histological specimen needs to be preserved appropriately and fixed adequately for microRNA detection.

For all the three techniques mentioned earlier, northern blotting, RT-PCR, and in situ hybridization, commercial kits and reagents are available.

4.2.3 Functional Analysis

To investigate the mechanisms by which changes in cellular microRNA result in disease, synthetic microRNA mimics and microRNA inhibitors are transfected into cells in vitro and in vivo. The method of in vitro transfection is similar to siRNAs targeting protein-coding mRNAs, either by lipofection or electroporation. Locked nucleic acid (LNA) oligonucleotides have an added advantage of being highly sensitive and specific. The ribose moiety of an LNA nucleotide is modified with an extra bridge connecting the 2′ oxygen and 4′ carbon resulting in increased affinity for the complementary strand compared to traditional DNA and RNA oligonucleotides.

Artificial microRNA sponges can also be used as effective microRNA inhibitors [94,95]. They are expressed from strong promoters and contain multiple binding sites for a particular microRNA of interest. This leads to derepression of targets of this particular microRNA as effectively as chemically modified antisense inhibitors.

The biological consequences of blocking a specific microRNA–mRNA interaction can be evaluated with the use of target site blockers. Target site blockers are synthetic antisense oligonucleotides that compete with the microRNA for a target site of a particular mRNA. The target site blocker hybridizes with the target mRNA and thus prevents the microRNA from interacting with the target mRNA of interest.

5. LIMITATIONS

Though the microRNA studies in IPF we discussed earlier have exciting in vitro and in vivo results, translating it to humans is largely an assumption. The widely used in vivo bleomycin model of IPF is not perfect; it does not mimic IPF in its entirety [96]. Bleomycin is associated with an initial phase of inflammation, which is not observed in IPF. The injury caused by bleomycin is reversible whereas IPF is slow and progressive. Secondly, microRNA manipulations are routinely performed in cells and in animal models to explore their mechanistic functions. Many experiments either have total knockdown of endogenous microRNAs or supplement exogenous microRNAs to unrealistic levels. This, most likely, saturates the cellular microRNA processing machinery allowing spurious results. Finally, most studies focus on changing microRNAs in isolation, a single microRNA at a time. While such studies are highly effective in determining the mechanistic effects of microRNAs, they do not mimic the situation in the human lung, where multiple microRNAs are changed, often in conflicting directions.

6. MICRORNA THERAPEUTICS

The realization that microRNAs play critical roles in cellular homeostasis and that their dysregulation causes a myriad diseases, considerable interest has developed in microRNA therapeutics. Targeting microRNAs has now become a viable option, with two clinical trials currently enrolling patients.

The first microRNA clinical trial (NCT01200420) [97] was initiated for the treatment of hepatitis C following positive results in chimpanzees [98]. Chronically infected chimpanzees administered an intravenous miR-122 inhibitor SPC3649 experienced long-lasting suppression of hepatitis C viremia. SPC3649 is an LNA-modified phosphorothioate oligonucleotide complementary to the 5′ end of miR-122 [98]. NCT01200420 was a multicenter Phase 2a study, in which patients with hepatitis C were administered 5 weekly doses of Miravirsen (SPC3649) over 29 days. They showed a dose-dependent reduction in hepatitis C virus RNA levels without evidence of viral resistance [98a]. A subsequent trial (NCT01727934) [99] is currently recruiting participants for a 12-week drug regimen.

Another phase I study (NCT01829971) [100] has been initiated for treating patients with primary liver cancer or liver metastasis from other cancers with miR-34 in a liposome-formulated compound (MRX34). The tumor suppressor miR-34 is significantly reduced or lost in both solid tumors and hematologic malignancies. MiR-34 targets oncogenes such as MYC, MET, BCL2, and β-catenin [101]. Intravenous administration of MRX34 to orthotopic mouse models of hepatocellular carcinoma showed increased survival [102,103].

Currently there are no active clinical trials testing microRNA therapeutics in IPF. However, results from in vivo mouse studies suggest miR-29 to be a potential therapeutic target. As discussed earlier, the miR-29 family is downregulated in IPF resulting in increase of extracellular matrix target genes [74,79]. Intravenous injection of synthetic miR-29 mimics decreased collagen expression in the extracellular matrix and reversed bleomycin-induced lung fibrosis in mice. An miR-29 mimic is currently in the pre-clinical phase for lung fibrosis by pharmaceutical companies [104].

There are currently still significant concerns about the therapeutic potential of microRNA-based interventions. The concerns have mostly to do with specificity, emerging resistance and drug delivery.

Specificity—Each microRNA targets numerous genes involving multiple pathways. Hence an important concern about microRNA therapeutics is the extent of off-target effects, especially in the potential for causing unexpected impacts on previously unrecognized targets or on unexpected cell types. A most frequent concern is the potential impact of administering chronically large amounts of nucleic acids and the potential for activation of innate immune and RNA sensing mechanisms. While many of these effects are not observed in animal models, they will need to be addressed carefully in microRNA-based interventions in humans.

Emerging resistance—Another concern is that of potential drug resistance mechanisms. There is a possibility that when an exogenous microRNA inhibitor is administered chronically, the endogenous microRNA biogenesis may be upregulated to counteract the intervention. Another mechanism of counteraction is that other microRNAs that target the same genes may change to negate the effect of the exogenous microRNA drug. While most of these effects are not usually observed in the relatively acute models used in animals, when considering chronic administration in humans such effects should be considered, and actively sought for before human studies are initiated.

Drug Delivery—One of the biggest challenges with microRNA-based interventions is delivery. Currently there is little experience in targeting such interventions to the lung. While intravenous delivery does result in significant retention in the lung, it requires very high doses, and aerosolized delivery needs to be assessed carefully for safety and effect. Thus, approaches for packaging and targeting microRNA-based interventions [105] are critical to make them viable in the clinical arena.

7. MICRORNA–LONG NONCODING RNA INTERACTION

Recent noncoding RNA studies have identified multiple transcriptional signals from genomic areas that were once classified as "junk DNA." Transcripts that are at least 200 base pairs in length and which do not have coding potential are known as long noncoding RNAs (lncRNAs). Majority of the lncRNAs are yet to be characterized and annotated. lncRNAs can be tissue-specific and cell-specific, with majority lncRNAs localized in the nucleus. They bind to DNA or RNA or proteins, some are precursors for small regulatory RNAs such as piwi RNAs. lncRNAs are involved in chromatin modification and transcriptional regulation. lncRNA transcripts are rapidly and extensively being investigated as targets for modification of abnormal cellular function.

The emerging role of lncRNAs in RNA–RNA cross talk introduces us to a new layer of gene regulation. lncRNAs act as natural sponges for microRNAs by binding to microRNAs with complementary sequences, thus relieving inhibition of other target transcripts [95]. For example, the lncRNA HULC sequesters miR-372 thus preventing repression of PRKACB, a target of miR-372 [106]. A second example is that of the lncRNA PTENP1 which acts as a sponge for microRNAs that target PTEN. PTENP1 shares a similar sequence with the 3′UTR of PTEN [107,108]. A third example is that of the lncRNA linc-MD1 which sponges miR-133 and miR-135 to regulate the expression of transcription factors that activate muscle-specific gene expression [109]. Such RNAs that compete for shared microRNAs are called competing endogenous RNAs. Similar microRNA–lncRNA interactions have not yet been identified in IPF, but there is a possibility of its existence.

lncRNAs having diagnostic potential have not been identified in lung diseases but have been identified in prostate and liver cancer. The lncRNA DD3 is prostate-specific and is a very specific marker of prostate cancer detectable in urine [110,111]. The lncRNA HULC is highly upregulated in hepatocellular carcinoma and detected in blood of patients [112]. The scope of lncRNAs as prognostic markers is currently limited. MALAT1 is the only lncRNA which is associated with high metastatic potential and poor prognosis in NSCLC [113].

Similar to microRNAs, the large numbers of lncRNAs have broad regulatory roles in cellular circuitry and hence poised as perfect candidates for cellular manipulations. An added advantage of lncRNAs is that due to their stage-specific and tissue-specific distribution, they may be easier therapeutic targets since the potential of off-target effects are considerably reduced. Therapeutic strategies would involve designing chemical compounds that interfere with lncRNA–protein interaction or those that prevent loading of the lncRNA onto its target genomic regions.

8. CONCLUSIONS AND FUTURE DIRECTIONS

The discovery that microRNAs are differentially expressed in IPF lungs, and the extent to which their altered expression affects both key regulators as well as downstream effectors of fibrosis provides us with a unique understanding of the complexity of transcriptional regulation in pulmonary fibrosis. It also allows us to identify key regulatory hubs that may act as master regulators of the lung phenotype in pulmonary fibrosis. Interestingly, some of them explain and seem to regulate fibrotic response in epithelial cells, others in fibroblasts, inflammatory cells and others regulate the interactions between the cells and the extracellular matrix. Thus it is possible that by using microRNA-based interventions we will be able to untangle the complex network of molecular events observed in IPF. While this may have a beneficial effect on its own, it will probably have a more significant impact in conjunction with other therapeutics and in fact the authors of this chapter believe that this is where microRNA-based interventions will find their niche in fibrosis—in synergism with other therapies, acting to induce a more favorable transcriptional environment for other drugs to exert their action. To maximally benefit from the huge potential of these therapeutics, significant efforts are needed in developing mimics and inhibitors and lung-targeting modalities. Most importantly, read outs that allow identification of off-target effects, target engagement, and potential surrogate markers of efficacy need to be initiated through collaborations of academia, industry, and the funding agencies.

LIST OF ACRONYMS AND ABBREVIATIONS

BMP7 Bone morphogenetic protein 7
CCL18 Chemokine (C–C motif) ligand 18
CDH1 E-cadherin
CDKN2B Cyclin-dependent kinase inhibitor 2B
ceRNA Competing endogenous RNA
CTGF Connective tissue growth factor
DLCO Diffusing capacity of the lung for carbon monoxide
DNMT1 DNA methyltransferase 1
ECM Extracellular matrix
EMT Epithelial mesenchymal transition
FVC Forced vital capacity
GATA3 GATA binding protein 3
HMGA2 High mobility group AT-hook 2
HRCT High resolution computed tomography
IGF1 Insulin-like growth factor 1

IGF1R Insulin-like growth factor 1 receptor
ILD Interstitial lung disease
KL-6 Mucin1, cell surface associated
KLF4 Kruppel-like factor 4
LNA Locked nucleic acid
lncRNA Long noncoding RNA
MET Mesenchymal epithelial transition
MMP7 Matrix metalloproteinase 7
NSCLC Non-small cell lung cancer
PFT Pulmonary function test
PTEN Phosphatase and tensin homolog
qRT-PCR Quantitative real-time polymerase chain reaction
RISC RNA-induced silencing complex
SP-A Surfactant protein A
SP-D Surfactant protein D
TGF-β Transforming growth factor beta
TJP1 Tight junction protein 1
TNF-α Tumor necrosis factor alpha
UIP Usual interstitial pneumonia
VEGF Vascular endothelial growth factor
WISP1 WNT1-inducible signaling pathway protein 1
ZEB1 Zinc finger E-box binding homeobox 1
ZEB2 Zinc finger E-box binding homeobox 2
α-SMA Alpha smooth muscle actin

REFERENCES

[1] Ozsolak F, Poling LL, Wang Z, Liu H, Liu XS, Roeder RG, et al. Chromatin structure analyses identify miRNA promoters. Genes Dev November 15, 2008;22(22): 3172–83.

[2] Ha M, Kim VN. Regulation of microRNA biogenesis. Nat Rev Mol Cell Biol August 2014;15(8):509–24.

[3] Lee Y, Kim M, Han J, Yeom KH, Lee S, Baek SH, et al. MicroRNA genes are transcribed by RNA polymerase II. EMBO J October 13, 2004;23(20):4051–60.

[4] Borchert GM, Lanier W, Davidson BL. RNA polymerase III transcribes human microRNAs. Nat Struct Mol Biol December 2006;13(12):1097–101.

[5] Lee Y, Ahn C, Han J, Choi H, Kim J, Yim J, et al. The nuclear RNase III Drosha initiates microRNA processing. Nature September 25, 2003;425(6956):415–9.

[6] Chiang HR, Schoenfeld LW, Ruby JG, Auyeung VC, Spies N, Baek D, et al. Mammalian microRNAs: experimental evaluation of novel and previously annotated genes. Genes Dev May 15, 2010;24(10):992–1009.

[7] Kozomara A, Griffiths-Jones S. miRBase: annotating high confidence microRNAs using deep sequencing data. Nucleic Acids Res January 2014;42(Database issue): D68–73.

[8] Ambros V, Bartel B, Bartel DP, Burge CB, Carrington JC, Chen X, et al. A uniform system for microRNA annotation. RNA March 2003;9(3):277–9.

[9] Lewis BP, Burge CB, Bartel DP. Conserved seed pairing, often flanked by adenosines, indicates that thousands of human genes are microRNA targets. Cell January 14, 2005;120(1):15–20.

[10] Krek A, Grun D, Poy MN, Wolf R, Rosenberg L, Epstein EJ, et al. Combinatorial microRNA target predictions. Nat Genet May 2005;37(5):495–500.

[11] Enright AJ, John B, Gaul U, Tuschl T, Sander C, Marks DS. MicroRNA targets in Drosophila. Genome Biol 2003;5(1):R1.

[12] John B, Enright AJ, Aravin A, Tuschl T, Sander C, Marks DS. Human MicroRNA targets. PLoS Biol November 2004;2(11):e363.

[13] Kertesz M, Iovino N, Unnerstall U, Gaul U, Segal E. The role of site accessibility in microRNA target recognition. Nat Genet October 2007;39(10):1278–84.

[14] Raghu G, Weycker D, Edelsberg J, Bradford WZ, Oster G. Incidence and prevalence of idiopathic pulmonary fibrosis. Am J Respir Crit Care Med October 1, 2006;174(7):810–6.

[15] Raghu G, Collard HR, Egan JJ, Martinez FJ, Behr J, Brown KK, et al. An official ATS/ERS/JRS/ALAT statement: idiopathic pulmonary fibrosis: evidence-based guidelines for diagnosis and management. Am J Respir Crit Care Med March 15, 2011;183(6): 788–824.

[16] Lee HL, Ryu JH, Wittmer MH, Hartman TE, Lymp JF, Tazelaar HD, et al. Familial idiopathic pulmonary fibrosis: clinical features and outcome. Chest June 2005;127(6):2034–41.

[17] Adkins JM, Collard HR. Idiopathic pulmonary fibrosis. Semin Respir Crit Care Med October 2012;33(5):433–9.

[18] Fernandez Perez ER, Daniels CE, Schroeder DR, St Sauver J, Hartman TE, Bartholmai BJ, et al. Incidence, prevalence, and clinical course of idiopathic pulmonary fibrosis: a population-based study. Chest January 2010;137(1):129–37.

[19] Ley B, Collard HR, King Jr TE. Clinical course and prediction of survival in idiopathic pulmonary fibrosis. Am J Respir Crit Care Med February 15, 2011;183(4): 431–40.

[20] Collard HR, Moore BB, Flaherty KR, Brown KK, Kaner RJ, King Jr TE, et al. Acute exacerbations of idiopathic pulmonary fibrosis. Am J Respir Crit Care Med October 1, 2007;176(7):636–43.

[21] Iwai K, Mori T, Yamada N, Yamaguchi M, Hosoda Y. Idiopathic pulmonary fibrosis. Epidemiologic approaches to occupational exposure. Am J Respir Crit Care Med September 1994;150(3):670–5.

[22] Nemery B, Nagels J, Verbeken E, Dinsdale D, Demedts M. Rapidly fatal progression of cobalt lung in a diamond polisher. Am Rev Respir Dis May 1990;141(5 Pt 1):1373–8.

[23] Pujol JL, Barneon G, Bousquet J, Michel FB, Godard P. Interstitial pulmonary disease induced by occupational exposure to paraffin. Chest January 1990;97(1):234–6.

[24] Vallyathan V, Bergeron WN, Robichaux PA, Craighead JE. Pulmonary fibrosis in an aluminum arc welder. Chest March 1982;81(3):372–4.

[25] Sherson D, Maltbaek N, Heydorn K. A dental technician with pulmonary fibrosis: a case of chromium-cobalt alloy pneumoconiosis? Eur Respir J November 1990;3(10):1227–9.

[26] Tsukamoto K, Hayakawa H, Sato A, Chida K, Nakamura H, Miura K. Involvement of Epstein-Barr virus latent membrane protein 1 in disease progression in patients with idiopathic pulmonary fibrosis. Thorax November 2000;55(11):958–61.

[27] Arase Y, Suzuki F, Suzuki Y, Akuta N, Kobayashi M, Kawamura Y, et al. Hepatitis C virus enhances incidence of idiopathic pulmonary fibrosis. World J Gastroenterol October 14, 2008;14(38):5880–6.

[28] Yonemaru M, Kasuga I, Kusumoto H, Kunisawa A, Kiyokawa H, Kuwabara S, et al. Elevation of antibodies to cytomegalovirus and other herpes viruses in pulmonary fibrosis. Eur Respir J September 1997;10(9):2040–5.

[29] Gross TJ, Hunninghake GW. Idiopathic pulmonary fibrosis. N Engl J Med August 16, 2001;345(7):517–25.

[30] Datta A, Scotton CJ, Chambers RC. Novel therapeutic approaches for pulmonary fibrosis. Br J Pharmacol May 2011;163(1):141–72.

[31] Andersson-Sjoland A, de Alba CG, Nihlberg K, Becerril C, Ramirez R, Pardo A, et al. Fibrocytes are a potential source of lung fibroblasts in idiopathic pulmonary fibrosis. Int J Biochem Cell Biol 2008;40(10):2129–40.

[32] Kage H, Borok Z. EMT and interstitial lung disease: a mysterious relationship. Curr Opin Pulm Med September 2012;18(5):517–23.

[33] Bartis D, Mise N, Mahida RY, Eickelberg O, Thickett DR. Epithelial-mesenchymal transition in lung development and disease: does it exist and is it important? Thorax August 2014;69(8):760–5.

[34] Katzenstein AL, Myers JL. Idiopathic pulmonary fibrosis: clinical relevance of pathologic classification. Am J Respir Crit Care Med April 1998;157(4 Pt 1):1301–15.

[35] du Bois RM, Weycker D, Albera C, Bradford WZ, Costabel U, Kartashov A, et al. Ascertainment of individual risk of mortality for patients with idiopathic pulmonary fibrosis. Am J Respir Crit Care Med August 15, 2011;184(4):459–66.

[36] Ley B, Ryerson CJ, Vittinghoff E, Ryu JH, Tomassetti S, Lee JS, et al. A multidimensional index and staging system for idiopathic pulmonary fibrosis. Ann Intern Med May 15, 2012;156(10):684–91.

[37] Zuo F, Kaminski N, Eugui E, Allard J, Yakhini Z, Ben-Dor A, et al. Gene expression analysis reveals matrilysin as a key regulator of pulmonary fibrosis in mice and humans. Proc Natl Acad Sci USA April 30, 2002;99(9):6292–7.

[38] Konishi K, Gibson KF, Lindell KO, Richards TJ, Zhang Y, Dhir R, et al. Gene expression profiles of acute exacerbations of idiopathic pulmonary fibrosis. Am J Respir Crit Care Med July 15, 2009;180(2):167–75.

[39] Selman M, Pardo A, Kaminski N. Idiopathic pulmonary fibrosis: aberrant recapitulation of developmental programs? PLoS Med March 4, 2008;5(3):e62.

[40] Barlo NP, van Moorsel CH, Ruven HJ, Zanen P, van den Bosch JM, Grutters JC. Surfactant protein-D predicts survival in patients with idiopathic pulmonary fibrosis. Sarcoidosis Vasc Diffuse Lung Dis July 2009;26(2):155–61.

[41] Kinder BW, Brown KK, McCormack FX, Ix JH, Kervitsky A, Schwarz MI, et al. Serum surfactant protein-A is a strong predictor of early mortality in idiopathic pulmonary fibrosis. Chest June 2009;135(6):1557–63.

[42] Satoh H, Kurishima K, Ishikawa H, Ohtsuka M. Increased levels of KL-6 and subsequent mortality in patients with interstitial lung diseases. J Intern Med November 2006;260(5):429–34.

[43] Prasse A, Probst C, Bargagli E, Zissel G, Toews GB, Flaherty KR, et al. Serum CC-chemokine ligand 18 concentration predicts outcome in idiopathic pulmonary fibrosis. Am J Respir Crit Care Med April 15, 2009;179(8):717–23.

[44] Ishii H, Mukae H, Kadota J, Kaida H, Nagata T, Abe K, et al. High serum concentrations of surfactant protein A in usual interstitial pneumonia compared with nonspecific interstitial pneumonia. Thorax January 2003;58(1):52–7.

[45] Ohnishi H, Yokoyama A, Kondo K, Hamada H, Abe M, Nishimura K, et al. Comparative study of KL-6, surfactant protein-A, surfactant protein-D, and monocyte chemoattractant protein-1 as serum markers for interstitial lung diseases. Am J Respir Crit Care Med February 1, 2002;165(3):378–81.

[46] Casoni GL, Ulivi P, Mercatali L, Chilosi M, Tomassetti S, Romagnoli M, et al. Increased levels of free circulating DNA in patients with idiopathic pulmonary fibrosis. Int J Biol Markers October–December 2010;25(4):229–35.

[47] Moeller A, Gilpin SE, Ask K, Cox G, Cook D, Gauldie J, et al. Circulating fibrocytes are an indicator of poor prognosis in idiopathic pulmonary fibrosis. Am J Respir Crit Care Med April 1, 2009;179(7):588–94.

[48] Borensztajn K, Crestani B, Kolb M. Idiopathic pulmonary fibrosis: from epithelial injury to biomarkers–insights from the bench side. Respiration 2013;86(6):441–52.

[49] King Jr TE, Bradford WZ, Castro-Bernardini S, Fagan EA, Glaspole I, Glassberg MK, et al. A phase 3 trial of pirfenidone in patients with idiopathic pulmonary fibrosis. N Engl J Med May 29, 2014;370(22):2083–92.

[50] Richeldi L, du Bois RM, Raghu G, Azuma A, Brown KK, Costabel U, et al. Efficacy and safety of nintedanib in idiopathic pulmonary fibrosis. N Engl J Med May 29, 2014;370(22):2071–82.

[51] Noble PW, Albera C, Bradford WZ, Costabel U, Glassberg MK, Kardatzke D, et al. Pirfenidone in patients with idiopathic pulmonary fibrosis (CAPACITY): two randomised trials. Lancet May 21, 2011;377(9779):1760–9.

[52] Richeldi L, Cottin V, Flaherty KR, Kolb M, Inoue Y, Raghu G, et al. Design of the INPULSIS trials: two phase 3 trials of nintedanib in patients with idiopathic pulmonary fibrosis. Respir Med July 2014;108(7):1023–30.

[53] Richeldi L, Costabel U, Selman M, Kim DS, Hansell DM, Nicholson AG, et al. Efficacy of a tyrosine kinase inhibitor in idiopathic pulmonary fibrosis. N Engl J Med September 22, 2011;365(12):1079–87.

[54] Pandit KV, Milosevic J, Kaminski N. MicroRNAs in idiopathic pulmonary fibrosis. Transl Res April 2011;157(4):191–9.

[55] Liu G, Friggeri A, Yang Y, Milosevic J, Ding Q, Thannickal VJ, et al. miR-21 mediates fibrogenic activation of pulmonary fibroblasts and lung fibrosis. J Exp Med August 2, 2010;207(8):1589–97.

[56] Milosevic J, Pandit K, Magister M, Rabinovich E, Ellwanger DC, Yu G, et al. Profibrotic role of miR-154 in pulmonary fibrosis. Am J Respir Cell Mol Biol December 2012;47(6):879–87.

[57] 6.2 Tv. TargetScan, Release 6.2. Available at: http://www.targetscan.org.

[58] Yang S, Cui H, Xie N, Icyuz M, Banerjee S, Antony VB, et al. miR-145 regulates myofibroblast differentiation and lung fibrosis. FASEB J June 2013;27(6):2382–91.

[59] Liu Y, Sinha S, Owens G. A transforming growth factor-beta control element required for SM alpha-actin expression in vivo also partially mediates GKLF-dependent transcriptional repression. J Biol Chem November 28, 2003;278(48):48004–11.

[60] Hinz B, Phan SH, Thannickal VJ, Galli A, Bochaton-Piallat ML, Gabbiani G. The myofibroblast: one function, multiple origins. Am J Pathol June 2007;170(6):1807–16.

[61] Lino Cardenas CL, Henaoui IS, Courcot E, Roderburg C, Cauffiez C, Aubert S, et al. miR-199a-5p is upregulated during fibrogenic response to tissue injury and mediates TGFbeta-induced lung fibroblast activation by targeting caveolin-1. PLoS Genet 2013;9(2):e1003291.

[62] Nho RS, Im J, Ho YY, Hergert P. MicroRNA-96 inhibits FoxO3a function in IPF fibroblasts on type I collagen matrix. Am J Physiol Lung Cell Mol Physiol October 15, 2014;307(8):L632–42.

[63] Lin H, Dai T, Xiong H, Zhao X, Chen X, Yu C, et al. Unregulated miR-96 induces cell proliferation in human breast cancer by downregulating transcriptional factor FOXO3a. PLoS ONE 2010;5(12):e15797.

[64] Nho RS, Hergert P, Kahm J, Jessurun J, Henke C. Pathological alteration of FoxO3a activity promotes idiopathic pulmonary fibrosis fibroblast proliferation on type i collagen matrix. Am J Pathol November 2011;179(5):2420–30.

[65] Boreddy SR, Pramanik KC, Srivastava SK. Pancreatic tumor suppression by benzyl isothiocyanate is associated with inhibition of PI3K/AKT/FOXO pathway. Clin Cancer Res April 1, 2011;17(7):1784–95.

[66] Gu TL, Tothova Z, Scheijen B, Griffin JD, Gilliland DG, Sternberg DW. NPM-ALK fusion kinase of anaplastic large-cell lymphoma regulates survival and proliferative signaling through modulation of FOXO3a. Blood June 15, 2004;103(12):4622–9.

[67] Roy SK, Chen Q, Fu J, Shankar S, Srivastava RK. Resveratrol inhibits growth of orthotopic pancreatic tumors through activation of FOXO transcription factors. PLoS ONE 2011;6(9):e25166.

[68] Bodempudi V, Hergert P, Smith K, Xia H, Herrera J, Peterson M, et al. miR-210 promotes IPF fibroblast proliferation in response to hypoxia. Am J Physiol Lung Cell Mol Physiol August 15, 2014;307(4):L283–94.

[69] Leone G, DeGregori J, Yan Z, Jakoi L, Ishida S, Williams RS, et al. E2F3 activity is regulated during the cell cycle and is required for the induction of S phase. Genes Dev July 15, 1998;12(14):2120–30.

[70] Zhang Z, Sun H, Dai H, Walsh RM, Imakura M, Schelter J, et al. MicroRNA miR-210 modulates cellular response to hypoxia through the MYC antagonist MNT. Cell Cycle September 1, 2009;8(17):2756–68.

[71] Pandit KV, Corcoran D, Yousef H, Yarlagadda M, Tzouvelekis A, Gibson KF, et al. Inhibition and role of let-7d in idiopathic pulmonary fibrosis. Am J Respir Crit Care Med July 15, 2010;182(2):220–9.

[72] Huleihel L, Ben-Yehudah A, Milosevic J, Yu G, Pandit K, Sakamoto K, et al. Let-7d microRNA affects mesenchymal phenotypic properties of lung fibroblasts. Am J Physiol Lung Cell Mol Physiol March 15, 2014;306(6):L534–42.

[73] Cushing L, Kuang PP, Qian J, Shao F, Wu J, Little F, et al. miR-29 is a major regulator of genes associated with pulmonary fibrosis. Am J Respir Cell Mol Biol August 2011;45(2):287–94.

[74] Parker MW, Rossi D, Peterson M, Smith K, Sikstrom K, White ES, et al. Fibrotic extracellular matrix activates a profibrotic positive feedback loop. J Clin Invest April 2014;124(4):1622–35.

[75] Fernandes DJ, Bonacci JV, Stewart AG. Extracellular matrix, integrins, and mesenchymal cell function in the airways. Curr Drug Targets May 2006;7(5):567–77.

[76] Pardo A, Selman M. Matrix metalloproteases in aberrant fibrotic tissue remodeling. Proc Am Thorac Soc June 2006;3(4):383–8.

[77] Strieter RM, Mehrad B. New mechanisms of pulmonary fibrosis. Chest November 2009;136(5):1364–70.

[78] Xiao J, Meng XM, Huang XR, Chung AC, Feng YL, Hui DS, et al. miR-29 inhibits bleomycin-induced pulmonary fibrosis in mice. Mol Ther June 2012;20(6):1251–60.

[79] Montgomery RL, Yu G, Latimer PA, Stack C, Robinson K, Dalby CM, et al. MicroRNA mimicry blocks pulmonary fibrosis. EMBO Mol Med October 2014;6(10):1347–56.

[80] Yang S, Banerjee S, de Freitas A, Sanders YY, Ding Q, Matalon S, et al. Participation of miR-200 in pulmonary fibrosis. Am J Pathol February 2012;180(2):484–93.

[81] Lamouille S, Xu J, Derynck R. Molecular mechanisms of epithelial-mesenchymal transition. Nat Rev Mol Cell Biol March 2014;15(3):178–96.

[82] Cho JH, Gelinas R, Wang K, Etheridge A, Piper MG, Batte K, et al. Systems biology of interstitial lung diseases: integration of mRNA and microRNA expression changes. BMC Med Genomics 2011;4:8.

[83] Liang H, Gu Y, Li T, Zhang Y, Huangfu L, Hu M, et al. Integrated analyses identify the involvement of microRNA-26a in epithelial-mesenchymal transition during idiopathic pulmonary fibrosis. Cell Death Dis 2014;5:e1238.

[84] Liang H, Xu C, Pan Z, Zhang Y, Xu Z, Chen Y, et al. The antifibrotic effects and mechanisms of microRNA-26a action in idiopathic pulmonary fibrosis. Mol Ther June 2014;22(6):1122–33.

[85] Wei C, Kim IK, Kumar S, Jayasinghe S, Hong N, Castoldi G, et al. NF-kappaB mediated miR-26a regulation in cardiac fibrosis. J Cell Physiol July 2013;228(7):1433–42.

[86] Das S, Kumar M, Negi V, Pattnaik B, Prakash YS, Agrawal A, et al. MicroRNA-326 regulates profibrotic functions of transforming growth factor-beta in pulmonary fibrosis. Am J Respir Cell Mol Biol May 2014;50(5):882–92.

[87] Dakhlallah D, Batte K, Wang Y, Cantemir-Stone CZ, Yan P, Nuovo G, et al. Epigenetic regulation of miR-17~92 contributes to the pathogenesis of pulmonary fibrosis. Am J Respir Crit Care Med February 15, 2013;187(4):397–405.

[88] Berschneider B, Ellwanger DC, Baarsma HA, Thiel C, Shimbori C, White ES, et al. miR-92a regulates TGF-beta1-induced WISP1 expression in pulmonary fibrosis. Int J Biochem Cell Biol August 2014;53:432–41.

[89] Konigshoff M, Kramer M, Balsara N, Wilhelm J, Amarie OV, Jahn A, et al. WNT1-inducible signaling protein-1 mediates pulmonary fibrosis in mice and is upregulated in humans with idiopathic pulmonary fibrosis. J Clin Invest April 2009;119(4):772–87.

[90] Izbicki G, Segel MJ, Christensen TG, Conner MW, Breuer R. Time course of bleomycin-induced lung fibrosis. Int J Exp Pathol June 2002;83(3):111–9.

[91] Gibson G. Microarray analysis: genome-scale hypothesis scanning. PLoS Biol October 2003;1(1):E15.

[92] Geiss GK, Bumgarner RE, Birditt B, Dahl T, Dowidar N, Dunaway DL, et al. Direct multiplexed measurement of gene expression with color-coded probe pairs. Nat Biotechnol March 2008;26(3):317–25.

[93] Moltzahn F, Hunkapiller N, Mir AA, Imbar T, Blelloch R. High throughput microRNA profiling: optimized multiplex qRT-PCR at nanoliter scale on the fluidigm dynamic arrayTM IFCs. J Vis Exp 2011;(54):e2552. http://dx.doi.org/10.3791/2552.

[94] Brown BD, Gentner B, Cantore A, Colleoni S, Amendola M, Zingale A, et al. Endogenous microRNA can be broadly exploited to regulate transgene expression according to tissue, lineage and differentiation state. Nat Biotechnol December 2007;25(12):1457–67.

[95] Ebert MS, Neilson JR, Sharp PA. MicroRNA sponges: competitive inhibitors of small RNAs in mammalian cells. Nat Methods September 2007;4(9):721–6.

[96] Moeller A, Ask K, Warburton D, Gauldie J, Kolb M. The bleomycin animal model: a useful tool to investigate treatment options for idiopathic pulmonary fibrosis? Int J Biochem Cell Biol 2008;40(3):362–82.

[97] Santaris Pharma A/S. Multiple ascending dose study of Miravirsen in treatment-naïve chronic hepatitis C subjects. Available from: http://clinicaltrials.gov/show/NCT01200420. NLM Identifier: NCT01200420.

[98] Lanford RE, Hildebrandt-Eriksen ES, Petri A, Persson R, Lindow M, Munk ME, et al. Therapeutic silencing of microRNA-122 in primates with chronic hepatitis C virus infection. Science January 8, 2010;327(5962):198–201
[a] Janssen HL, Reesink HW, Lawitz EJ, Zeuzem S, Rodriguez-Torres M, Patel K, et al. Treatment of HCV infection by targeting microRNA. N Engl J Med May 2, 2013;368(18):1685–94.

[99] Santaris Pharma A/S. Miravirsen study in null responder to pegylated interferon alpha plus ribavirin subjects with chronic hepatitis C. Available from: http://clinicaltrials.gov/show/NCT01727934. NLM Identifier: NCT01727934.

[100] Mirna Therapeutics I. A multicenter phase I study of MRX34, microRNA miR-RX34 liposome injectable suspension. Available from: http://clinicaltrials.gov/show/NCT01829971. NLM Identifier: NCT01829971.

[101] He L, He X, Lim LP, de Stanchina E, Xuan Z, Liang Y, et al. A microRNA component of the p53 tumour suppressor network. Nature June 28, 2007;447(7148):1130–4.

[102] Preclinical data of a microRNA-based therapy for hepatocellular carcinoma. In: Bader A, editor. Annual AACR Conference. 2012. Chicago (IL).

[103] The development of a miRNA-based therapeutic candidate for hepatocellular carcinoma. In: Daige C, editor. AACR-NCI-EORTC International Conference: Molecular Targets and Cancer Therapeutics. 2011. San Francisco (CA).

[104] Miragen Therapeutics. Available from: http://miragentherapeutics.com/pipeline/.

[105] Li Z, Rana TM. Therapeutic targeting of microRNAs: current status and future challenges. Nat Rev Drug Discov August 2014;13(8):622–38.

[106] Wang J, Liu X, Wu H, Ni P, Gu Z, Qiao Y, et al. CREB up-regulates long non-coding RNA, HULC expression through interaction with microRNA-372 in liver cancer. Nucleic Acids Res September 2010;38(16):5366–83.

[107] Poliseno L, Salmena L, Zhang J, Carver B, Haveman WJ, Pandolfi PP. A coding-independent function of gene and pseudogene mRNAs regulates tumour biology. Nature June 24, 2010;465(7301):1033–8.

[108] Alimonti A, Carracedo A, Clohessy JG, Trotman LC, Nardella C, Egia A, et al. Subtle variations in Pten dose determine cancer susceptibility. Nat Genet May 2010; 42(5):454–8.

[109] Cesana M, Cacchiarelli D, Legnini I, Santini T, Sthandier O, Chinappi M, et al. A long noncoding RNA controls muscle differentiation by functioning as a competing endogenous RNA. Cell October 14, 2011;147(2):358–69.

[110] Hessels D, Klein Gunnewiek JM, van Oort I, Karthaus HF, van Leenders GJ, van Balken B, et al. DD3(PCA3)-based molecular urine analysis for the diagnosis of prostate cancer. Eur Urol July 2003;44(1):8–15. Discussion-6.

[111] Tinzl M, Marberger M, Horvath S, Chypre C. DD3PCA3 RNA analysis in urine–a new perspective for detecting prostate cancer. Eur Urol August 2004;46(2):182–6. Discussion 7.

[112] Panzitt K, Tschernatsch MM, Guelly C, Moustafa T, Stradner M, Strohmaier HM, et al. Characterization of HULC, a novel gene with striking up-regulation in hepatocellular carcinoma, as noncoding RNA. Gastroenterology January 2007;132(1):330–42.

[113] Ji P, Diederichs S, Wang W, Boing S, Metzger R, Schneider PM, et al. MALAT-1, a novel noncoding RNA, and thymosin beta4 predict metastasis and survival in early-stage non-small cell lung cancer. Oncogene September 11, 2003;22(39):8031–41.

[114] Pandit KV, Milosevic J. MicroRNA regulatory networks in idiopathic pulmonary fibrosis. Biochem Cell Biol April 2015;93(2):129–37.

[115] van Rooij E, Sutherland LB, Thatcher JE, DiMaio JM, Naseem RH, Marshall WS, et al. Dysregulation of microRNAs after myocardial infarction reveals a role of miR-29 in cardiac fibrosis. Proc Natl Acad Sci USA September 2, 2008;105(35):13027–32.

CHAPTER 8

MicroRNA as Biomarkers of Malignant Mesothelioma

C.D. Hoang, R.A. Kratzke

Key Concepts

- Malignant pleural mesothelioma is an aggressive, relatively chemotherapy and radiation resistant disease without effective cure.
- MicroRNA (miRNA) display a specific pattern of derangement in mesothelioma and likely contribute to multiple mechanisms driving this malignancy.
- MiRNAs are a unique class of noncoding regulatory RNA that can simultaneously target multiple gene pathways.
- MiRNA in mesothelioma have diverse translational possibilities affecting pathogenetic, diagnostic, prognostic, and/ or therapeutic applications.

1. CLINICAL PROBLEM: MESOTHELIOMA

There is no effective cure or treatment for pleural asbestos–related diseases. Malignant mesothelioma is an aggressive neoplasm of the serosal lining of the pleural and peritoneal cavities arising from mesothelial cells. Clinically, the most common form is malignant pleural mesothelioma (MPM), which accounts for about 2500–3000 deaths per year in the United States [1]. Up to 80% of cases of MPM occur in patients 10–20 years after exposure to asbestos [2]. Due to this latency period between asbestos exposure and tumor development, the associated mortality rate in men, but not in women, continues to rise in industrialized countries at the rate of 5–10% per year, with a median survival time between 4 and 18 months [3]. This rising mortality rate is occurring despite the implementation of legislation limiting asbestos use and exposure in most industrialized countries. However, developing nations are increasing their asbestos imports and consumption [4]. Thus, the incidence of MPM continues to rise and is estimated to peak within the next 10–15 years in the western hemisphere, while cases worldwide are predicted to rise for up to 40 years [5,6].

Translating MicroRNAs to the Clinic
ISBN 978-0-12-800553-8

Complicating this trend in incidence rise, is the observation that MPM remains a relatively chemotherapy- and radiation-resistant cancer usually diagnosed in advanced stage. Regardless if patients are eligible for aggressive multimodal therapy to include surgical resection, the standard-of-care regimen based on pemetrexed and cisplatin has not significantly improved prognosis in over a decade of widespread use [7].

2. MOLECULAR PATHOLOGY

Aside from asbestos exposure, other factors like ionizing radiation or tumor DNA virus (simian virus-40) may act synergistically in MPM pathogenesis [8]. Acquired genetic lesions have been identified in MPM, including at the 9p21 locus (p16INK4a, p14ARF) and the 22q11-q13.1 locus (NF2) [9]. However, the molecular mechanisms controlling the malignant transformation of mesothelial cells remain poorly defined. This is underscored by the observation that these well-characterized etiologies incompletely account for the known incidence of MPM. About 10–20% of MPM occurrences have been documented in patients without previous exposure to asbestos, and only 60% of MPM tumors are known to harbor simian virus-40 viral DNA [9].

Accordingly, we and others have identified that multiple active oncogenic pathways are involved in MPM pathobiology [10,11]. It is likely that this multiplicity and complexity of pathogenic processes represent an ongoing barrier to translation of standard anticancer agents directed against mesothelioma. Similarly, targeted therapies aimed at specific oncogenes used alone or in combinations, would not be expected to exert much clinical impact on mesothelioma because of these interconnected, redundant pathogenic pathways. Successful interventions for MPM more likely require interference of multiple oncogenic pathways.

Furthermore, it is believed that the overall low survival statistics of thoracic cancers are (at least) partially attributable to diagnosis occurring at advanced stages. Despite extensive studies, there are no available biomarkers that could lead to earlier detection for MPM in routine clinical use, including the FDA-approved MESOMARK assay [12].

3. MICRORNAs AS ONCOGENES AND TUMOR SUPPRESSORS

MiRNAs are a family of short, noncoding regulatory RNA molecules expressed in a tissue-specific, developmentally regulated manner. Most miRNAs posttranscriptionally regulate the degradation or translation of target messenger RNAs (mRNAs), and may function as oncogenes or tumor suppressors [13,14]. miRNAs are frequently aberrantly expressed or mutated in cancer, and 50% of miRNA genes are located in cancer-associated genomic regions or fragile sites, suggesting a role in disease initiation and progression. Dysregulation of miRNA expression in various human cancers has been reported, including chronic lymphocytic leukemia, breast carcinoma, and lung carcinoma, for example [15,16].

Continuing work in this field has now confirmed at least 1996 unique human miRNA according to miRBase v21 released in 2014 [17]. However, rapid adoption of next-generation and deep sequencing methods have already suggested the existence of perhaps thousands more novel, yet to be characterized miRNA in the human genome that are tissue specific [18].

4. TRANSLATIONAL APPLICATIONS OF MICRORNA IN MESOTHELIOMA

4.1 Microarray Profiling Results

miRNA microarray profiling in human MPM has potential to rapidly identify relevant, disease-specific miRNA. This is a necessary endeavor to select specific miRNA and guide further mechanistic studies, since the regulation of mRNA targets by miRNA are highly complex.

Toward this end, Guled et al. [19], in 2009, were among the first to globally profile miRNA using a microarray from Agilent containing 723 human miRNA. They studied 17 MPM fresh-frozen samples (one sarcomatoid) wherein they identified 21 differentially expressed miRNA relative to normal tissue. Their findings were based on a comparison with a single sample of human pericardium as the normal control. Overall miRNA expression profiles of tumor tissue versus normal mesothelium: twelve of them, miR-let-7b*, -1228*, -195*, -30b*, -32*, -345, -483-3p, -584, -595, -615-3p, -934, and miR-885-3p, were highly expressed whereas the remaining nine, miR-let-7e*, miR-144*, miR-203, miR-340*, miR-34a*, miR-423, miR-582, miR-7-1*, and miR-9, were unexpressed or had severely reduced expression levels. Additional profiling revealed subsets of differentially expressed miRNAs that seemed to recapitulate the three distinct histopathologic subtypes of MPM. Epithelioid MPM preferentially expressed miR-135b, -181a-2*, -499-5p, -517b, -519d, -615-5p, and miR-624 in distinction to sarcomatoid MPM which displayed miR-301b, -433, and miR543. The biphasic subtype of MPM expressed more miR-218-2*, -346, -377*, -485-5p, and miR-525-3p. Furthermore, when the smoking status and asbestos exposure were taken into consideration, significantly differentially expressed miRNAs were identified in smokers versus non-smokers (miR-379, -301a, -299-3p, -455-3p, and miR-127-3p), but not in asbestos-exposed patients versus unexposed ones. Lastly, using inference from in silico, Web-based analysis, they identified likely target genes for these miRNAs such as CDKN2A, NF2, JUN, HGF, and PDGFA. Many of the altered miRNAs were located in chromosomal areas known to be deleted or gained in MPM such as 8q24, 1p36, and 14q32. Since this seminal study, a large body of literature has appeared to detail the roles of various specific miRNA in MPM, but surprisingly few comprehensive profiling reports.

The study by Busacca et al. [20] in 2010 revealed another set of differentially expressed miRNAs in MPM. The specimens used for profiling were the cell lines MPP-89 and REN compared against a control human mesothelial cells (HMC-telomerase reverse transcriptase). miRNA expression profiling was performed using locked nucleic acid–modified oligonucleotide microarrays manufactured by Exiqon. The MPM specimens revealed differential expression of 10 upregulated and 19 downregulated miRNAs when compared with the immortalized control cells. Further, they investigated miRNA expression on a panel of 24 MPM tissues, representative of the three histologic subtypes (epithelioid, biphasic, and sarcomatoid), by quantitative real-time PCR (qRT-PCR). The expression of miR-17-5p, -21, -29a, -30c, -30e-5p, -106a, and miR-143 was significantly associated with the histopathologic subtypes. Notably, the reduced expression of miR-17-5p and miR-30c correlated with better survival of patients with the MPM sarcomatoid subtype—the only report to date that has identified such miRNA specific to sarcomatoid outcome (see later discussion about prognostic miRNA in MPM for more examples). Similarly, Balatti et al. [21] reported in 2011 a comparative microarray analysis of five MPM cell lines (NCI-H2052, NCI-H28, MPP-89, MSTO-211H, and IST-MES2) versus five human normal pleural mesothelial short-term cultures. They used the Agilent microarray G4470A consisting of 470 human miRNA. Microarray profiling showed members of the oncomiR miR-17-92 cluster and its paralogs (miR-17-5p, -18a, -19b, -20a, -20b, -25, -92, -106a, -106b) were markedly upregulated, while miR-214 and miR-497 were found to be downregulated in MPM cells. However, conclusions drawn from cell line data, as in these two studies, do not always corroborate in a biologically relevant manner with data from human tumors, even if the findings are validated in a larger human data set, because of the inherent artificiality of in vitro systems [22].

Other miRNA microarray data sets related to human MPM remain relatively sparse. We could identify the study by Gee et al. [23] that included

15 nonhistologically defined MPM samples that were profiled against 10 lung adenocarcinomas to identify discriminant miRNAs. Therefore, their reported significant miRNAs are necessarily nonoverlapping with other microarray studies. They used a custom Affymetrix miRNA genechip (TGmirV1b520432f) which, at that time, contained 2564 human sequences of miRNA for profiling. Compared to lung adenocarcinoma, members of the miR-200 gene family were significantly downregulated in MPM, including miR-200a★, -200b, -200c, -141, -429, -203, and miR-205. Our own work derived miRNA signatures from the largest reported cohort of MPM tissue samples profiled using an Illumina miRNA BeadChip containing 1145 miRNA probes [24]. We analyzed 25 MPM tumors compared to an unmatched cohort of noncancerous, normal human pleura. We observed 49 overexpressed miRNA accompanied by 65 underexpressed miRNA. Specifically, miR-155★, -130b★, -21★, and miR-149 were significantly overexpressed in MPM while miR-551b, -1/206, -591, -483-5p, and miR-363★ were underexpressed in MPM.

Thus, in published reports, there are no consistent miRNAs and/or their gene targets that have been proposed to be critical to the malignant phenotype of MPM. As illustrated here, contributing reasons for nonoverlapping miRNA results include (1) diverse experimental designs with variability in specimen type and/or control normal(s), (2) using very different microarray platforms, (3) an ever-increasing database of confirmed unique miRNA, and not the least (4) a spectrum of nonstandardized bioinformatics algorithms. Further, with the multiple updates of the miRNA database (eg, www.mirbase.org), the nomenclature of some specific miRNA changed; so in older reports it is difficult to recognize synonymous labels for the same miRNA. In a larger context, beyond the scope of this discussion, is the complex nature of miRNA interactions with a multitude of mRNA targets via interactions with the 3´-untranslated region [25] as well as other specific genetic regions (eg, protein coding region of mRNAs [26]). With the rapid changes in our knowledge of the human miRNAome, updated miRNA profiling studies may yield fresh insights into the pathobiology of MPM.

4.2 Biomarkers

There are several proposed MPM biomarkers including soluble mesothelin-related peptide (SMRP), osteopontin, and megakaryocyte potentiation factor [27]. These biomarkers in MPM lack adequate sensitivity, specificity, and reproducibility [28] to be useful assays for clinical purposes. The problem is, at least, because of the increased complexity of the proteome which is subject to alternative splice variations, protease cleavages, and posttranslational modifications. These biologic processes produce active protein species numbering from 10^7 to 10^9 that need to be searched to find a useful biomarker [29], which represents a formidable undertaking. More recently, and of great promise, is the description of fibulin-3 protein levels in plasma of asbestos-exposed patients being able to identify those harboring MPM from those without MPM [30]. However, this biomarker still needs to be validated in prediagnostic and early diagnostic scenarios.

Alternatively, miRNA could obviate many of the technical challenges accompanying traditional molecules in biomarker development. miRNAs have high stability in serum and plasma; are low complexity molecules with a smaller range of possible molecules to search as compared to mRNA or proteins; easy to detect with highly sensitive PCR-based methods; and are indicative of biological processes underlying cancer(s) [11].

4.2.1 Pathobiology

Since miRNAs regulate many gene targets and hence biologic pathways, their profiling may reveal novel insights into disease processes. In MPM, identification of novel, disease-specific miRNA may potentially impact early detection, differential diagnosis, and/or prognosis. Recently, to delineate new insights into the pathogenesis of MPM, Ramirez-Salazar et al. [31] examined the role of deranged miRNA among the interrelated processes of pleural chronic inflammation and mesothelial hyperplasia which may be critical to the development of MPM. Using TaqMan PCR array, they identified the miRNAs expressed in pleural tissues diagnosed with MPM ($n = 5$ epithelioid), pleural chronic inflammation ($n = 4$), and mesothelial hyperplasia ($n = 5$), as well as in noncancerous/noninflammatory tissue as the normal control ($n = 5$). A total of 667 miRNAs were screened. Bioinformatics and network analysis (Cytoscape) of differentially expressed miRNAs to identify tumorigenesis-related miRNAs and their biological gene targets (networks) were performed. The targets of four downregulated miRNA in MPM (miR-181a-5p, -101-3p, -145-5p, and miR-212-3p), one in pleural chronic inflammation (miR-101-3p) and one in mesothelial hyperplasia (miR-494) were significantly enriched in "pathways in cancer" (based on KEGG and Reactome databases). Interactome gene networks revealed that >50% of downregulated miRNAs in MPM targeted the signaling activation molecule MAPK1, the transcription factor ETS1, and the mesenchymal transition-associated molecule FZDA, which have been associated with oncogenic function. Comparative analysis revealed that FZD4 was an overlapping gene target of downregulated miRNAs that were associated with "pathways in cancer" in MPM, pleural inflammation, and mesothelial hyperplasia. This network analysis revealed a potential combinatory effect of deregulated miRNAs in MPM pathogenesis and indicated potential molecular links between pleural inflammation and hyperplasia with tumorigenesis mechanisms in pleura.

4.2.2 Early Detection

Perhaps MPM is an ideal tumor example of the benefits of early detection. Namely, MPM has a long latency period of at least 20 years, is detected in advanced stages with current clinical practice(s), and is relatively resistant to available treatment modalities. Discovery of a practical early detection biomarker could directly address and improve these critical issues negatively impacting the prognosis of MPM.

Weber et al. [32] analyzed the cellular fraction of peripheral blood in patients with proven MPM of all histologic subtypes ($n = 23$ total) compared to asbestos-exposed subjects (controls). They used the mirVana oligonucleotide microarray containing 328 miRNAs from Ambion to profile for potential miRNA biomarkers. The best miRNA to pass their stringency criteria was miR-103 significantly downregulated in MPM. Next, they validated this result by assaying levels of miR-103 using qRT-PCR in the MPM patients compared to asbestos-exposed and cancer-free general population patients. MiR-103 had a sensitivity of 83% and specificity of 71% to discriminate MPM patients from asbestos-exposed controls, while a sensitivity of 78% and specificity of 76% to discriminate MPM patients from the general population. To further improve the performance of this marker and demonstrate better reliability, they suggested combining miR-103 profile with other blood markers such as SMRP or calretinin.

With a similar goal to identify an early biomarker for MPM, Santarelli et al. [33] systematically screened multiple MPM specimens to identify a potential early disease biomarker. A total of 88 characterized miRNAs involved in carcinogenesis were assessed by multiplex qRT-PCR in a cohort of fresh-frozen biopsies proven to be MPM compared against venous blood serum of healthy noncancer patients (control). Consistent with prior observations of multiple solid cancer types showing biased underexpression of miRNAs, possibly due to the loss of ability to process premiRNA structures [34], three from a group of eight miRNAs significantly downregulated were further confirmed in a set of formalin-fixed, paraffin-embedded specimens. MPM and adjacent noncancerous sample pairs were analyzed for expression levels of these three miRNAs (miR-335, -126, and miR-32 which all were downregulated by 15-fold in their screening). MiR-126 emerged as the most consistently downregulated miRNA, so it was selected as a biomarker in serum samples of asbestos-exposed and MPM patients both compared against controls. Using ROC analysis, cut-off values of miR-126 were determined to discriminate asbestos-exposed subjects from MPM patients ($\Delta C_T = -4.5$; sensitivity 73% and specificity 74%). Improved predictive performance for early detection of MPM was found with decreasing expression of miR-126, when correlated with increasing levels of SMRP.

In a follow-up study [35], this research group evaluated the accuracy and precision of detecting miR-126 in the serum of 45 treatment-naïve MPM patients compared against both non-small cell–lung cancer (NSCLC) and healthy control patients. MiR-126 was quantified in serum using endogenous and exogenous controls for normalization and both relative and absolute qRT-PCR methods. Circulating miR-126 detected in the serum by relative qRT-PCR was found lowly expressed in both malignancies, significantly differentiated MPM patients from healthy controls and NSCLC from MPM, but did not discriminate NSCLC patients from control subjects. Kaplan–Meier analysis revealed that a low level of circulating miR-126 in MPM patients was strongly associated with poor prognosis. Unfortunately, miR-126 lacks specificity as it is expressed in other cancers but perhaps could be used in combination with other biomarkers.

4.2.3 Diagnostics

Other malignancies involving the thorax confound the reliable diagnosis of MPM [36] based on conventional pathologic methods. Often, MPM can be difficult to distinguish from lung adenocarcinoma and metastatic epithelial cancers. It is also a challenge sometimes to differentiate MPM from reactive mesothelial proliferations. miRNAs are potentially effective novel platform of biomarkers because of their tissue-specific expression.

In a subsequent validation phase of the study by Gee et al. [23], they confirmed their miRNA microarray results by qRT-PCR on a different cohort of 100 MPM compared against 32 lung adenocarcinoma tissues. They observed significant downregulation (range of 6- to 42-fold) of miR-141, -200a★, -200b, -200c, -203, -205, and miR-429 in the MPM tissues. In a different study specifically aimed at exploiting miRNA for its diagnostic discrimination, Benjamin et al. [37] used a systematic approach using 7 MPM compared against 97 epithelial carcinomas. In their discovery phase, they observed a microarray (custom array of 747 DNA oligonucleotide probes) pattern of overexpressed miR-193-3p in MPM only, while miR-200c and miR-192 were overexpressed in lung adenocarcinomas and other tumors frequently metastasizing to the pleura. miRNA signatures by microarray were verified using qRT-PCR on a separate tissue cohort of 51 specimens. Next in a training set of 88 new specimens, they defined combinations of miRNAs (a classification rule based on "PCR score"), which can be used to discriminate MPM from other carcinomas. In a final validation phase, blinded tumor tissues were tested. All 14 MPM (all histologic subtypes) samples and 46 of 49 non-MPM samples were classified correctly based on miR-193-3p expression, yielding 100% sensitivity and 94% specificity. Also of interest is that both of these studies identified the miR-200 family with relative overexpression in lung adenocarcinoma in contradistinction to low expression in MPM.

In contradistinction to discovery from tissues, Kirschner et al. [38] investigated the ability of cell-free miRNAs in plasma and serum to serve as a diagnostic biomarker for MPM. They used the Human 8 × 15K miRNA Microarray Kit (V3, miRBase v12.0) from Agilent, which contained 854 unique miRNA probes, to profile plasma samples from five MPM (epithelioid and sarcomatoid) patients and three healthy controls. The 12 miRNAs with the most significantly and relatively different abundance between tumor patients and controls were validated in a larger series of MPM patients ($n = 15$) and in an independent, separate cohort of MPM patients ($n = 30$) using TaqMan qRT-PCR of plasma/serum. Receiver operating characteristics curve analysis showed that plasma miR-625-3p levels discriminated the between patient and control specimens with an accuracy of 82.4% (area under the curve, 0.824; CI, 0.669–0.979), a sensitivity of 73.33%, and a specificity of 78.57%. Logistic regression was used to investigate the performance of combinations of the three miRs (miR-29c★, -92a, and miR-625-3p) in terms of accuracy; however, miR-625-3p on its own provided the best and most simple classification. To confirm their findings, they assessed 18 MPM tissues and consistently observed miR-625-3p to be differentially upregulated by qRT-PCR.

A related diagnostic dilemma is the reliable differentiation of MPM from reactive mesothelial proliferations. Andersen et al. [39] recently screened with a real-time qRT-PCR-based platform, the expression of 742 miRNAs in formalin fixed, paraffin-embedded, preoperative diagnostic biopsy samples, surgically resected MPM specimens previously treated with chemotherapy, and corresponding nonneoplastic pleura, from five patients. They identified that miR-126, -143, -145, and miR-652 were significantly downregulated ($\geq$2-fold) in resected MPM and/or chemotherapy-naïve diagnostic tumor biopsy samples. The miRNA expression pattern was validated by qRT-PCR in a cohort of 40 independent MPM (epithelioid and biphasic). By performing binary logistic regression on the PCR data for the four miRNAs, their established four-miRNA classifier differentiated MPM from nonneoplastic pleura with high sensitivity and specificity (area under the curve, 0.96; 95% CI, 0.92–1.00). The classifier's optimal logit(P) value (see Ref. [39] for further explanation) of 0.62 separated nonneoplastic pleura from MPM samples with a sensitivity of 0.95 (95% CI, 0.89–1.00), a specificity of 0.93 (95% CI, 0.87–0.99), and an overall accuracy of 0.94 (95% CI, 0.88–1.00). Overall, these results indicate that these four miRNAs may be suitable biomarkers for distinguishing MPM from reactive mesothelial proliferations, and lend support for the concept of combinatorial sets of multiple miRNA are expected to be superior to single miRNA biomarkers. A few caveats to this study are the requirement of tissues obtained by invasive methods and that areas of tumors assayed by this PCR method had to contain >50% tumor-cell content to sustain accuracy. Ongoing refinement of this miRNA panel is in progress.

4.2.4 Prognosis

miRNAs are implicated in clinical outcomes of MPM. Pass et al. [40] provided one of the earliest examples of this concept in MPM. One of the unique aspects of their data set included accompanying clinical and outcomes information for each specimen along with molecular miRNA profiling. In 2004, they profiled 129 surgically resected MPM specimens with a custom-made miRNA microarray provided by Rosetta Genomics that contained approximately 900 unique probes applied to a training set of 37 patient tumors (epithelioid and nonepithelioid) and a test set of 92 patients (similar tumor histopathologic distribution). Using multivariate analysis, they identified a single miR independently and significantly associated with prognosis. They observed two groups in the test set with significantly different time to progression (or recurrence): one group had recurrence at 5.5 months versus the other group at 12.8 months ($P=$ .008). Similarly, patients in the survival test set separated into two distinct groups with observed median survival times of 9.1 months versus 21.6 months ($P=$ .0026). Notably, it was miR-29c* (updated currently in miRBase v21 as miR-29c-5p) expression that correlated with MPM histology. miR-29c* was differentially expressed between epithelioid subtype (being higher expression level) as compared to other histologies. miR-29c* could subclassify epithelioid MPM into short- versus long time for progression and overall survival times. Increased expression of miR-29c★ predicted a more favorable prognosis. Forcing overexpression of miR-29c★ by transfection into MPM cell lines with the premiR mimic demonstrated significantly decreased proliferation, migration, invasion, and colony formation. Additional prospective validation studies are required, however, to fully delineate the role of miR-29c★ (miR-29c-5p) in MPM as the study by Kirschner [38] was not able to demonstrate significance for miR-29c★ in their hands.

More recently, Kirschner et al. [41] have developed an MPM prognostic predictor model based on a group of six miRNAs. Candidate miRNAs from microarray profiling of tumor samples from eight long (median 53.7 months) and eight short (median 6.4 months) survivors following extrapleural pneumonectomy (EPP) were identified in the discovery phase. They used the same microarray platform based on an older Agilent production with 854 miRNA probes. The MPM patients all were epithelioid subtype without induction chemotherapy in clinical stages II or III. Tissues were dissected by laser capture microdissection prior to RNA analysis. After multiple stringency criteria were applied, some candidate miRNAs were validated by qRT-PCR in 48 additional EPP tumor samples. Kaplan–Meier log ranking was used to assess the association between miRNA expression and overall survival. Binary logistic regression constructed an miRNA signature (miR-Score, a weighted formula described in detail in Ref. 39) that was predictive of overall survival ≥20 months. Performance of their miR-Score was evaluated by receiver operating characteristic curve analysis and then validated in a separate cohort of 43 tumor samples from patients who underwent pleurectomy/decortication (P/D). The miR-Score (miR-21-5p, -23a-3p, -30e-5p, -221-3p, -222-3p, and miR-31-5p), predicted prolonged patient survival with an accuracy of 92.3% for those undergoing EPP versus 71.9% for those receiving P/D. Hazard ratios for score-negative patients were 4.12 (95% CI, 2.03–8.37) for EPP and 1.93 (95% CI, 1.01–3.69) for P/D. Adding their miR-Score to a set of clinical variables (histology, age, gender) increased predictive accuracy in the independent validation set from 76.3% for clinical factors only to 87.3%. During construction of this predictive miRNA model, the authors purposefully included previously reported miRNAs (miR-29c-5p, -31-5p, -106a-5p, -126-3p, -625-3p, -92a-3p, -23a-3p, and miR-24-3p). None but for the addition of miR-23a-3p passed to the final prognostic model of miR-Score.

This area of investigation is going to require larger, prospective investigations before any miRNA is identified as clinically relevant and prognostic. This remains a formidable challenge given the relative rarity of MPM patients who undergo multimodality treatments in an organized manner under study protocols.

4.2.5 Therapeutics and Targets

Inherent in studies of miRNAs that have relevance as biomarkers in MPM, is the potential to translate findings of the gene targets and networks into practical and/or novel anticancer therapies. Not all dysregulated miRNAs induce therapeutic phenotype changes when studied in MPM. So identifying the most suitable miRNA that may exert therapeutic effects in the broadest range of MPM types is actively under investigation. Beyond this, are significant biologic barriers that need to be addressed to specifically target and deliver such miRNA to MPM tumor cells. Following are some current examples of this emerging miRNA therapy prospects.

Ivanov et al. [42] were among early studies to show in MPM that miRNAs could exert beneficial phenotype effects when constitutively reexpressed in MPM cells. They reported that eight well-known MPM cell lines (eg, NCI-H2373, NCI-H2596, and NCI-H2461, etc.) failed to express miR-31, an miRNA linked with suppression of metastases in another solid tumor type. They explained that loss of miR-31 expression in MPM cells was due to homozygous deletion of the miR-31-encoding gene residing in chromosome region 9p21.3. They also noted that in clinical cohorts of MPM, those with 9p21.3 chromosome deletion showed a significantly shorter time to tumor recurrence. Functional assessment of miR-31 activity revealed its ability to inhibit proliferation, migration, invasion, and clonogenicity of those MPM cells tested. Reexpression of miR-31 by transfection of miRNA mimic suppressed the cell cycle and inhibited expression of multiple factors involved in maintenance of DNA replication and cell cycle progression, including pro-survival phosphatase PPP6C, which was known to facilitate chemotherapy and radiation therapy resistance, and maintenance of chromosomal stability. PPP6C mRNA, which has 3′-untranslated region with verified binding sites for miR-31, was consistently downregulated by miR-31 reexpression in MPM cells. Likewise, PPP6C mRNA was upregulated in clinical MPM specimens as compared with matched normal peritoneal tissues. These results suggest that the tumor suppressive propensity of miR-31 can be used for development of new therapies against mesothelioma (even possibly to other cancers that show loss of the 9p21.3 chromosome).

Similarly, our report of miR-1 [24], an miRNA controlling development of mesenchymal-derived tissues, revealed its strong effects consistent with a tumor suppressor role in MPM. We first screened and observed that the expression level of miR-1 was significantly lower in tumors (all histologic subtypes) as compared with normal pleural specimens. Subsequently, premiR of miR-1 was introduced into MPM cell lines to reintroduce forced overexpression of this miRNA. Phenotypic changes of these transfected cells were assayed. The cellular proliferation rate was significantly inhibited after overexpression of miR-1. Early and late apoptosis was increased markedly in miR-1-transfected cell lines. Our data suggested that overexpression of miR-1 induced a profound level of apoptosis in tested MPM cell lines, acting as a tumor suppressor. We confirmed our observations by assessing in the transduced MPM cells cell cycle-related, proapoptotic, and antiapoptotic genes by qRT-PCR, which all showed coordinated, significant changes characteristic of the apoptotic phenotype. An in silico analysis further supported miR-1 involvement in apoptotic gene pathways. Thus, our study suggests that at least miR-1 has potential pathogenic and therapeutic significance for MPM. Our unpublished results of the related miR-1 family member, miR-206 (significantly underexpressed in MPM tissues), also showed significant tumor suppressor function specific in MPM. The miR-1/206 family may represent an effective therapy against MPM by replacing underexpressed miR(s) dominant in the malignant phenotype.

The miR-34 family has been extensively reported in MPM. Kubo et al. [43] noted gene silencing of miR-34b/c but not of miR-34a using methylation-specific PCR. MiR-34 family members are direct targets of TP53 methylation-specific PCR screening of MPM tissues ($n = 47$) and cell lines. Aberrant methylation of miR-34b/c was observed in 6 of 6 MPM cell lines and in 40 of 47 MPM tissues, but not miR-34a. The consequences of miR-34b/c expression dysregulation were examined further. MiR-34b/c reexpression in cell lines inhibited colony formation. c-MET protein expression was strongly downregulated as well as other cell cycle proteins during this induction. G1 cell cycle arrest was evident in miR-34b/c transfectants. Cell migration, invasion, and motility were all significantly suppressed in miR-34b/c stable transfectants compared with control. Additional studies such as Maki et al. [44] showed that miR-34b/c could promote radiation-induced apoptosis in radiosensitized MPM cell lines. Tanaka et al. [45], conversely, confirmed the cell phenotype effects of miR-34b/c knockdown by RNA-based inhibitor transfection which increased malignant features in their tested mesothelial cell lines. MiR-34b/c knockdown in nonmalignant human mesothelial cell lines increased proliferation, migration, and invasion in transfectants. They confirmed upregulation of oncogenic proteins (eg, phospho-cMET). Lastly, Muraoka et al. [46] showed that miR-34b/c heavy methylation status correlated with a higher clinical stage of MPM disease. They assessed the methylation status in serum circulating DNA of patients with MPM versus healthy, noncancer control subjects. High methylation grade of miR-34b/c correlated with advanced MPM disease stage. Taken together, these studies indicate that restoring miR-34b/c expression (in MPM) could have therapeutic effects mimicking a tumor suppressor function.

In Japan, preclinical studies are now testing for feasibility and efficacy of therapeutic delivery of miR-34b/c as reported by Ueno et al. [47]. They examined the in vivo antitumor effects of miR-34b/c using adenovirus vector on MPM. Subcutaneously transplanted NCI-H290 human MPM cells in BALB/c mice (all xenografts >5 mm diameter) were treated with either an injected adenovirus vector expressing miR-34b/c, a luciferase driven by the cytomegalovirus promoter (Ad-miR-34b/c or Ad-Luc), or a PBS control. A statistically significant growth inhibition of the xenograft tumor volume was observed in the Ad-miR-34b/c group from day 6 onward compared to the Ad-Luc group. The inhibition rate of Ad-miR-34b/c, compared to the tumor volume treated with Ad-Luc, was 58.6% on day 10 and 54.7% on day 13. Thus, adenovirus-mediated miR-34b/c gene therapy could be useful for the clinical treatment of human MPM in the future. Similarly, Reid et al. [48] showed that expression of the miR-15/16 family was downregulated in MPM tumors and NCI MPM cell lines using TaqMan PCR assays. Using synthetic miRNA mimics to restore miR-16 expression led to growth inhibition, colony formation, and increased DNA fragmentation in treated MPM cells but not in control. These antitumor effects increased with corresponding miR-16 mimic concentration. Most interestingly, they used EGFR-targeted minicells packaged with miR-16 mimics. These were delivered intravenous to xenograft-bearing athymic nude mice. There was a dose-dependent inhibition of tumor growth. Thus, miR-16 has likely tumor suppressor function in MPM and represents yet another viable therapy option for further development. Currently, this concept has entered phase I clinical testing as the MesomiR 1 study registered with the NCI (NCT02369198).

There remains much opportunity, however, for ongoing refinement and therapeutic development of miRNA strategies. To date, in contrast to miRNA replacement therapy, no examples exist for specific miRNA knockdown that may have similar or unknown improved therapeutic effects. This strategy likely merits some consideration. Further, with the many gene targets of each miRNA, controlling or minimizing untoward effects on "good" genes downstream have yet to be elucidated. miRNA delivery in the human body and tumor-cell specific targeting remain significant challenges (eg, not all MPM tumors or and/or histologic subtypes exhibit EGFR to be susceptible to the minicell strategy of the study by Reid et al. [48]).

5. CONCLUSIONS AND FUTURE DIRECTIONS

The current body of literature concerning miRNA, a family of noncoding regulatory RNA molecules expressed in a tissue-specific manner, suggests that translational application(s) in MPM represent a promising field [13]. miRNAs negatively regulate the expression of tens to hundreds of target mRNAs, thereby controlling multiple cellular pathways simultaneously. The consequence of reduced expression or functioning of many endogenous miRNA induces oncogenic cellular transformation, suggesting that miRNA, as a group, act as tumor suppressors [7]. Similarly, a cellular protooncogene can escape miRNA-mediated repression and acquire gain of function independent of mutations to the protooncogene [49]. Reintroduction of such miRNAs into cancer cells with reduced expression might induce a therapeutic response by reactivation of antioncogenic pathways. Therefore, underexpressed miRNA specific to mesothelioma tissues may be most clinically relevant, so further discovery of underexpressed miRNA and investigation of their unique mechanisms are warranted.

This short review has discussed some of the larger studies of miRNA and their role in MPM. It is apparent that miRNAs hold great promise to improve our understanding of the underlying pathobiology of this poorly understood malignancy. Furthermore, judicious application of specific miRNA may improve early detection, differential diagnosis, and prognosis. Currently, miRNA-based therapeutics are in the early phases of investigation to bring in the clinic for use against human mesothelioma.

Undoubtedly, the miRNA database will continue to be refined and expected to increase in size and complexity. In fact, emerging data suggests there are many more lineage- and/or tissue-specific miRNAs still to be characterized, which promise many more new insights in the future.

LIST OF ACRONYMS AND ABBREVIATIONS

Ad Adenovirus
CI Confidence interval
DNA Deoxyribonucleic acid
eg, For example
EPP Extrapleural pneumonectomy
FDA US Food and Drug Administration

HMC Human mesothelial cell
KEGG Kyoto Encyclopedia of Genes and Genomes
Luc Luciferase
MPM Malignant pleural mesothelioma
mRNA Messenger RNA
miRNA MicroRNA
miR MicroRNA
mm Millimeters
NCI National Cancer Institute
NSCLC Non-small cell–lung carcinoma
n Number
PBS Phosphate-buffered saline
P/D Pleurectomy/decortication
PCR Polymerase chain reaction
qRT-PCR Quantitative real-time reverse transcriptase PCR
ROC Receiver operating curve (analysis)
RNA Ribonucleic acid
SMRP Soluble mesothelin-related peptide
US The United States of America
ΔCT Difference of cycle threshold (PCR)
= Equal to
> Greater than
≥ Greater than or equal
< Less than

ACKNOWLEDGMENTS

Chuong Hoang was previously supported by funds from the Mesothelioma Applied Research Foundation. The Kazan, McClain, Abrams, Fernandez, Lyons, Greenwood, Harley & Oberman Foundation, Inc. provided support for the purchase of dedicated scientific equipment used in miRNA research.

REFERENCES

[1] Teta MJ, Mink PJ, Lau E, Sceurman BK, Foster ED. US mesothelioma patterns 1973–2002: indicators of change and insights into background rates. Eur J Cancer Prev November 2008;17(6):525–34. PMID:18941374.
[2] Britton M. The epidemiology of mesothelioma. Semin Oncol February 2002;29(1): 18–25. PMID:11836665.
[3] Robinson BW, Lake RA. Advances in malignant mesothelioma. N Engl J Med October 13, 2005;353(15):1591–603. PMID:16221782.
[4] Brims FJ. Asbestos–a legacy and a persistent problem. J R Nav Med Serv 2009;95(1): 4–11. PMID:19425525.
[5] Ismail-Khan R, Robinson LA, Williams Jr CC, Garrett CR, Bepler G, Simon GR. Malignant pleural mesothelioma: a comprehensive review. Cancer Control October 2006;13(4):255–63. PMID:17075562.
[6] Murayama T, Takahashi K, Natori Y, Kurumatani N. Estimation of future mortality from pleural malignant mesothelioma in Japan based on an age-cohort model. Am J Ind Med January 2006;49(1):1–7. PMID:16362942.

[7] Vogelzang NJ, Rusthoven JJ, Symanowski J, Denham C, Kaukel E, Ruffie P, et al. Phase III study of pemetrexed in combination with cisplatin versus cisplatin alone in patients with malignant pleural mesothelioma. J Clin Oncol July 15, 2003;21(14):2636–44. PMID:12860938.

[8] Kaufman AJ, Pass HI. Current concepts in malignant pleural mesothelioma. Expert Rev Anticancer Ther February 2008;8(2):293–303. PMID:18279069.

[9] Carbone M, Kratzke RA, Testa JR. The pathogenesis of mesothelioma. Semin Oncol February 2002;29(1):2–17. PMID:11836664.

[10] Hoang CD, Zhang X, Scott PD, Guillaume TJ, Maddaus MA, Yee D, et al. Selective activation of insulin receptor substrate-1 and -2 in pleural mesothelioma cells: association with distinct malignant phenotypes. Cancer Res October 15, 2004;64(20):7479–85. PMID:15492273.

[11] Wittmann J, Jack HM. Serum microRNAs as powerful cancer biomarkers. Biochim Biophys Acta December 2010;1806(2):200–7. PMID:20637263.

[12] Ray M, Kindler HL. Malignant pleural mesothelioma: an update on biomarkers and treatment. Chest September 2009;136(3):888–96. PMID:19736192.

[13] Lu J, Getz G, Miska EA, Alvarez-Saavedra E, Lamb J, Peck D, et al. MicroRNA expression profiles classify human cancers. Nature June 9, 2005;435(7043):834–8. PMID:15944708.

[14] Volinia S, Calin GA, Liu CG, Ambs S, Cimmino A, Petrocca F, et al. A microRNA expression signature of human solid tumors defines cancer gene targets. Proc Natl Acad Sci USA February 14, 2006;103(7):2257–61. PMID:16461460.

[15] Calin GA, Dumitru CD, Shimizu M, Bichi R, Zupo S, Noch E, et al. Frequent deletions and down-regulation of micro- RNA genes miR15 and miR16 at 13q14 in chronic lymphocytic leukemia. Proc Natl Acad Sci USA November 26, 2002;99(24):15524–9. PMID:12434020.

[16] Zhang L, Huang J, Yang N, Greshock J, Megraw MS, Giannakakis A, et al. MicroRNAs exhibit high frequency genomic alterations in human cancer. Proc Natl Acad Sci USA June 13, 2006;103(24):9136–41. PMID:16754881.

[17] Kozomara A, Griffiths-Jones S. miRBase: annotating high confidence microRNAs using deep sequencing data. Nucleic Acids Res January 2014;42(Database issue):D68–73. PMID:24275495, PMCID:3965103.

[18] Londin E, Loher P, Telonis AG, Quann K, Clark P, Jing Y, et al. Analysis of 13 cell types reveals evidence for the expression of numerous novel primate- and tissue-specific microRNAs. Proc Natl Acad Sci USA March 10, 2015;112(10):E1106–15. PMID:25713380, PMCID:4364231.

[19] Guled M, Lahti L, Lindholm PM, Salmenkivi K, Bagwan I, Nicholson AG, et al. CDKN2A, NF2, and JUN are dysregulated among other genes by miRNAs in malignant mesothelioma -A miRNA microarray analysis. Genes Chromosom Cancer July 2009;48(7):615–23. PMID:19396864.

[20] Busacca S, Germano S, De Cecco L, Rinaldi M, Comoglio F, Favero F, et al. MicroRNA signature of malignant mesothelioma with potential diagnostic and prognostic implications. Am J Respir Cell Mol Biol March 2010;42(3):312–9. PMID:19502386.

[21] Balatti V, Maniero S, Ferracin M, Veronese A, Negrini M, Ferrocci G, et al. MicroRNAs dysregulation in human malignant pleural mesothelioma. J Thorac Oncol May 2011;6(5):844–51. PMID:21358347.

[22] Ross DT, Scherf U, Eisen MB, Perou CM, Rees C, Spellman P, et al. Systematic variation in gene expression patterns in human cancer cell lines. Nat Genet March 2000;24(3):227–35. PMID: 10700174.

[23] Gee GV, Koestler DC, Christensen BC, Sugarbaker DJ, Ugolini D, Ivaldi GP, et al. Downregulated microRNAs in the differential diagnosis of malignant pleural mesothelioma. Int J Cancer December 15, 2010;127(12):2859–69. PMID:21351265.

[24] Xu Y, Zheng M, Merritt RE, Shrager JB, Wakelee H, Kratzke RA, et al. miR-1 induces growth arrest and apoptosis in malignant mesothelioma. Chest November 2013;144(5):1632–43. PMID:23828229.

[25] Bartel DP, Chen CZ. Micromanagers of gene expression: the potentially widespread influence of metazoan microRNAs. Nat Rev Genet May 2004;5(5):396–400. PMID:15143321.

[26] Rigoutsos I. New tricks for animal microRNAS: targeting of amino acid coding regions at conserved and nonconserved sites. Cancer Res April 15, 2009;69(8):3245–8. PMID:19351814.

[27] Creaney J, Yeoman D, Demelker Y, Segal A, Musk AW, Skates SJ, et al. Comparison of osteopontin, megakaryocyte potentiating factor, and mesothelin proteins as markers in the serum of patients with malignant mesothelioma. J Thorac Oncol August 2008;3(8):851–7. PMID:18670302.

[28] Simpson RJ, Bernhard OK, Greening DW, Moritz RL. Proteomics-driven cancer biomarker discovery: looking to the future. Curr Opin Chem Biol February 2008;12(1):72–7. PMID:18295612.

[29] Ludwig JA, Weinstein JN. Biomarkers in cancer staging, prognosis and treatment selection. Nat Rev Cancer November 2005;5(11):845–56. PMID:16239904.

[30] Pass HI, Levin SM, Harbut MR, Melamed J, Chiriboga L, Donington J, et al. Fibulin-3 as a blood and effusion biomarker for pleural mesothelioma. N Engl J Med October 11, 2012;367(15):1417–27. PMID:23050525, PMCID: 3761217.

[31] Ramirez-Salazar EG, Salinas-Silva LC, Vazquez-Manriquez ME, Gayosso-Gomez LV, Negrete-Garcia MC, Ramirez-Rodriguez SL, et al. Analysis of microRNA expression signatures in malignant pleural mesothelioma, pleural inflammation, and atypical mesothelial hyperplasia reveals common predictive tumorigenesis-related targets. Exp Mol Pathol December 2014;97(3):375–85. PMID:25236577.

[32] Weber DG, Johnen G, Bryk O, Jockel KH, Bruning T. Identification of miRNA-103 in the cellular fraction of human peripheral blood as a potential biomarker for malignant mesothelioma—a pilot study. PLoS ONE 2012;7(1):e30221. PMID:22253921, PMCID: 3256226.

[33] Santarelli L, Strafella E, Staffolani S, Amati M, Emanuelli M, Sartini D, et al. Association of MiR-126 with soluble mesothelin-related peptides, a marker for malignant mesothelioma. PLoS ONE 2011;6(4):e18232. PMID:21483773, PMCID:3069972.

[34] Thomson JM, Newman M, Parker JS, Morin-Kensicki EM, Wright T, Hammond SM. Extensive post-transcriptional regulation of microRNAs and its implications for cancer. Genes Dev August 15, 2006;20(16):2202–7. PMID:16882971.

[35] Tomasetti M, Staffolani S, Nocchi L, Neuzil J, Strafella E, Manzella N, et al. Clinical significance of circulating miR-126 quantification in malignant mesothelioma patients. Clin Biochem May 2012;45(7–8):575–81. PMID:22374169.

[36] Allen TC. Recognition of histopathologic patterns of diffuse malignant mesothelioma in differential diagnosis of pleural biopsies. Arch Pathol Lab Med November 2005;129(11):1415–20. PMID:16253022.

[37] Benjamin H, Lebanony D, Rosenwald S, Cohen L, Gibori H, Barabash N, et al. A diagnostic assay based on microRNA expression accurately identifies malignant pleural mesothelioma. J Mol Diagn November 2010;12(6):771–9. PMID:20864637, PMCID: 2963911.

[38] Kirschner MB, Cheng YY, Badrian B, Kao SC, Creaney J, Edelman JJ, et al. Increased circulating miR-625-3p: a potential biomarker for patients with malignant pleural mesothelioma. J Thorac Oncol July 2012;7(7):1184–91. PMID:22617246.

[39] Andersen M, Grauslund M, Ravn J, Sorensen JB, Andersen CB, Santoni-Rugiu E. Diagnostic potential of miR-126, miR-143, miR-145, and miR-652 in malignant pleural mesothelioma. J Mol Diagn July 2014;16(4):418–30. PMID:24912849.

[40] Pass HI, Goparaju C, Ivanov S, Donington J, Carbone M, Hoshen M, et al. hsa-miR-29c* is linked to the prognosis of malignant pleural mesothelioma. Cancer Res March 1, 2010;70(5):1916–24. PMID:20160038.

[41] Kirschner MB, Cheng YY, Armstrong NJ, Lin RC, Kao SC, Linton A, et al. MiR-score: a novel 6-microRNA signature that predicts survival outcomes in patients with malignant pleural mesothelioma. Mol Oncol March 2015;9(3):715–26. PMID:25497279.

[42] Ivanov SV, Goparaju CM, Lopez P, Zavadil J, Toren-Haritan G, Rosenwald S, et al. Pro-tumorigenic effects of miR-31 loss in mesothelioma. J Biol Chem July 23, 2010;285(30):22809–17. PMID:20463022, PMCID:2906272.

[43] Kubo T, Toyooka S, Tsukuda K, Sakaguchi M, Fukazawa T, Soh J, et al. Epigenetic silencing of microRNA-34b/c plays an important role in the pathogenesis of malignant pleural mesothelioma. Clin Cancer Res August 1, 2011;17(15):4965–74. PMID:21673066.

[44] Maki Y, Asano H, Toyooka S, Soh J, Kubo T, Katsui K, et al. MicroRNA miR-34b/c enhances cellular radiosensitivity of malignant pleural mesothelioma cells. Anticancer Res November 2012;32(11):4871–5. PMID:23155254.

[45] Tanaka N, Toyooka S, Soh J, Tsukuda K, Shien K, Furukawa M, et al. Downregulation of microRNA-34 induces cell proliferation and invasion of human mesothelial cells. Oncol Rep June 2013;29(6):2169–74. PMID:23525472.

[46] Muraoka T, Soh J, Toyooka S, Aoe K, Fujimoto N, Hashida S, et al. The degree of microRNA-34b/c methylation in serum-circulating DNA is associated with malignant pleural mesothelioma. Lung Cancer December 2013;82(3):485–90. PMID:24168922.

[47] Ueno T, Toyooka S, Fukazawa T, Kubo T, Soh J, Asano H, et al. Preclinical evaluation of microRNA-34b/c delivery for malignant pleural mesothelioma. Acta Med Okayama 2014;68(1):23–6. PMID:24553485.

[48] Reid G, Pel ME, Kirschner MB, Cheng YY, Mugridge N, Weiss J, et al. Restoring expression of miR-16: a novel approach to therapy for malignant pleural mesothelioma. Ann Oncol December 2013;24(12):3128–35. PMID:24148817.

[49] Johnson SM, Grosshans H, Shingara J, Byrom M, Jarvis R, Cheng A, et al. RAS is regulated by the let-7 microRNA family. Cell March 11, 2005;120(5):635–47. PMID:15766527.

CHAPTER 9

MicroRNA, an Important Epigenetic Regulator of Immunity and Autoimmunity

R. Dai, S. Ansar Ahmed

Key Concepts

- MicroRNAs (miRNAs) are small, noncoding RNAs that regulate gene expression at the posttranscription level. miRNAs are now considered as key epigenetic regulators of almost all biological systems and processes including immune system development and responses.

- During normal immune cell development, differentiation, and immune responses, the miRNA expression/function is stringently regulated to ensure immune homeostasis. The induced miRNAs during specific immune responses may form positive or negative feedback loops with their target genes to either promote or resolve the innate and adaptive immune responses. Therefore, disruption of miRNA expression/function impairs immune cell development, causes abnormal immune responses, and even induces inflammatory autoimmunity.

- Of the numerous miRNAs involved in the immunity and autoimmunity, miR-146a and miR-155 are the two most well-studied miRNAs, which regulate immune system development and function at multiple levels. Dysregulated miR-146a and miR-155 expression has been identified in many human autoimmune diseases including systemic lupus erythematosus (SLE).

- Various dysregulated miRNAs have been identified in distinct sample sources in human and murine lupus, a number of which correlate with lupus onset/severity, and contribute to lupus pathogenesis by regulating DNA methylation signaling pathway, T- and B-cell function, and the production of inflammatory mediators.

- The dysregulation of miRNA during lupus pathogenesis is contributed by multiple factors including genetic, epigenetic, hormonal, and environmental factors.

- Despite challenges that remain in the miRNA field, there is a tremendous potential of using miRNAs as novel biomarkers and therapeutic targets for SLE and other autoimmune disorders.

Translating MicroRNAs to the Clinic
ISBN 978-0-12-800553-8

1. INTRODUCTION

The quantitative and qualitative regulation of immune responses are critical to effectively ward off infections and other assaults, and to prevent inflammatory and autoimmune diseases. The induction, temporal maintenance, and resolution of immune responses are complex processes that warrant the coordination and fine-tuning of genetics and epigenetics. Epigenetics is the study of heritable gene expression or physiological trait changes that are not caused by any alteration in genomic DNA sequences. Prior to the discovery of miRNA, DNA methylation and histone modification are the two major epigenetic mechanisms, which regulate gene expression mainly at the transcriptional level and are critically involved in the modulation of the immune system in both physiological and pathological conditions [1]. miRNAs are small noncoding RNAs that regulate genome expression at the posttranscriptional level, thus providing a new layer of epigenetic regulation of diverse biological systems including the immune system [2].

The first miRNA, *lin-4*, was identified in the nematode *Caenorhabditis elegans* at 1993 [3], a seminal finding that was not well appreciated until the identification of the second miRNA, *let-7*, in *C. elegans* at 2000 [4]. Currently, the miRBase database contains a total of 28,645 miRNA precursor entries producing 35,828 mature miRNAs in 223 species (http://www.mirbase.org; version 21, June 2014) [5]. A single miRNA may regulate hundreds to over thousands of target genes, albeit the effect may be mild or moderate [6,7]. Thus far, there are 1881 human miRNAs in the database, which regulate more than 30% of human genes, and play a crucial role in the maintenance of the cellular homeostasis and normal function [8–11]. The association between aberrant miRNA expression/function and diverse human diseases, including cancers, metabolic disease, neurodegenerative, infectious, and autoimmune diseases, has garnered a widespread attention of biomedical researchers [12–16]. This chapter summarizes the current knowledge regarding the regulatory role of miRNA in immunity and autoimmunity, with an emphasis on SLE, a prototypical autoimmune disease.

2. MICRORNA BIOGENESIS

It is estimated that nearly 80% of mammalian miRNA genes are located at intergenic or intronic regions of either protein-coding or nonprotein-coding transcripts, whereas only a small number of miRNA genes (20%) are located in the exon region of noncoding RNAs [17,18]. Some miRNAs could be either exonic or intronic miRNAs, depending on the alternative splicing pattern of host genes. As illustrated in the Fig. 9.1, miRNAs are generated by two pathways: canonical and noncanonical (mirtron) pathways [2]. In both pathways, miRNA genes are transcribed by RNA polymerase II and III into primary miRNA transcripts (primiRNAs) containing either a single or clusters of hairpin loop structures. In the canonical miRNA biogenesis pathway, the primiRNAs are cleaved in the nucleus by a microprocessor complex comprising the nuclear RNase III enzyme, Drosha and the double-stranded-RNA-binding protein, DiGeorge syndrome critical region protein 8 (DGCR8) to produce precursor miRNAs (premiRNAs) [19,20]. In contrast, due to the lack of the lower stem-loop structure and the flanking single-stranded segments that are crucial for the binding and processing by the DGCR8/Drosha microprocessor, in the noncanonical pathway, the mirtrons bypass Drosha processing in the nucleus. The branched pre-mirtrons are spliced out from the primary transcripts by spliceosome, and then go through lariat-mediated debranching to generate premiRNAs [21].

After being exported into the cytoplasm by the nucleocytoplasmic shuttle protein exportin 5, the premiRNAs are further processed by the RNAIII enzyme, Dicer, to yield ~22 nucleotides (nts) long, imperfect matched miRNA/miRNA* duplexes. The duplexes are then loaded into the Argonaute (Ago) protein to generate RNA-induced silencing complex (RISC) where the guided strand remains as a mature miRNA to interact with the 3' untranslated region (UTR) of its target messenger RNA (mRNA), leading to either translational repression or cleavage and degradation of target mRNA [8,22]. The complementary strand (the passenger strand miRNA*) is degraded in the RISC [23]. Although miRNAs usually regulate gene expression negatively through interacting with 3' UTR, in certain circumstances, they could also stimulate gene expression via directly binding to the target gene or indirectly by suppressing the suppressor of gene expression [24]. Also, miRNAs may interact with the 5' UTR of target mRNA to promote or repress target gene expression [25].

3. MICRORNA IN NORMAL IMMUNE SYSTEM DEVELOPMENT AND FUNCTION

It has now been unequivocally established that specific miRNAs are involved in the development of lymphoid cell subsets, and functioning of the innate and adaptive immune responses [2,10,26,27].

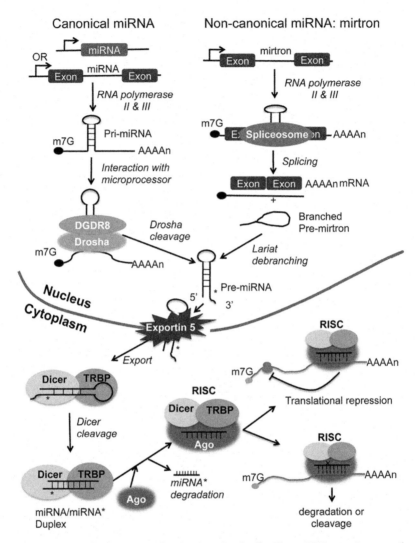

Figure 9.1 *miRNA biogenesis and action in animal cells.* The miRNA gene is transcribed by RNA polymerase II and III into primiRNA, which is processed sequentially in the nucleus and cytoplasm to produce mature, single-stranded miRNA. Canonically the primiRNA is processed by DGCR8/Drosha microprocessor in the nucleus to generate premiRNA. The noncanonical miRNA, mirtron, bypasses Drosha processing and is spliced by the spliceosome to generate premiRNA. After it is exported into the cytoplasm by Exportin 5, the premiRNA is further processed by Dicer to produce short miRNA/miRNA* duplex, which is then loaded into Argonaute protein to form RNA-induced silencing complex (RISC) with other cofactors such as Dicer and TAR RNA binding protein (TRBP). Depending on the complementary degree between miRNA and the 3′UTR sequence of target gene, miRNA regulates target gene expression through either translational inhibition or mRNA degradation/cleavage. *Modified from the Fig. 1 of our previous review article published at* Translational Research *[2].*

3.1 MicroRNA in the Regulation of Innate Immunity

3.1.1 MicroRNA Regulation of Innate Immune Cell Development

The innate immune system comprises cells such as neutrophils and mono-cyte-derived macrophages to form the first line of cellular defense against invading pathogens. Granulocytic neutrophils are the most abundant white blood cells and are essential components of the innate immune system because of their potent phagocytic role. Although conflicting results were reported with different research systems, miR-223, a myeloid-lineage spe-cific miRNA, is undoubtedly a key miRNA in the regulation of neutrophil differentiation and function (Fig. 9.2) [28]. MiR-223 was initially reported to promote granulocyte differentiation in acute promyelocytic leukemia (APL) cells by forming a regulatory minicircuitry comprised of miR-223, transcription factors CCAAT enhancer binding protein(C/EBP)α, and nuclear factor 1 A-type (NFI-A) [29]. In undifferentiated APL cells, NFI-A binds to the miR-223 promoter and maintains a low level of miR-223. Fol-lowing all-trans-retinoic acid (ATRA)-induced granulocytic differentia-tion, C/EBPα was activated to induce miR-223, which in turn targeted and repressed NFI-A, resulting the replacement of NFI-A by C/EBPα on the miR-223 promoter to further boost miR-223 production and promote

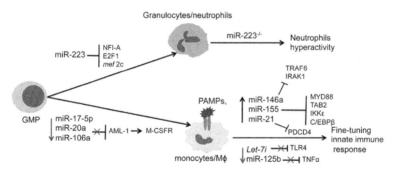

Figure 9.2 *miRNA regulation of innate immunity.* This figure illustrates selected miRNAs that play a key role in the development and function of neutrophils and monocyte/mac-rophage (Mφ). MiR-223 has been reported to target different genes such as NFI-A, E2F1, and Mef2c to regulate granulocytic differentiation and function. The decreased expres-sion of miR-17-5p, miR-20a, and miR-106a during monocytopoiesis releases their suppres-sive effect on AML-1 to promote monocyte–macrophage differentiation and maturation. Diverse miRNAs are altered in response to toll-like receptor (TLR) activation, and these miRNAs in turn target distinct signaling components of TLR pathways to fine-tune the innate immune responses. Please see the text for a detailed explanation. *GMP,* Granulo-cyte–monocyte progenitor; *PAMPs,* pathogen-associated molecular patterns. *Updated from the Fig. 2 of our previous review article published at* Translational Research *[2].*

granulocyte differentiation. In addition, miR-223 was shown to target cell cycle regulator, E2F transcription factor (E2F)1 to promote granulocytic differentiation in activated acute myeloid leukemia (AML) cells [30]. Nevertheless, an in vivo study with miR-223 knockout mice revealed that miR-223 acted as a negative regulator of granulocyte differentiation by targeting myocyte-specific enhancer factor 2C (Mef2c), a transcription factor promoting myeloid progenitor differentiation [31]. The depletion of miR-223 led to the upregulation Mef2c, which caused a cell-autonomous expansion of granulocyte progenitors, resulting in an increased number of granulocytes in mice with miR-223 deficiency. Furthermore, the miR-223 knockout mice spontaneously developed inflammatory lung pathology due to the hyperactivity of neutrophils, suggesting a role of miR-223 in the regulation of neutrophil function. In addition to miR-223, several other miRNAs such as miR-34a, Let-7c, miR-29a, miR-142-3p were reported to promote ATRA-induced granulocytic differentiation in AML cells [32–34].

Monocyte-derived macrophages play critical roles in innate immune responses by engulfing invading pathogens, presenting antigens, and releasing inflammatory signaling molecules to activate adaptive immune cells. Monocytopoiesis is regulated by a circuitry loop consisting of miR-17-5p, miR-20a, miR-106a, acute myeloid leukemia-1 (AML-1), and macrophage colony-stimulating factor receptor (M-CSFR) [35]. During monocytopoiesis, the expression of miR-17-5p, miR-20a, and miR-106a was decreased. In parallel, their target gene, AML-1, was upregulated to promote the expression of M-CSFR, which plays a pivotal role in monocyte–macrophage differentiation and maturation (Fig. 9.2). On the other hand, AML-1 also negatively regulated the expression of miR-17-5p-20a-106a by binding to the promoter of the miR-17-92 and miR-106a-92 cluster, suggesting a mutual negative feedback regulation loop in monocytopoiesis [35].

3.1.2 MicroRNA Regulation of Innate Immune Responses

The distinct toll-like receptors (TLRs), located on either the cell surface or inside cellular compartments of innate immune cells, such as macrophages, play a key role in the innate inflammatory responses. The TLRs recognize and bind to specific microbial products called pathogen-associated molecular patterns (PAMPs) and then trigger downstream signaling pathways to initiate inflammatory responses for eliminating invading pathogens [36]. In normal physiological conditions, the TLR signaling is tightly controlled in vivo by different classes of negative regulators to prevent overwhelming inflammation [37]. A number of miRNAs are reported to alter dramatically in response to PAMPs or inflammatory cytokines stimulation. These induced miRNAs target distinct signaling components of TLR signaling pathways to fine-tune the innate immune response, thereby adding a new layer of negative feedback regulation of TLR signaling (Fig. 9.2) [26,38–40]. As they have been well-reviewed recently, the TLR-regulated miRNAs feedback regulate TLR signaling at multiple levels by targeting TLRs themselves, TLR signaling molecules such as tumor necrosis factor receptor-associated factor 6 (TRAF6), IL-1 receptor-associated kinase (IRAK), myeloid differentiation primary response protein 88 (MyD88), TGF-beta activated kinase 1/MAP3K7 binding protein (TAB)2, and IκB kinases (IKK), TLR-induced transcription factors such as C/EBPβ, TLR-associated regulatory molecules such as programmed cell death protein 4 (PDCD4), and also TLR-induced functional cytokines such as tumor necrosis factor (TNF)α, Interleukin (IL)-6, Interferon (IFN)α/β/γ, and IL-10 [26,40]. It is noteworthy that diverse miRNAs may be induced in a specific cell sequentially following TLR activation to control the strength, duration, and resolution of inflammatory responses [26].

MiR-146a, arguably one of the best-studied miRNAs, was the first miRNA that was identified as a negative feedback regulator of TLR signaling [41]. MiR-146a was rapidly upregulated in monocytes in response to the activation of TLR4, TLR2, and TLR5 signaling pathways or the stimulation of inflammatory cytokines such as TNFα and IL-1β. The induced miR-146a suppressed TLR-mediated inflammatory response by targeting TRAF6 and IRAK1 (Fig. 9.2) [41]. MiR-155, another miRNA that has been well investigated, was induced in macrophages/monocytes following the activation of TLR4, TLR2, TLR3, or TLR9 signaling pathway, and it could negatively regulate the inflammatory responses by targeting MyD88, TAB2, IKKε, and/or other components in the TLR signaling pathways [26,42]. Lipopolysaccharide (LPS)-induced miR-21 negatively regulated TLR4 signaling by targeting tumor suppressor PDCD4, leading to reduced NF-κB activity and promotion of suppressor cytokine IL-10 [43]. Some miRNAs such as Let7i and miR-125b were decreased following TLR activation to tweak the innate immune responses to pathogens [44,45]. LPS stimulation and parasite *Cryptosporidium parvum* infection decreased *let7i* expression in cultured human cholangiocytes, leading to the upregulation of TLR4 protein and enhanced immune responses. While inhibition of *let7i* decreased parasite infection, enhanced *let7i* expression increased parasite burden in the infected cells [44]. LPS-mediated reduction of miR-125b expression may contribute to elevated TNFα level in activated macrophages as miR-125b was shown to target TNFα mRNA (Fig. 9.2) [45]. Overall, the above studies exemplify that miRNAs play an integral role in calibrating innate immunity to effectively combat infections and prevent devastating inflammation.

3.2 MicroRNA in the Regulation of Adaptive Immunity

Innate immune cells such as antigen-presenting cells interact with T and B lymphocytes, the major cellular components of adaptive immunity, to induce specific immunity. MiRNAs are critical for the functioning of the adaptive immunity as evidenced by studies where depletion of Dicer, Drosha, and Ago (key components of miRNA biogenesis pathways) impaired T and B lymphocyte development, differentiation, and immune responses [2,46–49]. Fig. 9.3 illustrates the specific miRNAs that are critically involved in the regulation of T and B cells.

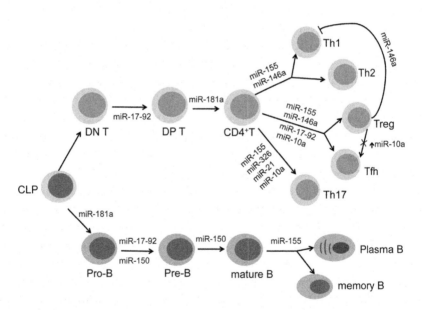

Figure 9.3 *miRNA regulation of adaptive immunity.* This figure depicts selected key miRNAs such as miR-155, miR-146a, and miR-17-92 that are critically involved in the development, differentiation, and function of T cells (upper part of the figure), and B cells (lower part of the figure). Please see the text for a detailed explanation. *CLP,* Common lymphoid progenitor; *DN,* CD4⁻CD8⁻ double negative; *DP,* CD4⁺CD8⁺ double positive. *Updated from the Fig. 3 of our previous review article published at* Translational Research *[2].*

3.2.1 MiR-155 Regulation of Adaptive Immunity

MiR-155 is a dominant regulator of lymphocyte development, differentiation, and function (Fig. 9.3). MiR-155 is upregulated in both T and B cells upon activation, and plays a key role in maintaining lymphocyte homeostasis and normal immune responses by targeting different genes including c-maf, Pu.1, and activation-induced cytidine deaminase, and suppressor of cytokine signaling 1 (SOCS1) [50–53]. It is noteworthy that even though miR-155-deficient mice seemingly had normal lymphocyte development, defective functioning of T and B cells was evident in these mice with increased Th2 polarization, Th2 cytokine production, and impaired B-cell responses [50]. Moreover, miR-155 is intrinsically required for B-cell development and function since miR-155 deficiency led to reduced germinal center (GC) size, decreased GC responses, low numbers of plasma and memory B cells, and impaired production of class-switched high affinity IgG antibodies [50–52,54].

MiR-155 has a critical role in the development of regulatory T cells (Tregs), which are central cell subsets in suppressing immune responses including autoimmune inflammatory responses to maintain self-tolerance. MiR-155-deficient mice had a reduced number of Tregs in both thymus and peripheral lymphoid tissues. But interestingly, the miR-155-deficient Tregs had competent suppressive activity in vivo [55,56]. This suggests that miR-155 is critical for Tregs development, but may not be essential for Treg-suppressive functions. Nevertheless, another study indicated an indirect involvement of miR-155 in Treg-suppressive function. Since miR-155 inhibition led to increased cell sensitivity, there was enhanced natural Treg (nTreg)-mediated suppression in human and mouse CD4$^+$ Th effector cells with miR-155 inhibition [57].

MiR-155 also regulates differentiation of Th17 cells, which are potent inflammatory cells. Compared to wild-type mice, miR-155-deficient mice displayed a remarkable resistance to myelin oligodendrocyte glycoprotein$_{35-55}$ (MOG$_{35-55}$)-induced experimental autoimmune encephalomyelitis (EAE), resulting from a significant reduction Th-17 cell numbers and IL-17 cytokine levels in the miR-155 null mice [58]. MiR-155 promotes the differentiation of another important T-cell subset, the follicular T (Tfh) cells, which interact with B cells to help B-cell activation during humoral responses [59]. There was a spontaneous Tfh cell accumulation in aged miR-146a knockout mice, which was mediated by miR-155 since depletion of miR-155 prevented the accumulation of Tfh and inflammation in these mice. Further investigation revealed that miR-155 targeted multiple genes including pellino E3 ubiquitin protein ligase (Peli)1, IKKε, and FOS-like Antigen 2 (Fosl2) to regulate Tfh cell development [59].

3.2.2 MiR-146a Regulation of Adaptive Immunity

Besides its important role in the regulation of innate immune responses, miR-146a is also critically involved in adaptive immunity (Fig. 9.3). MiR-146a expression is dynamically regulated during CD4$^+$ T cell differentiation. MiR-146a expressed at low levels in the naïve CD4$^+$ T cells, and its expression was upregulated in differentiated Th1 cells and abolished in differentiated Th2 cells [60]. T-cell receptor stimulation activated NF-κB signaling to highly induce miR-146a, which in turn negatively regulated NF-κB activity to dampen T cell–mediated inflammation [61]. Therefore, because of the lack of miR-146a-mediated negative feedback regulation, miR-146a-deficient T cells were hyperactive to both acute antigen stimulation and chronic inflammatory responses. The mice with T-cell deficiency of miR-146a developed T cell–associated autoimmunity, characterized by the accumulation of activated T cells and multiorgan inflammation [61].

MiR-146a also plays a role in regulation of Treg and Tfh cell development and function. MiR-146a is highly expressed in Tregs, and it targets STAT-1 to regulate Treg-mediated suppression of IFNγ-dependent Th1 responses and inflammation [62]. Ablation of miR-146a specifically in Tregs led to elevated production of Th1 cytokine IFNγ and development of severe Th1-mediated lesions in mice [62]. Depletion of miR-146a in conventional T cells caused a bias of Tfh cell differentiation, which was associated with the development of low-grade chronic and systemic inflammation in the mice [59]. While it was suggested previously that the promotion of Tfh cells in miR-146a-deficient mice was miR-155 mediated, the direct role of miR-146a on Tfh cells has emerged recently. A recent report revealed that miR-146a is highly expressed in Tfh cells, and that loss of miR-146a led a cell-intrinsic accumulation of Tfh cells and GC B cells resulting from enhanced inducible T cell costimulator (ICOS)-ICOS ligand (ICOSL) signaling [63]. This observation suggests a direct involvement of miR-146a in limiting Tfh cell and GC responses.

3.2.3 Other Key MicroRNAs Involved in Adaptive Immunity

In addition to miR-155 and miR-146a, there are many other miRNAs that are involved in the adaptive immunity (Fig. 9.3). In immature T cells, miR-181a is highly expressed, which fine-tunes T cell–receptor signaling and sensitivity by targeting multiple phosphatases [64]. Inhibition of miR-181a expression in immature T cells reduced T-cell sensitivity and impaired both positive and negative selection during T-cell development [64]. MiR-181a also affects B-cell lineage. Overexpressing of miR-181a in hematopoietic progenitor cells promoted the development of CD19$^+$ B cells [65].

MiR-150 is selectively expressed in mature resting B and T cells, but not in their progenitors [60,66]. MiR-150 modulates B-cell development by targeting transcription factor *c-myb* [66]. MiR-150 overexpression in hematopoietic stem cells blocked B-cell development at pro- to pre-B stages, while miR-150 depletion enhanced B1 cell development and antibody responses [66,67].

MiR-17-92, in contrast to miR-150, is highly expressed in the lymphocyte precursors and decreased following maturation. While overexpression of miR-17-92 in lymphocyte progenitors promoted T and B lymphocyte proliferation and lymphoma development by targeting proapoptotic proteins, *Bim* and phosphatase and tensin homolog (PTEN), depletion of miR-17−92 inhibited B-cell development at the pro-B to pre-B transition due to the apoptotic death of pro-B cells, resulting from the upregulation of its target protein *Bim* [68,69]. Recent studies have shown that miR-17−92 regulates Treg and Tfh cells [27]. Overexpression miR-17−92 suppressed Foxp3$^+$ iTreg cell differentiation in vitro, suggesting a suppressive role of miR-17-92 on iTreg differentiation [70]. MiR-17-92 is also critical for robust Tfh cell differentiation and function since depletion of miR-17-92 in T cells led to compromised Tfh differentiation, germinal center formation and antibody responses, and failure in controlling chronic viral infection [71]. The important role of miR-17-92 in Tfh cell is further supported by the finding that transgenic overexpression of miR-17-92 in T cells led to a spontaneous Tfh accumulation and the development of fatal immune-related disorder in mice [71]. While PTEN is also partially involved, PH domain leucine-rich repeat protein phosphatase (PHLPP)2 is the main target of miR-17-92 by which miR-17-92 controls the migration of CD4$^+$ T cells to B-cell follicles and promotion of Tfh differentiation [71].

Several miRNAs, in addition to miR-155, are critically involved in the Th17 cell differentiation and production of IL-17 (Fig. 9.3). MiR-326 targets Ets-1, a transcription factor that negatively regulates Th17 development. Increased expression miR-326 in the human patients with multiple sclerosis (MS) and EAE mice caused a repression of Ets-1 and consequently a promotion of Th17 differentiation and IL-17 production [72]. A latest report indicated that miR-21 was elevated in Th17 cells and it promoted Th17 cell differentiation by targeting mothers against decapentaplegic homolog 7

(SMAD7), a negative regulator of TGF-β signaling [73]. The mice lacking miR-21 in T cells had reduced number of Th17 cells and were resistant to EAE [73]. In contrast, miR-10a was highly induced in retinoic acid–activated T cells, and it limited Th17 cell differentiation accompanying with a promotion of Treg cell differentiation [74]. Also, induced miR-10a expression in Tregs suppressed the conversion of iTreg into Tfh cells by targeting B-cell lymphoma-6 and nuclear receptor corepressor 2, suggesting a role of miR-10a in the regulation of Th cell plasticity [74].

In summary, the above studies demonstrate that the expression and functions of specific miRNAs described earlier are critical in T and B lymphocyte development, differentiation, and function. For the maintenance of immune homeostasis, the expression and function of miRNAs must be tightly regulated. Conceivably, dysregulation of miRNA in immune cells can induce autoimmunity.

4. MICRORNA IN THE DEVELOPMENT AND PREVENTION OF AUTOIMMUNITY

Activation and proliferation of autoreactive cells has serious consequences that can lead to autoimmune diseases. Therefore, physiologically, autoreactive cells are either developmentally eliminated or silenced in the peripheral lymphoid organs. However, autoreactive cells are capable of escaping stringent checkpoints in central and peripheral lymphoid organs and can be activated by poorly described extrinsic and intrinsic triggers [75]. Once activated, the autoreactive cells are able to mount a devastating attack on self-tissues, leading to the development of autoimmune disease. It is conceivable that the genetic defects of any gene involved in maintenance of self-tolerance can contribute to the development of autoimmune disease. Alternatively, epigenetic modulation of genes involved in self-tolerance control by histone modification, DNA methylation, and miRNAs has gained much attention in recent years for understanding autoimmunity [76]. While histone modification and DNA methylation are crucial epigenetic regulators of immunity and autoimmunity, they are not the focus of this chapter. Here, we comprehensively discuss the important role of miRNA in autoimmunity. The vital contribution of miRNA to autoimmunity is emphasized by the findings that disruption of either whole miRNA biogenesis pathway or individual miRNAs in immune cells leads to the development of autoimmune-related disorders [2,48,68,77–79].

As stated earlier, miR-146a is a dominant negative regulator of innate and T cell–mediated inflammation. Due to the loss of miR-146a-mediated resolution of inflammatory responses, aged miR-146a knock-out mice died prematurely as the result of the development of a spontaneous autoimmune disorder, characterized by splenomegaly, lymphadenopathy, and multiorgan inflammation [79]. The chronic inflammation and autoimmunity in miR-146a knockout mice was due to the accumulation of Tfh cells and elevation of GC B cells and autoantibodies in the mice [59]. MiR-146a has also been linked with autoimmunity by regulating Treg cell function via targeting STAT-1. Ablation of miR-146a in Treg cells resulted a selective loss of Treg-mediated suppression of Th1 inflammatory responses, leading to elevated IFNγ production and the development of autoimmune diseases in mice [62]. Interestingly, while the mice with T cell depletion of miR-146a developed autoimmune disorders at late life due to unresolved T cell responses [61,79], the miR-146a overexpression transgenic mice also developed autoimmune lymphoproliferative syndrome starting at an early age due to the downregulation of Fas in GC B cells [80]. This suggests that a specific miRNA has diverse roles in distinct cell types via targeting different genes to control autoimmunity. In addition to miR-146a, disruption of other specific miRNAs such as miR-17-92 in lymphocytes also induces autoimmunity [68].

So far, numerous dysregulated miRNAs have been identified in human patients with different autoimmune disorders such as SLE, rheumatoid arthritis (RA), MS, psoriasis, and Sjogren's syndrome [2,81,82]. It is interesting to notice that a specific miRNA is differentially dysregulated in distinct autoimmune disease. For example miR-146a was downregulated in the peripheral blood mononuclear cells (PBMCs) of lupus patients, but was upregulated in the PBMCs of RA patients [83,84], suggesting the specificity of miRNA function in certain disease microenvironments. In this chapter, we summarize the recent finding with regard to the pathogenic contribution of miRNA dysregulation specifically to SLE.

5. MICRORNA IN SYSTEMIC LUPUS ERYTHEMATOSUS

SLE is a multisystemic autoimmune disorder that afflicts kidneys, joints, vascular tissues, skin, the central nervous system, and other tissue types. The hallmark of lupus is the induction of autoantibodies to a variety of nuclear components and phospholipids. In both humans and a variety of animal models, females are more susceptible to the disease than males. Multiple immunological abnormalities are evident in lupus that include impairments in B- and T-lymphocyte development/functions, aberrant type I IFN, TLR, and NF-κB signaling pathways, and abnormal interactions among immune cells [2,85]. Relevant to this chapter, signature dysregulated miRNAs have been identified in both human and murine lupus. Evidence is emerging to show that miRNAs are vital epigenetic contributors to lupus pathogenesis by modulating autoimmune-related genes and signaling pathways [2,82,86,87].

In the first report of miRNA dysregulation in human lupus, 16 lupus-related miRNAs were found to be altered in the PBMCs from patients with lupus, but not with idiopathic thrombocytopenic purpura (an organ-specific autoimmune disease) [88]. However, the initial report did not address any functional relevance of these miRNA changes. Subsequent to this initial report, dysregulated miRNA expression profiles have been iden-tified in various types of immune cells including PBMC, CD4$^+$ T cells, B cells, and also lupus target organs, such as kidneys from both human and murine lupus [2,82,86]. It is noteworthy that in different human lupus miRNA studies, there were some discrepancies with regard to the dysregu-lation of a specific miRNA. The precise reasons for these discrepancies is unclear but potentially could be due to the intrinsic genetic differences related to the ethnicity of the patients, diversity in the medical history, and disease severity of the patients with lupus, distinct sample sources, cellular heterogeneity of the samples, and also possibly the sensitivity of detection methods. Also, while there is a broad agreement between human and murine lupus miRNA data; not all dysregulated miRNAs identified in murine lupus were evident in human lupus [2,82]. Nevertheless, the critical patho-genic contribution of specific dysregulated miRNAs to lupus is increasingly appreciated in recent studies, which are discussed later.

5.1 MicroRNA Regulation of DNA Methylation in Systemic Lupus Erythematosus T Cells

DNA methylation is one of the best-described epigenetic mechanisms so far. The critical involvement of CD4$^+$ T cell DNA hypomethylation in lupus etiology has been reported a long time ago, and is recently reinforced with the advanced genome-wide DNA methylation profiling technology. Several miRNAs have been reported to contribute to lupus etiology via targeting DNA methylation signaling pathway (Fig. 9.4). MiR-21 and miR-148a were upregulated in CD4$^+$ T cells from both MRL/MpJ-Fas^{lpr}/J (MRL-*lpr*) mice and human patients with lupus [89]. While miR-148a targeted DNA

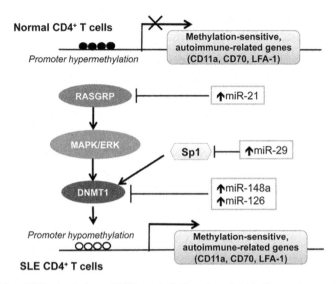

Figure 9.4 *miRNA regulation of DNA methylation signaling pathways in systemic lupus erythematosus (SLE) CD4⁺ T cells.* In normal CD4⁺ T cells, the expression of methylation-sensitive, autoimmune-related genes is blocked since the promoters are methylated (black circle). In SLE CD4⁺ T cells, increased miR-148a, miR-126, miR-21, and miR-29 target DNMT1 directly or indirectly, which reduce the DNMT1 expression level, leading to the promoter hypomethylation (open circle) and then induction of autoimmune-related genes such as *CD11a, CD70*, and *LFA-1*. Please see text for a detailed explanation.

methyltransferase (DNMT)1 directly, miR-21 indirectly downregulated DNMT1 by targeting its upstream regulator, RAS guanyl nucleotide-releasing protein 1 (RASGRP1). Overexpression of miR-148a and miR-21 in CD4⁺ T cells led to DNA hypomethylation and subsequently resulted in increased expression of autoimmune-associated methylation-sensitive genes, CD70 and lymphocyte function-associated antigen 1 (LFA-1) [89]. Another study reported that miR-126a was upregulated in lupus CD4⁺ T cells, and its expression level inversely correlated with the DNMT1 level since miR-126 targeted DNMT1 directly in CD4⁺ T cells [90]. In addition, miR-29 was upregulated in lupus T cells, which contributed to DNA hypomethylation in lupus CD4⁺ T cells via targeting Sp1, a positive regulator of DNMT1 [91]. The overexpression of miR-126 or miR-29 in normal CD4⁺ T cells caused DNA demethylation and upregulation of genes encoding CD11a and CD70, leading T cell and B cell hyperactivity, while the inhibition of miR-126 or miR-29 in lupus CD4⁺ T cells had an opposite effect [90,91]. In summary these data strongly suggest the critical pathogenic involvement of miRNA regulation of DNA methylation in lupus.

5.2 MicroRNA Regulation of Inflammatory Mediators and T-Cell Function in Systemic Lupus Erythematosus

Dysregulated miRNAs have also been implicated in lupus pathogenesis by affecting the production of lupus-associated inflammatory cytokines and chemokines in PBMCs and T cells (Fig. 9.5). Increased type I IFN production and signaling is a hallmark of human lupus. In one study, 42 differentially expressed miRNAs were identified in PBMCs from human patients when compared to healthy controls [83]. Among these dysregulated miRNAs, miR-146a, a key regulator of both immunity and autoimmunity, was profoundly decreased (more than six-fold) in the PBMCs of patients with lupus versus healthy controls. Further investigation revealed that miR-146a negatively regulated the type I IFN pathway by targeting interferon regulatory factor (IRF)-5 and STAT-1. Overexpression of miR-146a in normal PBMCs greatly reduced TLR7 induction of IFNα/β, and inhibition of miR-146a increased the production of IFNα/β [83]. These data suggest that decreased miR-146a in PBMCs may contribute to the enhanced type I IFN signaling in human lupus.

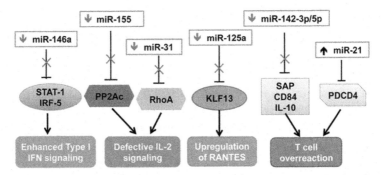

Figure 9.5 *miRNA regulation of inflammatory mediators and T-cell function in systemic lupus erythematosus.* The upregulated miRNA reduces the expression of its specific target gene(s) and the downregulated miRNA increases the expression of its target gene(s) by releasing its inhibitory effect to cause abnormal T-cell function and inflammatory mediators in lupus. Decreased miR-146a level leads to the upregulation of IRF5 and STAT-1, and consequently enhanced type I IFN signaling in lupus. The downregulation of miR-155 and miR-31 causes defective IL-2 signaling in lupus by upregulation of their target gene *PP2Ac* and *RhoA*, respectively. Decreased miR-125a increases expression of transcription factor KLF13, leading to the promotion of its regulated chemokine RANTES in lupus T cells. The downregulation of miR-142-3p/5p causes the upregulation of several target genes such as *SAP*, *CD84* and *IL-10*, leading to T cell overreaction in lupus. The increased miR-21 in lupus T cells causes T-cell overexpression by reducing the expression of PDCD4.

Impaired IL-2 production and signaling has been implicated in lupus pathogenesis. MiR-31 is a positive regulator of IL-2 production in activated T cells via targeting Ras homolog gene family member A (RhoA) [92]. MiR-31 was markedly reduced in human lupus T cells, which caused the upregulation of RhoA and then the reduction of IL-2 in human lupus T cells. Ectopic expression of miR-31 in lupus T cells restored the expression of IL-2 [92]. While a previous study has shown the upregulation of miR-155 in the PBMCs of human patients with lupus [93], a most recent study revealed that there was reduced expression of miR-155 in the PBMCs of juvenile SLE patients, which correlated with disease severity and proteinuria. Ectopic expression of miR-155 led to decreased expression of protein phosphatase 2A (PP2Ac) and enhanced IL-2 expression. This suggests a potential role of miR-155 in juvenile lupus by affecting IL-2 production [94].

MiR-125a is mainly expressed in T cells, and it targets Kruppel-like factor 13 (KLF13) gene, a critical transcription factor in the regulation of the chemokine RANTES [95]. Reduced expression of miR-125a in the PBMCs from patients with lupus resulted in the upregulation of KLF13, which in turn contributed to the elevation of the inflammatory chemokine RANTES in lupus T cells [95]. miR-142-3p/5p was significantly downregulated in lupus CD4$^+$ T cells, which led to the increase of their target genes such as signaling lymphocytic activation molecule-associated protein (SAP), CD84, and IL-10, leading to T and B cell overactivation. The inhibition of miR-142-3p/5p in healthy donor CD4$^+$ T cells caused T-cell overactivation and B-cell hyperstimulation, whereas overexpression of miR-142-3p/5p in lupus CD4$^+$ T cells had an opposite effect [96]. MiR-21 was upregulated in both human and murine lupus and its expression level positively correlated with disease activity [93,97]. An in vitro study suggested that increased miR-21 expression in lupus T cells repressed PDCD4, leading to aberrant T cell responses in lupus, characterized as enhanced proliferation, IL-10 production, CD40L expression and their capacity to drive B-cell maturation [93]. Overall, these data implicate the role of miRNA in regulation of lupus-related inflammatory mediators and T-cell responses (Fig. 9.5).

5.3 MicroRNA Regulation of B-Cell Function in Systemic Lupus Erythematosus

Production of autoantibodies against a wide range of self-antigens is one of the hallmarks of SLE, suggesting a critical involvement of B-cell mediated humoral responses in lupus etiology. The B cell–depleting medicine such as Benlysta, which depletes B cells to reduce the production of autoantibodies, is seemingly an effective treatment for human lupus. The direct involvement of miRNA in the regulation B-cell function in the lupus setting has been evidenced recently (Fig. 9.6).

MiR-155 is a multifaceted regulator of immunity. The expression of miR-155 in B cells is intrinsically required for normal B-cell development and function. The upregulation of miR-155 has been observed in both murine and human lupus B cells [93,98]. To understand the direct role of miR-155 in autoimmunity, Thai et al. crossed the B6.MRL- Fas^{lpr}/J (B6-lpr) lupus mice, which carry the mutation on Fas gene (Fas^{lpr}), with bic/miR-155$^{-/-}$ mice to generate a B6-lpr model with miR-155 deficiency (miR-155$^{-/-}$ Fas^{lpr}). Significantly, depletion of miR-155 alleviated the autoimmune splenomegaly, and reduced anti-dsDNA IgG autoantibody production in miR-155$^{-/-}$ Fas^{lpr} mice. The mitigating effect of miR-155 depletion on lupus-related symptoms such as B-cell hyperactivation, proliferation, and

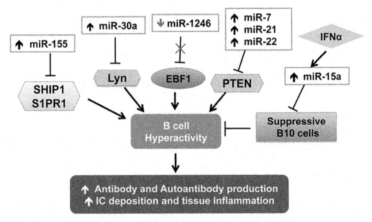

Figure 9.6 *miRNA regulation of B cell function in SLE.* Several miRNAs are reported to be dysregulated in lupus B cells, which target different signaling molecules that regulate B-cell development and activation, leading to B-cell hyperactivity and production of autoantibodies. The increased miR-155 expression in lupus B cells may reduce SHIP1 and S1PR1 to cause B-cell hyperactivity. The upregulation of miR-30a, miR-7, miR-21, and miR22 in lupus B cells leads to abnormal B-cell activity via targeting Lyn and PTEN, two negative regulators of B-cell development and function, respectively. On the other hand, reduced miR-1246 level in lupus B cells releases its suppressive effect on EBF1, a positive regulator of B-cell development and activation to induce B-cell hyperactivity. Increased miR-15a plays a role in IFNα-mediated promotion of B-cell hyperactivity and autoimmune responses via suppressing IL-10 producing regulatory B-10 cells.

autoimmune responses in B6-*lpr* mice is mediated by the derepression of miR-155 target gene Src homology-2 domain-containing inositol 5-phosphatase 1 (SHIP1) [99].The autoimmune amelioration effect of depleting miR-155 in B6-*lpr* was also reported in a recent independent study [100]. Compared to regular B6-*lpr* mice, miR-155$^{-/-}$ Fas^{lpr} mice had reduced renal inflammation as the result of less deposition of total IgA, IgM, and IgG, and less infiltration of inflammatory cells in the kidney. Sphingosine-1-phosphate receptor 1 (S1PR1) was identified as a new target of miR-155, by which miR-155 regulates autoinflammation [100].

Lyn, a member of the Src family of nonreceptor-type protein tyrosine kinase, is a negative regulator of B-cell activation. Mice with Lyn deficiency display spontaneous B-cell hyperactivity and develop a lupus-like autoimmune disease [97]. It was reported that miR-30a was significantly increased in human lupus B cells, which targeted Lyn directly, leading to decreased expression of Lyn in lupus B cells and then B-cell hyperactivity [101].The promotion effect of ectopic miR-30a expression on B-cell proliferation and the production of IgG antibodies could be abolished by inducing overexpression of Lyn [101].

The most current study identified that decreased miR-1246 expression in B cells from human patients with lupus contributed to enhanced B-cell activation via targeting early B-cell factors (EBF1). EBF1 is a key regulator of B-cell development, activation, and proliferation through activation of AKT signaling pathway [102]. Inhibition of miR-1246 in healthy B cells upregulated EBF1 expression and promoted AKT phosphorylation, leading to B-cell responsiveness. Deliberate increase of miR-1246 in lupus B cells reduced the B-cell responses by inhibiting EBF1 expression.This study also demonstrated a positive feedback loop comprised of AKT signaling pathway, p53, miR-1246, and EBF1 in the acceleration of B-cell responses in lupus.The initial B-cell activation led to phosphorylation of AKT and then reduction of p53, which consequently downregulated miR-1246, leading the upregulation of EBF1 and further promotion of B-cell responsiveness [102]. Another study has shown that miR-7, miR-21, and miR-22 were increased in lupus B cells, which inversely correlated with PTEN expression level in lupus B cells [103]. Inhibition of miR-7 rescued defective PTEN-related B-cell hyperresponsiveness, suggesting the contribution of miR-7 to B-cell abnormalities in lupus [103].

IFNα has been shown to promote lupus disease in murine lupus models such as NZB/W$_{F1,}$ presumably involving the upregulation of miR-15a [104]. IFNα treatment upregulated the expression of miR-15a in both splenic cells and plasma, which correlated with autoantibody production in NZB/W$_{F1}$ mice [104]. MiR-15a promotion of lupus is believed to be mediated through B-10 cells, an immunosuppressive subset of B cells producing IL-10 [104]. Together, the above findings demonstrate the significant contribution of altered miRNAs in B cells to in lupus pathogenesis (Fig. 9.6).

5.4 MicroRNA Regulation of Renal Inflammation in Systemic Lupus Erythematosus

The kidney is the principal target organ of autoimmune attacks in SLE, resulting in lupus nephritis (LN). LN is a leading cause of morbidity and mortality in lupus. Immune-complexes get deposited in the renal glomeruli to initiate a type III hypersensitivity immune attack causing attraction of damaging inflammatory cells including neutrophils, macrophages, and Th17 cells, culminating in LN. The dysregulated miRNAs have been identified in kidney cells of human patients with LN and also murine lupus, which correlated with local tissue inflammation [105].

A murine model of chronic kidney disease, B6.MRLc1, has been utilized to decipher the contribution of miRNAs to renal inflammation [106]. MiR-146a was significantly increased in the kidneys of B6.MRLc1 mice when compared to control mice. The increase of miR-146a correlated with the expression of the inflammatory-associated genes NF-κBp65, IRAK1, IL1β, IL10, and chemokine (C-X-C motif) ligand (CXCL)s and the development of chronic kidney inflammation [106]. Of relevance, the glomerulus of human patients with LN also demonstrated increased miR-146a compared to healthy controls. The expression levels of miR-146a seemingly correlated with LN [107]. Further studies are needed to decisively determine the causative role of miR-146a in LN.

In normal C57BL/6 mice, miR-26a is one of the most abundant miRNAs in glomeruli, which in striking contrast was significantly decreased in B6.MRLc1 mice, correlating with the progression of podocyte injury in autoimmune glomerulonephritis [108]. Notably, decreased glomerular miR-26a levels were also reported in human patients with LN and IgA nephropathy when compared to healthy controls [108]. Progressive LN culminates in renal fibrosis, which is measured by chronicity index (CI), a semiquantitative score of chronic kidney injury. The upregulation of miR-150 in kidneys of LN patients positively correlated with CI score and profibrotic protein expression [109]. Further study revealed that miR-150 promoted profibrotic protein expression indirectly via downregulation of

SOCS1, an antifibrotic protein. Moreover, miR-150 inhibition reversed TGF-β1-induced suppression of SOCS1 to promote profibrotic protein in tubular cells and podocytes, suggesting the involvement of miR-150 in TGF-β1-mediated profibrotic effect [109].

MiR-23 was commonly downregulated in the inflammatory lesions in different human autoimmune diseases including lupus, RA, and MS and their correlated murine models [110]. The downregulation of miR-23b in the local inflammatory cells, such as primary kidney cells in lupus, was associated with the increase of IL-17 during autoimmune pathogenesis. MiR-23b targets TAB2, TAB3, and IKKα, thereby playing an important role in the regulation of IL-17-, TNFα-, and IL-1β-induced activation of NF-κB signaling and production of inflammatory cytokines. The upregulation of IL-17 suppressed miR-23b in the target tissues, and subsequently caused enhanced inflammation in resident cells and inflammatory lesions. The mice with transgenic overexpression of miR-23b were resistant to the development of autoimmune disease, such as collagen-induced arthritis and EAE, due to the suppression of IL-17-mediated inflammation. This study strongly suggests a key role of miR-23b in the regulation of IL-17-associated autoimmune inflammation in localized tissues for different autoimmune diseases, including LN [110].

6. CELL-FREE CIRCULATING MICRORNAs AS BIOMARKERS IN SYSTEMIC LUPUS ERYTHEMATOSUS

The detection of miRNAs in cell-free body fluid (such as serum, plasma, and urinary sediments) and the correlation of cell-free miRNAs with lupus make miRNAs attractive, noninvasive biomarkers for lupus prognosis, diagnosis, and monitoring therapy. Compared to healthy controls, the serum levels of miR-146a and miR-155 were reduced, while the urinary level of miR-146a was increased in human patients with SLE [111]. Of note, serum miR-146a level inversely correlated with proteinuria and SLE disease index (SLEDAI). Both serum miR-146a and miR-155 correlated with lymphocyte count and estimated glomerular filtration rate (eGFR). Further, calcitriol treatment, which has a beneficial effect on LN, significantly increased serum miR-146a [111]. This study suggested the potential use of serum miR-155 and miR-146a as biomarkers of SLE. In addition to miR-146a and miR-155, most likely with the same cohort of lupus and healthy control subjects, the same research group reported alteration of other miRNAs, including the miR-200 family, miR-205 and miR-192 in serum and urinary sediments in human SLE [112]. The serums levels of miR-200a/b/c, miR-429, miR205, and miR-192, and the urinary levels of miR200a/c, miR-141, miR-429, and miR-192 were lower in patients with lupus when compared to healthy controls. While the eGFR correlated with serum miR-200b/c, miR-429, miR-205, and miR-192, the proteinuria inversely correlated with serum miR-200a/c. The serum miR-200a also inversely correlated with SLEDAI [112].

By profiling miRNA expression in a Chinese cohort of healthy controls, patients with SLE, and patients with RA, Wang et al. reported that miR-126 was specifically increased in the plasma of patients with SLE, but not RA. The upregulation of miR-21, miR-451, miR-223, and miR-16, and the downregulation of miR-125a-3p, miR-155, and miR-146a in the plasma were commonly observed in patients with SLE and RA [113]. These data suggest that miR-126 can potentially serve as specific diagnostic biomarkers for SLE. In a recent study of circulating miRNAs, Carlsen et al. profiled the expression of 45 miRNAs in the plasma samples from a primary Denmark cohort, and then validated the finding of dysregulated miRNAs in an independent Sweden cohort as well as a separate disease control cohort [114]. Seven miRNAs were significantly altered in plasma from SLE patients. MiR-142-3p and miR-181a were increased, whereas miR-106a, miR-17, miR-20a, miR-203, and miR-92a were decreased in plasma from the patients with SLE. Of note, there was no specific dysregulated miRNA common to both cohorts correlated with SLEDAI. Also, there was no correlation of the above dysregulated miRNA with disease activity in individual patients. However, when analyzing the two consecutive data sets as a whole, they did observe a significant correlation between the SLEDAI score and altered expression of miR-142-3p and miR-181a in the plasma [114]. There exists urgent need for further innovation and improvements in sensitivity for the techniques used in profiling body fluid miRNA for successful clinical application of noncellular miRNAs as biomarkers.

7. REGULATION OF MICRORNA EXPRESSION IN SYSTEMIC LUPUS ERYTHEMATOSUS

With the emerging recognition of the involvement of miRNA in autoimmune disease, it is vital to understand the cause of miRNA dysregulation in lupus and other autoimmune disorders. Such an understanding will form a sound basis for better clinical management of patients and even potential prevention of these disorders. It is highly likely that miRNA dysregulation in SLE may be caused by multiple interacting factors including genetic, epigenetic, environmental, and hormonal factors.

7.1 Genetic and Epigenetic Regulation of MicroRNA Expression in Lupus

Although the precise reasons for miRNA dysregulation in human disease are not comprehensively understood, several plausible mechanisms have been proposed [2,115]. The miRNA dysregulation in human diseases may be caused by the direct deletion, amplification, and mutation of miRNA coding regions, or the genetic and epigenetic defects in transcription factors and signaling molecules that regulate specific miRNAs [115]. Since miR-146a has been shown to play a key role in immunity and autoimmunity, the association between miR-146a genetic single nucleotide polymorphisms and lupus susceptibility has been extensively investigated in different ethnic populations. By sequencing the potential regulatory regions of miR-146a, a novel genetic variant (rs57095329 A/G) in the miR-146a promoter was identified in multiple cohorts of Asian lupus patients, which was strongly associated with lupus susceptibility. The risk-associated G allele had decreased binding to transcription factor Ets-1, leading to lower miR-146a levels in patients with SLE [116]. In a case control study, the genetic variant located at an intergenic region between the pituitary tumor-transforming 1 (PTTG1) and miR-146a gene, rs2431697 was found to be genetically associated with SLE in Europeans. The risk allele in rs2431697 correlated with the downregulation of miR-146a in the patients with lupus [117].

While miRNA contributes to lupus pathogenesis by directly or indirectly targeting DNA methylation pathways, the interaction between DNA methylation and miRNA in physiological and pathological conditions is bidirectional. The overexpression of X chromosome–linked miRNAs such as miR-98, miR-188, miR-421, and miR-503 in women with lupus is likely contributed by X-chromosome demethylation [118]. A genome-wide DNA methylation, mRNA, and miRNA expression analysis revealed that 36 upregulated miRNAs such as miR-181 in human lupus CD4$^+$ T cells were located near CpG sites and were hypomethylated when compared with normal controls, and eight downregulated miRNAs were hypermethylated in lupus CD4$^+$ T cells. Moreover, we have reported that a large cluster of genomic imprinting miRNAs form DLK1–Dio3 region such as miR-127 and miR-379 were substantially upregulated in murine lupus splenocytes [98]. Further investigation revealed the upregulation of DLK1–Dio3 miRNAs in MRL-*lpr* splenic cells was associated with the DNA hypomethylation [119]. Together, these newly emerged data strongly suggest the involvement of genetics and epigenetics in the miRNA dysregulation in SLE.

7.2 Hormonal and Environmental Regulation of MicroRNA Expression in Lupus

Compared to males, females are predominantly affected by many autoimmune diseases including SLE [120,121]. The female to male ratio for SLE is 9–13:1. The sex differences in the immune capability and autoimmune disease susceptibility are attributed to multiple factors, including the intrinsic genome difference between males and females, sex hormones, sex-specific gene regulation in response to environmental stimuli, and also sex-related epigenome profile [121].

Estrogen regulates miRNA expression via activation of estrogen receptors (ERs) in different biological systems, including the immune system. Estrogen-activated ER, estrogen-regulated miRNAs (such as miR-17-92), and protein coding genes (such as c-Myc and E2F) form a regulatory circuitry in estrogen action, thereby fine-tuning the estrogen-mediated cellular responses [121]. It is clear that estrogen has profound influences on immune system development and function. Estrogen has been shown to exert both antiinflammatory and proinflammatory immune responses in different studies, which in large part is due to variations in dose and duration of estrogen employed [2,122,123]. We have reported that estrogen regulated miRNA expression in splenic cells, and that selective estrogen-regulated miRNAs such as miR-146a and miR-223 regulated LPS induced IFNγ production in splenocytes from estrogen-treated mice [124]. These data suggest a novel, miRNA-related mechanism in the estrogen modulation of immune function.

While the detrimental effect of estrogen in human lupus is seemingly equivocal, the lupus promotion effect of estrogen has been clearly documented in murine lupus models, especially in the NZB/W$_{F1}$ model [2,125]. Of note, estrogen-downregulated miRNAs, such miR-146a, miR-125a, and miR-143, were decreased in human lupus, while estrogen-upregulated miR-148a was also increased in human lupus [2,124]. This observation suggests that estrogen might contribute to the sex bias of lupus and affect disease course by regulating lupus-related miRNA expression. We recently reported the sex differences in the expression of murine lupus-related miRNAs including the miR-182-96-183 cluster, miR-155, miR-31, miR148a, miR-127, and miR-379 in NZB/W$_{F1}$ mice, which were markedly evident after the onset of lupus [126]. Moreover, estrogen treatment promoted the levels of aforementioned lupus-related miRNAs (with exception of miR-155) in orchidectomized male NZB/W$_{F1}$ mice to a similar level as that in the age-matched female mice. Together, we surmised that the dysregulation of miRNAs in lupus may be contributed by both lupus manifestation and female hormone estrogen [126].

Environmental factors such as tobacco smoke, infectious agents, radiation, ultraviolet (UV) light, chemical compounds, and diet are critically associated with lupus onset and flare in genetically susceptible people. While different mechanisms are involved, increased evidence suggests that environmental factors have profound influences on autoimmune disease–related epigenome profiles [86,127]. In a recent review report, Vrijens et al. critically reviewed the signature miRNA expression in response to all types of environmental exposures including estrogenic endocrine disrupting chemicals, to which humans can be exposed to during a lifetime [128]. Many environmental exposure-regulated miRNAs such as miR-21, miR-146a, miR-155 have been implicated in lupus pathogenesis. Nevertheless, further investigation is warranted to determine the direct involvement of miRNA in the environmental effect on lupus in a disease setting.

8. CONCLUSION AND PERSPECTIVE

Even though only two decades ago miRNAs were identified as key epigenetic regulators of mammalian genome, there has been a stunning pace of publications documenting the broad physiological and pathological roles of miRNA in the immune development, functions, and immune-related disorders. A single miRNA exerts diverse immunoregulatory functions, which may be specific at distinct developmental stages, in a specific immune cell type, or during effector function. As summarized earlier, miR-146a and miR-155 are two well-studied multifaceted miRNAs, which play important roles in the regulation of immune cell development, differentiation, function, and in the shaping of both the innate and adaptive immune responses. Not surprisingly, dysregulated miR-146a and miR-155 expression/function has been widely associated with different autoimmune diseases such as SLE, RA, and MS. Although miR-146a was decreased in PBMC from human patients with lupus [83], it was increased in PBMC and RASFs from human patients with RA and in murine lupus T cells [84,98]. Thus, the distinctive role of individual miRNA must be recognized based on the cell type and disease phenotype, and therefore cannot be generalized.

The signature lupus-related miRNA expression profile has been identified in different cellular sources and also cell-free body fluid. The relative stability of miRNAs and the correlation of dysregulated miRNAs with lupus onset and progress render miRNAs as valuable potential biomarkers for diagnosis as well as therapeutic targets for lupus treatment. Given that a specific lupus-related gene may contain targeting sites for different miRNAs [129], it is plausible that multiple disease-associated miRNAs, rather than a single miRNA, synergistically act together to disrupt self-tolerance and to promote autoimmunity development. Therefore, targeting a single miRNA may not be sufficient for therapy efficacy in a particular human pathological condition. Despite the challenges of correcting miRNA dysregulation in vivo in specific cell types/tissues, there is still a huge potential of using miRNAs as gene therapy targets in vivo to treat cancer and other human diseases [130–132]. SPC3649, an inhibitor of miR-122 is currently under phase II clinical trial for treatment of hepatitis C virus infection [133]. Although thus far, effective approaches for targeting specific miRNA in distinct immune cell type or location is a technical challenge, the therapeutic potential of miRNA in autoimmune disease such as SLE is still promising. Given the encouraging data in murine lupus models that depletion of miR-155 significant alleviated lupus-related symptoms in B6-*lpr* mice, miRNA-based therapy is a plausible alternative, or adjunct, therapeutic strategy for SLE [99,100]. Further, a study with a different murine lupus model B6.Sle123 revealed that systemic silencing miR-21 in vivo significantly ameliorated autoimmune-related splenomegaly, suggesting the therapeutic potential of targeting miR-21 [97]. We anticipate that in the near future, to treat autoimmune diseases, novel effective miRNA-based gene therapies will be developed to replace the traditional immune-suppressive therapies, which are usually lifelong with undesirable side effects.

GLOSSARY

Adaptive immunity Specific immune responses by the generation of specific antibodies and cell-mediated responses to a specific antigen.

Autoimmunity The immune system misdirects immune responses to self-antigens. Technically, autoimmunity occurs during physiological response such as elimination of senescent cells by autoantibodies.

Autoimmune disease A pathological condition, where autoreactive cells and/or autoantibodies launch a deleterious attack on the body's own tissues, resulting in functional or structural damage.

B6.MRL- Fas*lpr***/J (B6-lpr)** A congenic murine lupus model developed by introducing Fas gene mutation (Fas*lpr*) from MRL-*lpr* mice to C57BL/6 mice.

DNA methylation A biochemical process that adds a methyl (CH3) group to the 5′ cytosine within a CpG dinucleotide context.

Epigenetics A study of heritable gene expression and cellular phenotype changes that are not caused by any alteration in genetic DNA sequences.

Follicular B helper T cell (Tfh) A subpopulation of T cells that are located at B-cell follicles and provide help for B-cell development, differentiation, and function.

Germinal center (GC) A structure developed within the B-cell follicles of secondary lymphoid organs for helping B-cell proliferation and differentiation into high affinity, class-switched plasma, and memory B cells. The GC arises once the follicular B cells are exposed to and activated by T cell–dependent antigens.

Helper T cell (Th) A type of $CD4^+$ cells that helps other immune cells in the immune response by releasing cytokines.

Histone modification A variety of biochemical processes that modify histone protein to alter chromatin structure and gene expression, which include, but are not limited to methylation, acetylation, and phosphorylation.

Immune response The reactions of the body's immune system to defend against foreign substances or its own tissues.

Inflammation A body's protective response to harmful stimuli such as damaged cells and pathogens. Either compromised or overwhelming inflammation will cause the development of disease.

Innate immunity Nonspecific immune defense mechanism that kills pathogens once they invade the body.

Interferon A group of cytokine proteins that are critically involved in host immune responses to distinct pathogens. Based on the type of receptor they interact with, the interferon cytokines are classified to type I (eg, $IFN\alpha/\beta$), type II (eg, $IFN\gamma$), and type III (eg, $IFN\lambda$).

Lymphocyte A type of mononuclear, nonphagocytic white blood cell found in the blood and lymphoid organs. T and B lymphocytes are the two major types of lymphocytes.

Macrophage A type of mononuclear white blood cell that has phagocytic function.

MicroRNA (miRNA) Small, single strand, noncoding ribonucleic acid oligonucleotide that regulate gene expression at the posttranscription level.

MRL/MpJ-Fas^{lpr}/J (MRL-lpr) Mice with a spontaneous mutation on Fas gene (Fas^{lpr}) in MRL/MpJ genetic background that develop a variety of autoimmune diseases, notably lupus.

Neutrophil A type of granulocyte characterized by lobed nucleus and granules containing inflammatory molecules. In most species, neutrophils are the predominant leukocyte, and their numbers in peripheral blood are enhanced following acute infections or inflammation.

NZB/W$_{F1}$ The F1 hybrid of the New Zealand black and New Zealand white mouse, which is a classical murine model for human lupus.

Peripheral blood mononuclear cell (PBMC) Any blood cell having a round nucleus, ie, lymphocytes, monocytes, macrophages, and dendritic cells.

Regulatory T cell (Treg) A subpopulation of T cells that generally suppress the immune responses of other T cells.

RNA-induced silencing complex (RISC) A multiprotein complex where miRNAs or other regulatory RNAs (such as small interfering RNAs) recognize and bind to complementary messenger RNA (mRNA) to silence gene expression.

Systemic lupus erythematosus (SLE) A prototypic autoimmune disease that affects the skin, joints, kidneys, brain, and other organs. SLE predominantly occurs in women.

Toll like receptor A class of proteins that recognize and bind to specific microbial components to trigger innate inflammatory responses.

LIST OF ACRONYMS AND ABBREVIATIONS

AID Activation-induced cytidine deaminase
Ago Argonaute
AML-1 Acute myeloid leukemia-1
APL Acute promyelocytic leukemia
ATRA All-trans-retinoic acid
BCL B-cell lymphoma
B6-*lpr* B6.MRL- Fas^{lpr}/J
C/EBP CCAAT enhancer binding protein
CXCL Chemokine (C-X-C motif) ligand
DGCR8 DiGeorge syndrome critical region protein 8
DNMT DNA methyltransferase
EAE Experimental autoimmune encephalomyelitis
E2F1 E2F transcription factor 1
eGFR Estimated glomerular filtration rate
ER Estrogen receptor
Fosl2 FOS-like antigen 2
GC Germinal center
ICOS Inducible T-cell costimulator
IFN Interferon
IKK IκB kinases
IL Interleukin
IRAK Interleukin-1 receptor-associated kinase 1
IRF Interferon regulatory factor
KLF13 Kruppel-like factor 13
LPS Lipopolysaccharide
MCP-1 Monocyte chemotactic protein-1
M-CSFR Macrophage colony-stimulating factor
Mef2c Myocyte-specific enhancer factor 2C
miRNA MicroRNA
MRL-*lpr* MRL/MpJ-Fas^{lpr}/J
MOG$_{35-55}$ Myelin oligodendrocyte glycoprotein$_{35-55}$
mRNA Messenger RNA
MS Multiple sclerosis
MyD88 Myeloid differentiation primary response protein 88
NCOR2 Nuclear receptor corepressor 2
NFI-A Nuclear factor 1 A-type
NF-κB Nuclear factor-κB
Nt Nucleotide
PAMPs Pathogen-associated molecular patterns
PBMC Peripheral blood mononuclear cell

PHLPP2 PH domain leucine-rich repeat protein phosphatase 2
PP2Ac Protein phosphatase 2A
primiRNA Primary miRNA transcripts
premiRNA Precursor miRNA
PTEN Phosphatase and tensin homolog
PTTG1 Pituitary tumor-transforming 1
RA Rheumatoid arthritis
RASF Synovial fibroblasts
RASGRP1 RAS guanyl nucleotide-releasing protein 1
RhoA Ras homolog gene family member A
RISC RNA-induced silencing complex
RNA Ribonucleic acid
SAP Signaling lymphocytic activation molecule-associated protein
SHIP1 Src homology-2 domain-containing inositol 5-phosphatase 1
SMAD7 Mothers against decapentaplegic homolog 7
SLE Systemic lupus erythematosus
SLEDAI Systemic lupus erythematosus disease index
SNP Single nucleotide polymorphism
SOCS1 Suppressor of cytokine signaling 1
STAT Signal transducers and activators of transcription
TAB2 TGF-beta activated kinase 1/MAP3K7 binding protein 2
TCR T cell receptor
Th T helper
TNF Tumor necrosis factor
TLR Toll-like receptor
TRAF-6 Tumor necrosis factor receptor associated factor 6
TRBP TAR RNA binding protein
Treg Regulatory T cell
UTR Untranslated region

ACKNOWLEDGMENTS

We thank the support from Lupus Foundation of America, VMRCVM Intramural Research Competition Grant (IRC), American Autoimmune Related Diseases Association (AARDA), and Alliance for Lupus Research (ALR). We also thank Dr. Michael Edwards for critical reviewing of the manuscripts.

REFERENCES

[1] Obata Y, Furusawa Y, Hase K. Epigenetic modifications of the immune system in health and disease. Immunol Cell Biol 2015;93(3):226–32.
[2] Dai R, Ahmed SA. MicroRNA, a new paradigm for understanding immunoregulation, inflammation, and autoimmune diseases. Transl Res 2011;157(4):163–79.
[3] Lee RC, Feinbaum RL, Ambros V. The *C. elegans* heterochronic gene lin-4 encodes small RNAs with antisense complementarity to lin-14. Cell 1993;75(5):843–54.
[4] Reinhart BJ, Slack FJ, Basson M, Pasquinelli AE, Bettinger JC, Rougvie AE, et al. The 21-nucleotide let-7 RNA regulates developmental timing in *Caenorhabditis elegans*. Nature 2000;403(6772):901–6.

[5] Griffiths-Jones S, Grocock RJ, van Dongen S, Bateman A, Enright AJ. miRBase: microRNA sequences, targets and gene nomenclature. Nucleic Acids Res 2006; 34(Database issue):D140–4.

[6] Baek D, Villen J, Shin C, Camargo FD, Gygi SP, Bartel DP. The impact of microRNAs on protein output. Nature 2008;455(7209):64–71.

[7] Selbach M, Schwanhausser B, Thierfelder N, Fang Z, Khanin R, Rajewsky N. Widespread changes in protein synthesis induced by microRNAs. Nature 2008;455(7209):58–63.

[8] Lewis BP, Burge CB, Bartel DP. Conserved seed pairing, often flanked by adenosines, indicates that thousands of human genes are microRNA targets. Cell 2005;120(1): 15–20.

[9] Bernstein E, Kim SY, Carmell MA, Murchison EP, Alcorn H, Li MZ, et al. Dicer is essential for mouse development. Nat Genet 2003;35(3):215–7.

[10] Baltimore D, Boldin MP, O'Connell RM, Rao DS, Taganov KD. MicroRNAs: new regulators of immune cell development and function. Nat Immunol 2008;9(8): 839–45.

[11] Ambros V. The functions of animal microRNAs. Nature 2004;431(7006):350–5.

[12] Zhang B, Farwell MA. MicroRNAs: a new emerging class of players for disease diagnostics and gene therapy. J Cell Mol Med 2008;12(1):3–21.

[13] Zhang B, Pan X, Cobb GP, Anderson TA. MicroRNAs as oncogenes and tumor suppressors. Dev Biol 2007;302(1):1–12.

[14] Krutzfeldt J, Stoffel M. MicroRNAs: a new class of regulatory genes affecting metabolism. Cell Metab 2006;4(1):9–12.

[15] Eacker SM, Dawson TM, Dawson VL. Understanding microRNAs in neurodegeneration. Nat Rev Neurosci 2009;10(12):837–41.

[16] Pauley KM, Cha S, Chan EK. MicroRNA in autoimmunity and autoimmune diseases. J Autoimmun 2009;32(3–4):189–94.

[17] Rodriguez A, Griffiths-Jones S, Ashurst JL, Bradley A. Identification of mammalian microRNA host genes and transcription units. Genome Res 2004;14(10A):1902–10.

[18] Kim VN, Nam JW. Genomics of microRNA. Trends Genet 2006;22(3):165–73.

[19] Gregory RI, Yan KP, Amuthan G, Chendrimada T, Doratotaj B, Cooch N, et al. The microprocessor complex mediates the genesis of microRNAs. Nature 2004;432(7014): 235–40.

[20] Lee Y, Ahn C, Han J, Choi H, Kim J, Yim J, et al. The nuclear RNase III Drosha initiates microRNA processing. Nature 2003;425(6956):415–9.

[21] Ruby JG, Jan CH, Bartel DP. Intronic microRNA precursors that bypass Drosha processing. Nature 2007;448(7149):83–6.

[22] Bartel DP. MicroRNAs: genomics, biogenesis, mechanism, and function. Cell 2004;116(2):281–97.

[23] Kim VN, Han J, Siomi MC. Biogenesis of small RNAs in animals. Nat Rev Mol Cell Biol 2009;10(2):126–39.

[24] Vasudevan S, Tong Y, Steitz JA. Switching from repression to activation: microRNAs can up-regulate translation. Science 2007;318(5858):1931–4.

[25] Lee I, Ajay SS, Yook JI, Kim HS, Hong SH, Kim NH, et al. New class of microRNA targets containing simultaneous 5'-UTR and 3'-UTR interaction sites. Genome Res 2009;19(7):1175–83.

[26] O'Neill LA, Sheedy FJ, McCoy CE. MicroRNAs: the fine-tuners of Toll-like receptor signalling. Nat Rev 2011;11(3):163–75.

[27] Baumjohann D, Ansel KM. MicroRNA-mediated regulation of T helper cell differentiation and plasticity. Nat Rev 2013;13(9):666–78.

[28] Montagner S, Deho L, Monticelli S. MicroRNAs in hematopoietic development. BMC Immunol 2014;15:14.

[29] Fazi F, Rosa A, Fatica A, Gelmetti V, De Marchis ML, Nervi C, et al. A minicircuitry comprised of microRNA-223 and transcription factors NFI-A and C/EBPalpha regulates human granulopoiesis. Cell 2005;123(5):819–31.

[30] Pulikkan JA, Dengler V, Peramangalam PS, Peer Zada AA, Muller-Tidow C, Bohlander SK, et al. Cell-cycle regulator E2F1 and microRNA-223 comprise an autoregulatory negative feedback loop in acute myeloid leukemia. Blood 2010; 115(9):1768–78.

[31] Johnnidis JB, Harris MH, Wheeler RT, Stehling-Sun S, Lam MH, Kirak O, et al. Regulation of progenitor cell proliferation and granulocyte function by microRNA-223. Nature 2008.

[32] Pelosi A, Careccia S, Lulli V, Romania P, Marziali G, Testa U, et al. miRNA let-7c promotes granulocytic differentiation in acute myeloid leukemia. Oncogene 2013;32(31):3648–54.

[33] Pulikkan JA, Peramangalam PS, Dengler V, Ho PA, Preudhomme C, Meshinchi S, et al. C/EBPalpha regulated microRNA-34a targets E2F3 during granulopoiesis and is down-regulated in AML with CEBPA mutations. Blood 2010;116(25):5638–49.

[34] Wang XS, Gong JN, Yu J, Wang F, Zhang XH, Yin XL, et al. MicroRNA-29a and microRNA-142-3p are regulators of myeloid differentiation and acute myeloid leukemia. Blood 2012;119(21):4992–5004.

[35] Fontana L, Pelosi E, Greco P, Racanicchi S, Testa U, Liuzzi F, et al. MicroRNAs 17-5p-20a-106a control monocytopoiesis through AML1 targeting and M-CSF receptor upregulation. Nat Cell Biol 2007;9(7):775–87.

[36] Takeda K, Akira S. Toll-like receptors in innate immunity. Int Immunol 2005;17(1):1–14.

[37] Liew FY, Xu D, Brint EK, O'Neill LA. Negative regulation of toll-like receptor-mediated immune responses. Nat Rev 2005;5(6):446–58.

[38] Gantier MP, Sadler AJ, Williams BR. Fine-tuning of the innate immune response by microRNAs. Immunol Cell Biol 2007;85(6):458–62.

[39] Taganov KD, Boldin MP, Baltimore D. MicroRNAs and immunity: tiny players in a big field. Immunity 2007;26(2):133–7.

[40] He X, Jing Z, Cheng G. MicroRNAs: new regulators of Toll-like receptor signalling pathways. BioMed Res Int 2014;2014:945169.

[41] Taganov KD, Boldin MP, Chang KJ, Baltimore D. NF-kappaB-dependent induction of microRNA miR-146, an inhibitor targeted to signaling proteins of innate immune responses. Proc Natl Acad Sci USA 2006;103(33):12481–6.

[42] O'Connell RM, Taganov KD, Boldin MP, Cheng G, Baltimore D. MicroRNA-155 is induced during the macrophage inflammatory response. Proc Natl Acad Sci USA 2007;104(5):1604–9.

[43] Sheedy FJ, Palsson-McDermott E, Hennessy EJ, Martin C, O'Leary JJ, Ruan Q, et al. Negative regulation of TLR4 via targeting of the proinflammatory tumor suppressor PDCD4 by the microRNA miR-21. Nat Immunol 2010;11(2):141–7.

[44] Chen XM, Splinter PL, O'Hara SP, LaRusso NF. A cellular micro-RNA, let-7i, regulates Toll-like receptor 4 expression and contributes to cholangiocyte immune responses against *Cryptosporidium parvum* infection. J Biol Chem 2007;282(39):28929–38.

[45] Tili E, Michaille JJ, Cimino A, Costinean S, Dumitru CD, Adair B, et al. Modulation of miR-155 and miR-125b levels following lipopolysaccharide/TNF-alpha stimulation and their possible roles in regulating the response to endotoxin shock. J Immunol 2007;179(8):5082–9.

[46] Lee HM, Nguyen DT, Lu LF. Progress and challenge of microRNA research in immunity. Front Genet 2014;5:178.

[47] Cobb BS, Nesterova TB, Thompson E, Hertweck A, O'Connor E, Godwin J, et al. T cell lineage choice and differentiation in the absence of the RNase III enzyme Dicer. J Exp Med 2005;201(9):1367–73.

[48] Chong MM, Rasmussen JP, Rudensky AY, Littman DR. The RNAseIII enzyme Drosha is critical in T cells for preventing lethal inflammatory disease. J Exp Med 2008;205(9):2005–17.

[49] O'Carroll D, Mecklenbrauker I, Das PP, Santana A, Koenig U, Enright AJ, et al. A slicer-independent role for Argonaute 2 in hematopoiesis and the microRNA pathway. Genes & Dev 2007;21(16):1999–2004.

[50] Rodriguez A, Vigorito E, Clare S, Warren MV, Couttet P, Soond DR, et al. Requirement of bic/microRNA-155 for normal immune function. Sci (New York, NY) 2007;316(5824):608–11.

[51] Thai TH, Calado DP, Casola S, Ansel KM, Xiao C, Xue Y, et al. Regulation of the germinal center response by microRNA-155. Sci (New York, NY) 2007;316(5824):604–8.

[52] Vigorito E, Perks KL, Abreu-Goodger C, Bunting S, Xiang Z, Kohlhaas S, et al. MicroRNA-155 regulates the generation of immunoglobulin class-switched plasma cells. Immunity 2007;27(6):847–59.

[53] Dorsett Y, McBride KM, Jankovic M, Gazumyan A, Thai TH, Robbiani DF, et al. MicroRNA-155 suppresses activation-induced cytidine deaminase-mediated Myc-Igh translocation. Immunity 2008;28(5):630–8.

[54] Calame K. MicroRNA-155 function in B cells. Immunity 2007;27(6):825–7.

[55] Kohlhaas S, Garden OA, Scudamore C, Turner M, Okkenhaug K, Vigorito E. Cutting edge: the Foxp3 target miR-155 contributes to the development of regulatory T cells. J Immunol 2009;182(5):2578–82.

[56] Lu LF, Thai TH, Calado DP, Chaudhry A, Kubo M, Tanaka K, et al. Foxp3-dependent microRNA155 confers competitive fitness to regulatory T cells by targeting SOCS1 protein. Immunity 2009;30(1):80–91.

[57] Stahl HF, Fauti T, Ullrich N, Bopp T, Kubach J, Rust W, et al. miR-155 inhibition sensitizes CD4+Th cells for TREG mediated suppression. PLoS ONE 2009;4(9):e7158.

[58] O'Connell RM, Kahn D, Gibson WS, Round JL, Scholz RL, Chaudhuri AA, et al. MicroRNA-155 promotes autoimmune inflammation by enhancing inflammatory T cell development. Immunity 2010.

[59] Hu R, Kagele DA, Huffaker TB, Runtsch MC, Alexander M, Liu J, et al. miR-155 promotes T follicular helper cell accumulation during chronic, low-grade inflammation. Immunity 2014;41(4):605–19.

[60] Monticelli S, Ansel KM, Xiao C, Socci ND, Krichevsky AM, Thai TH, et al. MicroRNA profiling of the murine hematopoietic system. Genome Biol 2005; 6(8):R71.

[61] Yang L, Boldin MP, Yu Y, Liu CS, Ea CK, Ramakrishnan P, et al. miR-146a controls the resolution of T cell responses in mice. J Exp Med 2012;209(9):1655–70.

[62] Lu LF, Boldin MP, Chaudhry A, Lin LL, Taganov KD, Hanada T, et al. Function of miR-146a in controlling Treg cell-mediated regulation of Th1 responses. Cell 2010;142(6):914–29.

[63] Pratama A, Srivastava M, Williams NJ, Papa I, Lee SK, Dinh XT, et al. MicroRNA-146a regulates ICOS-ICOSL signalling to limit accumulation of T follicular helper cells and germinal centres. Nat Commun 2015;6:6436.

[64] Li QJ, Chau J, Ebert PJ, Sylvester G, Min H, Liu G, et al. miR-181a is an intrinsic modulator of T cell sensitivity and selection. Cell 2007;129(1):147–61.

[65] Chen CZ, Li L, Lodish HF, Bartel DP. MicroRNAs modulate hematopoietic lineage differentiation. Sci (New York, NY) 2004;303(5654):83–6.

[66] Xiao C, Calado DP, Galler G, Thai TH, Patterson HC, Wang J, et al. MiR-150 controls B cell differentiation by targeting the transcription factor c-Myb. Cell 2007; 131(1):146–59.

[67] Zhou B, Wang S, Mayr C, Bartel DP, Lodish HF. miR-150, a microRNA expressed in mature B and T cells, blocks early B cell development when expressed prematurely. Proc Natl Acad Sci USA 2007;104(17):7080–5.

[68] Xiao C, Srinivasan L, Calado DP, Patterson HC, Zhang B, Wang J, et al. Lymphoproliferative disease and autoimmunity in mice with increased miR-17-92 expression in lymphocytes. Nat Immunol 2008;9(4):405–14.

[69] Ventura A, Young AG, Winslow MM, Lintault L, Meissner A, Erkeland SJ, et al. Targeted deletion reveals essential and overlapping functions of the miR-17 through 92 family of miRNA clusters. Cell 2008;132(5):875–86.

[70] Jiang S, Li C, Olive V, Lykken E, Feng F, Sevilla J, et al. Molecular dissection of the miR-17-92 cluster's critical dual roles in promoting Th1 responses and preventing inducible Treg differentiation. Blood 2011;118(20):5487–97.

[71] Kang SG, Liu WH, Lu P, Jin HY, Lim HW, Shepherd J, et al. MicroRNAs of the miR-17 approximately 92 family are critical regulators of T(FH) differentiation. Nat Immunol 2013;14(8):849–57.

[72] Du C, Liu C, Kang J, Zhao G, Ye Z, Huang S, et al. MicroRNA miR-326 regulates TH-17 differentiation and is associated with the pathogenesis of multiple sclerosis. Nat Immunol 2009;10(12):1252–9.

[73] Murugaiyan G, da Cunha AP, Ajay AK, Joller N, Garo LP, Kumaradevan S, et al. MicroRNA-21 promotes Th17 differentiation and mediates experimental autoimmune encephalomyelitis. J Clin Invest 2015;125(3):1069–80.

[74] Takahashi H, Kanno T, Nakayamada S, Hirahara K, Sciume G, Muljo SA, et al. TGF-beta and retinoic acid induce the microRNA miR-10a, which targets Bcl-6 and constrains the plasticity of helper T cells. Nat Immunol 2012;13(6):587–95.

[75] Goodnow CC. Multistep pathogenesis of autoimmune disease. Cell 2007;130(1): 25–35.

[76] Hewagama A, Richardson B. The genetics and epigenetics of autoimmune diseases. J Autoimmun 2009;33(1):3–11.

[77] Zhou X, Jeker LT, Fife BT, Zhu S, Anderson MS, McManus MT, et al. Selective miRNA disruption in T reg cells leads to uncontrolled autoimmunity. J Exp Med 2008;205(9):1983–91.

[78] Liston A, Lu LF, O'Carroll D, Tarakhovsky A, Rudensky AY. Dicer-dependent microRNA pathway safeguards regulatory T cell function. J Exp Med 2008; 205(9):1993–2004.

[79] Boldin MP, Taganov KD, Rao DS, Yang L, Zhao JL, Kalwani M, et al. miR-146a is a significant brake on autoimmunity, myeloproliferation, and cancer in mice. J Exp Med 2011;208(6):1189–201.

[80] Guo Q, Zhang J, Li J, Zou L, Zhang J, Xie Z, et al. Forced miR-146a expression causes autoimmune lymphoproliferative syndrome in mice via downregulation of Fas in germinal center B cells. Blood 2013;121(24):4875–83.

[81] Deng X, Su Y, Wu H, Wu R, Zhang P, Dai Y, et al. The role of microRNAs in autoimmune diseases with skin involvement. Scand J Immunol 2015;81(3):153–65.

[82] Shen N, Liang D, Tang Y, de Vries N, Tak PP. MicroRNAs–novel regulators of systemic lupus erythematosus pathogenesis. Nat Rev Rheumatol 2012;8(12): 701–9.

[83] Tang Y, Luo X, Cui H, Ni X, Yuan M, Guo Y, et al. MicroRNA-146A contributes to abnormal activation of the type 1 interferon pathway in human lupus by targeting the key signaling proteins. Arthritis Rheum 2009;60(4):1065–75.

[84] Pauley KM, Satoh M, Chan AL, Bubb MR, Reeves WH, Chan EK. Upregulated miR-146a expression in peripheral blood mononuclear cells from rheumatoid arthritis patients. Arthritis Res Ther 2008;10(4):R101.

[85] Peng SL. Altered T and B lymphocyte signaling pathways in lupus. Autoimmun Rev 2009;8(3):179–83.

[86] Zan H, Tat C, Casali P. MicroRNAs in lupus. Autoimmunity 2014;47(4):272–85.

[87] Yan S, Yim LY, Lu L, Lau CS, Chan VS. MicroRNA regulation in systemic lupus erythematosus pathogenesis. Immune Netw 2014;14(3):138–48.

[88] Dai Y, Huang YS, Tang M, Lv TY, Hu CX, Tan YH, et al. Microarray analysis of microRNA expression in peripheral blood cells of systemic lupus erythematosus patients. Lupus 2007;16(12):939–46.

[89] Pan W, Zhu S, Yuan M, Cui H, Wang L, Luo X, et al. MicroRNA-21 and microRNA-148a contribute to DNA hypomethylation in lupus CD4+ T cells by directly and indirectly targeting DNA methyltransferase 1. J Immunol 2010;184(12):6773–81.

[90] Zhao S, Wang Y, Liang Y, Zhao M, Long H, Ding S, et al. MicroRNA-126 regulates DNA methylation in CD4+ T cells and contributes to systemic lupus erythematosus by targeting DNA methyltransferase 1. Arthritis Rheum 2011;63(5):1376–86.

[91] Qin H, Zhu X, Liang J, Wu J, Yang Y, Wang S, et al. MicroRNA-29b contributes to DNA hypomethylation of CD4+ T cells in systemic lupus erythematosus by indirectly targeting DNA methyltransferase 1. J Dermatol Sci 2013;69(1):61–7.

[92] Fan W, Liang D, Tang Y, Qu B, Cui H, Luo X, et al. Identification of microRNA-31 as a novel regulator contributing to impaired interleukin-2 production in T cells from patients with systemic lupus erythematosus. Arthritis Rheum 2012;64(11): 3715–25.

[93] Stagakis E, Bertsias G, Verginis P, Nakou M, Hatziapostolou M, Kritikos H, et al. Identification of novel microRNA signatures linked to human lupus disease activity and pathogenesis: miR-21 regulates aberrant T cell responses through regulation of PDCD4 expression. Ann Rheum Dis 2011;70(8):1496–506.

[94] Lashine YA, Salah S, Aboelenein HR, Abdelaziz AI. Correcting the expression of miRNA-155 represses PP2Ac and enhances the release of IL-2 in PBMCs of juvenile SLE patients. Lupus 2015;24(3):240–7.

[95] Zhao X, Tang Y, Qu B, Cui H, Wang S, Wang L, et al. MicroRNA-125a contributes to elevated inflammatory chemokine RANTES via targeting KLF13 in systemic lupus erythematosus. Arthritis Rheum 2010.

[96] Ding S, Liang Y, Zhao M, Liang G, Long H, Zhao S, et al. Decreased microRNA-142-3p/5p expression causes CD4+ T cell activation and B cell hyperstimulation in systemic lupus erythematosus. Arthritis Rheum 2012;64(9):2953–63.

[97] Garchow BG, Bartulos Encinas O, Leung Y T, Tsao PY, Eisenberg RA, Caricchio R, et al. Silencing of microRNA-21 in vivo ameliorates autoimmune splenomegaly in lupus mice. EMBO Mol Med 2011;3(10):605–15.

[98] Dai R, Zhang Y, Khan D, Heid B, Caudell D, Crasta O, et al. Identification of a common lupus disease-associated microRNA expression pattern in three different murine models of lupus. PLoS ONE 2010;5(12):e14302.

[99] Thai TH, Patterson HC, Pham DH, Kis-Toth K, Kaminski DA, Tsokos GC. Deletion of microRNA-155 reduces autoantibody responses and alleviates lupus-like disease in the Fas(lpr) mouse. Proc Natl Acad Sci USA 2013;110(50):20194–9.

[100] Xin Q, Li J, Dang J, Bian X, Shan S, Yuan J, et al. miR-155 deficiency ameliorates autoimmune inflammation of systemic lupus erythematosus by targeting S1pr1 in Faslpr/lpr mice. J Immunol 2015.

[101] Liu Y, Dong J, Mu R, Gao Y, Tan X, Li Y, et al. MicroRNA-30a promotes B cell hyperactivity in patients with systemic lupus erythematosus by direct interaction with Lyn. Arthritis Rheum 2013;65(6):1603–11.

[102] Luo S, Liu Y, Liang G, Zhao M, Wu H, Liang Y, et al. The role of microRNA-1246 in the regulation of B cell activation and the pathogenesis of systemic lupus erythematosus. Clin Epigenet 2015;7(1):24.

[103] Wu XN, Ye YX, Niu JW, Li Y, Li X, You X, et al. Defective PTEN regulation contributes to B cell hyperresponsiveness in systemic lupus erythematosus. Sci Transl Med 2014;6(246):246ra99.

[104] Yuan Y, Kasar S, Underbayev C, Vollenweider D, Salerno E, Kotenko SV, et al. Role of microRNA-15a in autoantibody production in interferon-augmented murine model of lupus. Mol Immunol 2012;52(2):61–70.

[105] Chafin CB, Reilly CM. MicroRNAs implicated in the immunopathogenesis of lupus nephritis. Clin Dev Immunol 2013;2013:430239.

[106] Ichii O, Otsuka S, Sasaki N, Namiki Y, Hashimoto Y, Kon Y. Altered expression of microRNA miR-146a correlates with the development of chronic renal inflammation. Kidney Int 2012;81(3):280–92.

[107] Lu J, Kwan BC, Lai FM, Tam LS, Li EK, Chow KM, et al. Glomerular and tubulointerstitial miR-638, miR-198 and miR-146a expression in lupus nephritis. Nephrology 2012;17(4):346–51.

[108] Ichii O, Otsuka-Kanazawa S, Horino T, Kimura J, Nakamura T, Matsumoto M, et al. Decreased miR-26a expression correlates with the progression of podocyte injury in autoimmune glomerulonephritis. PLoS ONE 2014;9(10):e110383.

[109] Zhou H, Hasni SA, Perez P, Tandon M, Jang SI, Zheng C, et al. miR-150 promotes renal fibrosis in lupus nephritis by downregulating SOCS1. J Am Soc Nephrol 2013; 24(7):1073–87.

[110] Zhu S, Pan W, Song X, Liu Y, Shao X, Tang Y, et al. The microRNA miR-23b suppresses IL-17-associated autoimmune inflammation by targeting TAB2, TAB3 and IKK-alpha. Nat Med 2012;18(7):1077–86.

[111] Wang G, Tam LS, Li EK, Kwan BC, Chow KM, Luk CC, et al. Serum and urinary cell-free MiR-146a and MiR-155 in patients with systemic lupus erythematosus. J Rheumatol 2010;37(12):2516–22.

[112] Wang G, Tam LS, Li EK, Kwan BC, Chow KM, Luk CC, et al. Serum and urinary free microRNA level in patients with systemic lupus erythematosus. Lupus 2011;20(5): 493–500.

[113] Wang H, Peng W, Ouyang X, Li W, Dai Y. Circulating microRNAs as candidate biomarkers in patients with systemic lupus erythematosus. Transl Res 2012;160(3): 198–206.

[114] Carlsen AL, Schetter AJ, Nielsen CT, Lood C, Knudsen S, Voss A, et al. Circulating microRNA expression profiles associated with systemic lupus erythematosus. Arthritis Rheum 2013;65(5):1324–34.

[115] Croce CM. Causes and consequences of microRNA dysregulation in cancer. Nat Rev 2009;10(10):704–14.

[116] Luo X, Yang W, Ye DQ, Cui H, Zhang Y, Hirankarn N, et al. A functional variant in microRNA-146a promoter modulates its expression and confers disease risk for systemic lupus erythematosus. PLoS Genet 2011;7(6):e1002128.

[117] Lofgren SE, Frostegard J, Truedsson L, Pons-Estel BA, D'Alfonso S, Witte T, et al. Genetic association of miRNA-146a with systemic lupus erythematosus in Europeans through decreased expression of the gene. Genes Immun 2012;13(3): 268–74.

[118] Hewagama A, Gorelik G, Patel D, Liyanarachchi P, McCune WJ, Somers E, et al. Overexpression of X-linked genes in T cells from women with lupus. J Autoimmun 2013;41:60–71.

[119] Dai R, Lu R, Ahmed SA. The upregulation of genomic imprinted DLK1-Dio3 miRNAs in murine lupus is associated with global DNA hypomethylation. PLoS ONE 2016;11(4):e0153509.

[120] Ahmed SA, Talal N. Sex hormones and the immune system–Part 2. Animal data, Baillieres Clin Rheumatol 1990;4(1):13–31.

[121] Dai R, Ahmed SA. Sexual dimorphism of miRNA expression: a new perspective in understanding the sex bias of autoimmune diseases. Ther Clin Risk Manag 2014;10: 151–63.

[122] Straub RH. The complex role of estrogens in inflammation. Endocr Rev 2007; 28(5):521–74.

[123] Lang TJ. Estrogen as an immunomodulator. Clin Immunol Orl Fla 2004;113(3): 224–30.

[124] Dai R, Phillips RA, Zhang Y, Khan D, Crasta O, Ahmed SA. Suppression of LPS-induced IFN{gamma} and nitric oxide in splenic lymphocytes by select estrogen-regulated miRNA: a novel mechanism of immune modulation. Blood 2008;112(12): 4591–7.

[125] Roubinian J, Talal N, Siiteri PK, Sadakian JA. Sex hormone modulation of autoimmunity in NZB/NZW mice. Arthritis Rheum 1979;22(11):1162–9.

[126] Dai R, McReynolds S, Leroith T, Heid B, Liang Z, Ahmed SA. Sex differences in the expression of lupus-associated miRNAs in splenocytes from lupus-prone NZB/WF1 mice. Biol Sex Differ 2013;4(1):19.

[127] Javierre BM, Hernando H, Ballestar E. Environmental triggers and epigenetic deregulation in autoimmune disease. Discov Med 2011;12(67):535–45.

[128] Vrijens K, Bollati V, Nawrot TS. MicroRNAs as potential signatures of environmental exposure or effect: a systematic review. Environ Health Perspect 2015;123(5): 399–411.

[129] Vinuesa CG, Rigby RJ, Yu D. Logic and extent of miRNA-mediated control of autoimmune gene expression. Int Rev Immunol 2009;28(3–4):112–38.

[130] Esquela-Kerscher A, Trang P, Wiggins JF, Patrawala L, Cheng A, Ford L, et al. The let-7 microRNA reduces tumor growth in mouse models of lung cancer. Cell Cycle 2008;7(6):759–64.

[131] Wiggins JF, Ruffino L, Kelnar K, Omotola M, Patrawala L, Brown D, et al. Development of a lung cancer therapeutic based on the tumor suppressor microRNA-34. Cancer Res 2010;70(14):5923–30.

[132] van Rooij E, Purcell AL, Levin AA. Developing microRNA therapeutics. Circulation Res 2012;110(3):496–507.

[133] Janssen HL, Reesink HW, Lawitz EJ, Zeuzem S, Rodriguez-Torres M, Patel K, et al. Treatment of HCV infection by targeting microRNA. N Engl J Med 2013; 368(18):1685–94.

CHAPTER 10

Microrna-Linked Heart Disease and Therapeutic Potential

K. Ono

Key Concepts

In this review, the current understanding of miRNA function in the pathogenesis of cardiovascular diseases will be summarized. There will also be a focus on the miRNA-based therapeutics in this field.

1. INTRODUCTION

Despite recent progress in diagnosis and treatment, cardiovascular disease is the leading cause of morbidity and mortality in developed countries. Therefore, understanding the underlying molecular mechanisms in the pathogenesis of cardiovascular disease may spur the development of novel therapeutic strategies. Generally, pathological processes in the cardiovascular system are associated with a characteristic cascade of altered intracellular signaling and gene expression [1]. Thus, the various signaling pathways that underlie pathological changes have been the subject of intense investigation. It is known that the regulation of cardiac gene expression is complex, with individual genes controlled by multiple transcription factors associated with their regulatory enhancer/promoter sequences to activate gene expression [2]. Moreover, epigenetic regulation and alternative splicing mechanisms also further complicate the patterns of gene expression.

Recently, microRNAs (miRNAs; miRs) have reshaped our view of how gene expression is controlled by adding another layer of regulation at the posttranscriptional level. miRNAs are a class of short (~22 nt), endogenous, noncoding RNAs that mediate posttranscriptional gene silencing [3]. Since the discovery of the first miRNAs, lin-4 and let-7, in *Caenorhabditis elegans* [4,5], 28,645 mature miRNAs have been annotated in 206 species according to the miRBase Sequence Database released on June 21, 2014 (http://www.mirbase.org/index.shtml). Most miRNAs are transcribed by RNA

Translating MicroRNAs to the Clinic
ISBN 978-0-12-800553-8

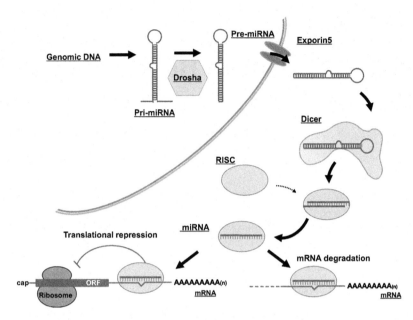

Figure 10.1 MicroRNA biogenesis pathway.

polymerase II from intergenic, intronic, or polycistronic loci to long primary transcripts called primiRNAs (Fig. 10.1). PrimiRNAs are processed in the nucleus by the Drosha-DGCR8 complex to approximately 70 nt premiRNA hairpin structures and these then move to the cytoplasm through exportin 5 [6,7]. Then, the miRNA hairpin structures are processed in the cytoplasm by the Dicer–TRBP complex to approximately 22 nt miRNA duplexes (Fig. 10.1) [8]. In addition to the canonical miRNA biogenesis pathway, many Drosha-DGCR8-independent pathways can produce premiRNAs. The most common alternative pathway involves short intronic hairpins, termed mirtrons, that are spliced and debranched to form premiRNA hairpins [9,10]. It is also known that maturation of miR-451 is independent of Dicer. Dicer knockout cells can generate mature miR-451 but not other miRNAs. Drosha generates a short premir-451 hairpin that is directly cleaved by Ago2 followed by resection of its 3′ terminus [11]. In the cytoplasm, miRNA duplexes are incorporated into an Argonaute protein to form miRNA-induced silencing complexes (miRISCs), followed by unwinding of the duplex and retention of the mature miRNA strand (guide strand) in the miRISC, while the complementary strand (passenger strand) is released and degraded [12]. In some cases, both of the strands are incorporated into the miRISC and suppress target mRNAs. Recent studies have revealed that miRNAs have both a powerful and fine regulatory function. In fact, a single miRNA can suppress the expression of several different genes, and multiple miRNAs can target the same genes. Moreover, some miRNAs are located within the same region as a cluster, thus making their regulatory potential even larger. It is estimated that 30–50% of all protein-coding genes are controlled by miRNAs. Thus, it is not surprising that miRNAs regulate cardiovascular diseases and can be potential therapeutic targets. Cardiovascular diseases encompassed many pathologies including cardiac hypertrophy, myocardial ischemia/myocardial infarction (MI), cardiac fibrosis, and vascular diseases that will be discussed in more detail in the following sections. The implications of miRNA-derived regulation in cardiovascular pathology have only been recognized recently, and research on miRNAs in relation to such diseases has now become a rapidly evolving field. In this review, I will summarize the current understanding of miRNA function in the pathogenesis of cardiovascular diseases. I will also focus on the miRNA-based therapeutics in this field.

2. CARDIAC HYPERTROPHY

Because cardiac hypertrophy, an increase in heart size, is associated with nearly all forms of heart failure (HF), it is of great clinical importance to understand the mechanisms responsible for cardiac hypertrophy. Therefore, the regulation and functions of hypertrophy-associated genes has attracted great interest from many researchers.

Pathological hypertrophy is mainly caused by hypertension, loss of myocytes following ischemic damage, and genetic alterations that cause cardiomyopathy. Moreover, metabolic abnormality or neurohormonal activation can also lead to hypertrophy [13]. Thus, pathological hypertrophy has been most studied in relation to miRNAs in the heart to date.

2.1 Mir-1

MiR-1 was first discovered in 2001 by northern analysis; it is strongly expressed in the human adult heart but not in human brain, liver, kidney, lung, or colon [14]. In fact, it is a cardiac- and skeletal-muscle-specific miRNA, and it is probably one of the most abundant miRNAs in the heart. Actually, miR-1 accounted for nearly 40% of all miRNA transcripts in cardiomyocytes [15]. Targeted deletion of miR-1 in mice revealed that it has numerous functions in the heart, including regulation of cardiac morphogenesis, electrical conduction, and cell-cycle control [16]. Several groups reported that miR-1 regulates genes involved in cardiac hypertrophy [17–20]. Many miRNAs are either up- or downregulated after transverse aortic constriction (TAC)-induced hypertrophy in a mouse model. MiR-1 was downregulated as early as day 1, persisting through day 7 after TAC. Overexpression experiments showed that miR-1 inhibited its in silico-predicted, growth-related targets, including Ras GTPase-activating protein, cyclin-dependent kinase 9, fibronectin, and Ras homolog enriched in brain. As a result, miR-1 reduced protein synthesis and cell size. It was also reported that miR-1 targets a cytoskeletal regulatory protein, twinfilin 1 (Twf1), which binds to actin monomers and prevents their assembly into filaments [20]. Downregulation of miR-1 is induced by hypertrophic stimuli, such as TAC or α-adrenergic stimulation with phenylephrine, which results in increased Twf1 expression, and overexpression of Twf1 is sufficient to induce cardiac hypertrophy. Other targets of miR-1 is insulin-like growth factor (IGF-1), IGF-1 receptor [19], calmodulin 1 and 2, Mef2a [18], and sodium calcium exchanger [21]. MiR-1 is downregulated in patients with aortic stenosis [22] and acromegaly-associated cardiac hypertrophy [19]. Recently, it was shown that administration of miR-1 by adenoviral vector regressed cardiac hypertrophy in rats [23]. Thus, restoration of miR-1 gene expression is a potential novel therapeutic strategy to reverse pressure-induced cardiac hypertrophy and prevent maladaptive cardiac remodeling.

2.2 MiR-133

MiR-133 is one of the most abundant miRNAs in human myocardium and is also a muscle-specific miRNA. MiR-1-1/miR-133a-2, miR-1-2/miR-133a-1, and miR-206/miR-133b form clusters in three different chromosomal regions in the human genome. Downregulation of both miR-133 and miR-1 was observed in mouse and human models of cardiac hypertrophy. In vitro overexpression of miR-133 inhibited cardiac hypertrophy. In contrast, suppression of miR-133 by complementary RNA to miR-133 induced hypertrophy and ventricular dilation in vivo. RhoA, a GDP–GTP exchange protein regulating cardiac hypertrophy; Cdc42, a signal transduction kinase implicated in hypertrophy; and Nelf-A/WHSC2, a nuclear factor involved in cardiogenesis have been identified as miR-133-specific targets. Thus, miR-133 is also a key regulator of cardiac hypertrophy and can be used for future therapeutic applications [24]. However, work in other groups suggested more complex results concerning miR-133 in the regulation of cardiac hypertrophy [25]. They reported that miR-133a was downregulated by TAC and isoproterenol-induced hypertrophy, but not in two genetic hypertrophy models. Overexpression of miR-133 driven by the MYH6 promoter had no effect on postnatal cardiac development assessed by measures of structure, function, and mRNA profile. However, transgenic expression of miR-133a prevented TAC-associated miR-133a downregulation and improved myocardial fibrosis and diastolic function without affecting the extent of hypertrophy. These results suggest that forced overexpression of miR-133 may be cardioprotective in certain models of pathological hypertrophy.

2.3 Mir-208

It is known that miR-208a, -208b, and -499 are located within the introns of myosin heavy chain (MHC) genes Myh6, Myh7, and Myh7b, respectively. During embryogenesis, there is a switch from fetal β-MHC to α-MHC with a concurrent increase in the expression of miR-208a and reduction in miR-208b [26]. It was reported that miR-208a$^{-/-}$ mice showed reduced cardiac hypertrophy in response to pressure overload [27]. Targets of miR-208a include thyroid hormone receptor-associated protein 1 [26,27], suggesting that miR-208a initiates cardiomyocyte hypertrophy by regulating triiodothyronine-dependent repression of β-MHC expression. Overexpression of miR-208a was sufficient to upregulate Myh7 and to elicit cardiac hypertrophy, resulting in systolic dysfunction [26]. Therapeutic inhibition of miR-208a by subcutaneous delivery of antimiR-208a was evaluated during hypertension-induced HF in Dahl hypertensive rats [28]. AntimiR-208a dose-dependently prevented pathological myosin switching and cardiac remodeling while improving cardiac function, overall health, and survival. Although miR-208a is required for cardiac hypertrophy, the role of miR-208b in these pathologic conditions remains to be elucidated. MiR-499 is considered likely to play a role in myosin gene regulation [29,30].

3. CARDIAC FIBROSIS

Cardiac fibrosis is a major aspect of myocardial remodeling and an important contributor to the development of cardiac dysfunction in diverse pathologic conditions, such as MI, in ischemic, dilated, and hypertrophic cardiomyopathies, and HF [31–36]. The extracellular deposition of collagen by fibroblasts contributes to this adverse remodeling. Cardiac fibrosis leads to an increased mechanical stiffness, initially causing diastolic dysfunction, and eventually resulting in systolic dysfunction and overt HF. In addition, fibrosis can also disturb the electrical continuity between cardiomyocytes, leading to conduction slowing and hence an increase in the risk of arrhythmias. It is also possible that the enhanced diffusion distance for cardiac substrates and oxygen to cardiac myocytes, caused by fibrosis, negatively influences the myocardial balance between energy demand and supply [33,34]. Recent findings suggest that in addition to the resident cardiac fibroblasts circulating bone-marrow-derived fibroblast (fibrocytes) may contribute to cardiac fibrogenesis [37].

3.1 MiR-29

In an effort to identify miRNAs involved in post-MI remodeling, miRNA expression profiles were compared between mouse hearts in the border zone of an infarcted region and the noninfarcted (remote) myocardium 3 and 14 days after MI. Interestingly, all three members of the miR-29 family, miR-29a, -29b, and -29c, were downregulated in response to MI. It was shown that the miR-29 family, which is highly expressed in fibroblasts, targets mRNAs encoding a multitude of extracellular matrix–related proteins involved in fibrosis, including col1a1, col3a1, elastin, and fibrillin [38]. Because the miR-29 family was dramatically repressed in the border zone flanking the infracted area in a mouse model of MI, it is tempting to speculate that upregulation of miR-29 may be a therapeutic option for MI. Regulation and function of miR-29 was also studied in Duchenne muscular dystrophy (DMD) [39]. It was demonstrated that miR-29 expression was downregulated in dystrophic muscles of mdx mice, a model of DMD. Restoration of its expression by intramuscular and intravenous injection improved dystrophy pathology by both promoting regeneration and inhibiting fibrogenesis. Mechanistic studies revealed that loss of miR-29 in myoblasts promotes their transdifferentiation into myofibroblasts through targeting extracellular molecules including collagens and microfibrillar-associated protein 5. Therefore, miR-29 replacement therapy may be a promising treatment approach for DMD.

3.2 MiR-21

MiR-21 is expressed in all cell types of the cardiovascular system, most prominently in cardiac fibroblasts but rather weakly in cardiomyocytes. Furthermore, miR-21 is among the most strongly upregulated miRNAs in response to a variety of forms of cardiac stress [38,40,41]. Thum et al. showed that miR-21 is selectively upregulated in cardiac fibroblasts in the failing heart, where it represses the expression of Sprouty homolog 1

(SPRY1), a negative regulator of the extracellular signal-regulated kinase/ mitogen-activated protein kinase (ERK-MAPK) signaling pathway [42]. Upregulation of miR-21 in response to cardiac injury was shown to enhance ERK-MAPK signaling, leading to fibroblast proliferation and fibrosis. Importantly, in vivo silencing of miR-21 by a specific antagomir in a mouse pressure-overload-induced disease model reduced cardiac ERK-MAP kinase activity, inhibited interstitial fibrosis, and attenuated cardiac dysfunction. MiR-21-dependent targeting of SPRY1 and PDCD4 was also shown to promote the fibroblastoid phenotype in the epicardial-to-mesenchymal transition [43]. Phosphatase and tensin homologue (PTEN) has also been demonstrated to be a direct target of miR-21 in cardiac fibroblasts [44]. Previous reports characterized PTEN as a suppressor of matrix metalloprotease-2 (MMP-2) expression [45,46]. Ischemia-reperfusion (I/R) injury in the heart induced miR-21 in cardiac fibroblasts in the infracted region. Thus, I/R injury-induced miR-21-limited PTEN function and caused activation of the Akt pathway and increased MMP-2 expression in cardiac fibroblasts.

3.3 Mir-24

MiR-24 was shown to be downregulated in a murine model of MI [47]. The change in miR-24 expression was closely related to ECM remodeling. It was also demonstrated that upregulation of miR-24 through intramyocardial injection of lentiviruses could improve heart function and attenuate fibrosis in the infarct border zone of the heart 2 weeks after MI. In vitro experiments showed that TGF-β (a pathological mediator of fibrotic disease) increased miR-24 expression, and overexpression of miR-24 reduced TGF-β secretion and Smad2/3 phosphorylation in cardiac fibroblasts. The authors identified the protease furin as an miR-24 target by a bioinformatics analysis and showed that inhibition of furin was sufficient to inhibit downstream TGF-β signaling. However, extra caution is needed when applying miR-24 augmentation in the treatment for fibrosis after MI. Lorenzen et al. indicated that miR-24 also regulated heme oxygenase 1 and the H2A histone family, member X, in the kidney in vivo. They showed that silencing of miR-24 in mice before I/R injury resulted in a significant improvement in survival and kidney function, a reduction of apoptosis, improved histologic tubular epithelial injury, and less infiltration of inflammatory cells [48]. These results indicate that miR-24 promotes renal ischemic injury by stimulating apoptosis in endothelial and tubular epithelial cell, and similar results may occur when miR-24 is overexpressed in the heart.

4. MYOCARDIAL ISCHEMIA AND CELL DEATH

Cardiomyocyte death/apoptosis is a key cellular event in ischemic hearts. There are miRNAs that have been shown to exert proapoptotic effects by targeting key cardioprotective proteins. It is known that early reperfusion of the ischemic heart remains the most effective intervention for improving clinical outcomes after an MI. However, abnormal increases in intracellular Ca^{2+} during myocardial reperfusion can cause cardiomyocyte death, known as I/R injury. Cardiac I/R injury is also accompanied by dynamic changes in the expression of miRNAs.

4.1 MiR-320

MiR-320 expression was significantly decreased in the hearts on I/R in vivo and ex vivo [49]. Overexpression of miR-320 enhanced cardiomyocyte death and apoptosis, whereas knockdown was cytoprotective, after simulated I/R. Furthermore, transgenic mice with cardiac-specific overexpression of miR-320 revealed an increased extent of apoptosis and infarction size in hearts on I/R in vivo and ex vivo relative to wild-type controls. In contrast, in vivo suppression of miR-320 reduced infarction size relative to the controls. Heat-shock protein 20 (HSP20), a known cardioprotective protein, was identified as a target of miR-320. Therefore, knockdown of endogenous miR-320 can provide protection against cardiomyocyte apoptosis through the upregulation of HSP20.

4.2 MiR-34

MiR-34a and -34b/c were found to be direct, conserved p53 target genes that presumably mediate induction of apoptosis, cell-cycle arrest, and senescence [50]. MiR-34a, -34b, and -34c were upregulated in the heart in response to stress. Therapeutic inhibition of all three miR-34 family members using a subcutaneously delivered seed-targeting 8-mer locked nucleic acid (LNA)-modified antimiRNA (LNA-antimiR-34) attenuated ischemia-induced remodeling and improved cardiac recovery [51]. In contrast, inhibition of miR-34a alone with LNA-antimiR-34a (15-mer) provided no benefit in the MI model. These data underscore the utility of seed-targeting 8-mer LNA-antimiRs in the development of new therapeutic approaches for pharmacologic inhibition of disease-implicated miRNA seed families. Other researchers indicated that miR-34a is induced in the ageing heart, and that in vivo silencing or genetic deletion of miR-34a reduces age-associated cardiomyocyte cell death [52]. Moreover, miR-34a inhibition reduced cell death and fibrosis following acute MI and improved recovery of myocardial function in their experiments. Mechanistically, PNUTS (also known as PPP1R10), which reduces telomere shortening, DNA damage responses, and cardiomyocyte apoptosis, was identified as a novel direct miR-34a target. Thus, miR-34 inhibition improves functional recovery after acute MI.

4.3 MiR-214

MiR-214 is upregulated during ischemic injury and HF. It was shown that genetic deletion of miR-214 in mice caused a loss of cardiac contractility, increased apoptosis, and excessive fibrosis in response to I/R injury [53]. The cardioprotective roles of miR-214 during I/R injury were attributed to repression of the mRNA encoding sodium/calcium exchanger 1, a key regulator of Ca^{2+} influx; and to repression of several downstream effectors of Ca^{2+} signaling that mediate cell death. These results suggested a pivotal role for miR-214 as a regulator of cardiomyocyte Ca^{2+} homeostasis and survival during cardiac injury. It is known that overexpression of NCX1 and intracellular Ca^{2+} overload induce cardiac hypertrophy and failure. Therefore, it is likely that miR-214 plays a cardioprotective role in a variety of stress settings.

5. VASCULAR DISEASES

miRNAs are also important in vascular development, physiology, and diseases. Profiling of endothelially expressed miRNAs has been performed using human umbilical vein endothelial cells. These results revealed high expression levels of miR-221/222, miR-21, let-7 family, miR-17-92 cluster, miR-23-24 cluster, and miR-126 in vascular endothelial cells. Among them, miR-126 is the only miRNA considered to be expressed specifically in endothelial cells [54]. On the other hand, miR-145 and -143 show a smooth-muscle-specific expression pattern. The most common vascular disease is atherosclerosis. Recent studies have shown the involvement of miRNAs in atherosclerosis and lipid homeostasis. Therefore, not only vascular-specific miRNAs but also metabolism-related miRNAs are involved in vascular diseases.

5.1 MiR-17-92 Cluster

The miR-17-92 cluster is one of the most important miRNAs for the regulation of angiogenesis. It encodes six miRNAs (miR-17, -18a, -19a, -20a, -19b-1, and -92-1), which are tightly grouped within an 800-bp region, and it is transcriptionally regulated by c-Myc [55]. Some of these miRNAs promote angiogenesis but others do not. In particular, miR-19 is primarily responsible for thrombospondin-1 downregulation and miR-18 for CTGF downregulation in response to Myc to promote tumor angiogenesis [56]. On the other hand, miR-92a inhibits the growth of new blood vessels (angiogenesis) [57]. Forced overexpression of miR-92a in endothelial cells blocked angiogenesis, and systemic inhibition of miR-92a led to enhanced blood vessel growth and functional recovery of damaged tissue in mouse models of limb ischemia and MI. Therefore, miR-92a may serve as a valuable therapeutic target in the setting of ischemic disease.

5.2 MiR-126

MiR-126 is an abundant, endothelial cell-enriched miRNA that is encoded in the second intron of an endothelial-cell-specific gene, *Egfl7*. It was shown that mechano-sensitive zinc finger transcription factor Klf2a induced miR-126 expression to activate vascular endothelial growth factor signaling [58]. This work described a novel genetic mechanism, in which an miRNA-facilitated integration of a physiological stimulus with growth factor signaling in endothelial cells to guide angiogenesis. MiR-126 also has an antiinflammatory effect. Actually, transfection of endothelial cells with an oligonucleotide that decreased miR-126 levels permitted an increase in tumor necrosis factor-α-stimulated vascular adhesion molecule 1 expression and increased leukocyte adherence to endothelial cells [59]. The apparent role of miR-126 in angiogenesis has led to increasing interest in miR-126 overexpression as a therapeutic approach. It has been reported that systemic delivery of miR-126 by miRNA-loaded bubble liposomes improved blood flow and may be useful for the treatment of hind-limb ischemia [60].

5.3 MiR-145/143

The miR-143 and -145 encoding genes are highly conserved and lie in close proximity to each other on murine chromosome 18 and human chromosome 5. MiR-145 and -143 are cotranscribed in multipotent murine cardiac progenitors before becoming localized to smooth muscle cells. It was reported that miR-145 is necessary for myocardin-induced reprogramming of adult fibroblasts into smooth muscle cells and sufficient to induce differentiation of multipotent neural crest stem cells into vascular smooth muscle [61]. In adult rats, miR-145 is selectively expressed in vascular smooth muscle cells of the vascular wall, and its expression was significantly downregulated in vascular walls with neointimal lesion formation [62]. The target of miR-145 is KLF5 and its downstream signaling molecule myocardin. Restoration of miR-145 in balloon-injured arteries via adenovirus-inhibited neointimal growth and might be used for treatment of a variety of proliferative vascular diseases.

5.4 Mir-33a/b

Recent studies have indicated that miR-33a/b, located in the intron of sterol regulatory element binding proteins (SREBPs), control cholesterol homeostasis [63–67]. In humans, miR-33a and -33b are encoded in the introns of *SREBF2* and *SREBF1*, respectively [64,68], whereas in rodents, there is a deletion in part of the miR-33b encoding lesion and miR-33b cannot be expressed. MiR-33a and -33b share the same seed sequence and differ in only two nucleotides. Thus, most of the targets of these miRNAs are supposed to be the same [69]. MiR-33a and -33b are considered to be cotranscribed and regulate lipid homeostasis with their host genes.

Several groups, including ours, reported that miR-33a targets *Abca1* and *Abcg1* in vivo, using either antisense technology or by generating miR-33a knockout mice [63–66]. ABCA1 mediates the efflux of cholesterol to lipid-poor apolipoprotein A-I (apoA-I) and forms nascent high density lipoprotein (HDL). ABCG1 plays a critical role in the process of cholesterol efflux, which exports cellular cholesterol to large HDL particles. As a result, plasma HDL levels increased to 35–50%, without affecting other lipoproteins in antimiR-33a-treated mice [63–65]. MiR-33a knockout mice also showed an increase in the levels of ABCA1 and ABCG1, and a 25–40% increase in serum HDL [66]. In contrast, HDL-C levels in miR-33b knock-in (KI) mice, which had miR-33b in the same intron as in humans, were reduced by almost 35%, even in miR-33b KI hetero mice compared with control mice [69].

It has already been proven that antisense inhibition of miR-33a resulted in regression of atherosclerotic plaque in low density lipoprotein receptor-deficient mice by promoting reverse cholesterol transport [70]. Moreover, miR-33a-deficient reduced the progression of atherosclerosis in apoE-deficient mice (Fig. 10.2) [71]. Therefore, miR-33a/b are good targets for the prevention of atherosclerosis.

6. HEART FAILURE

Because all of the previously described pathologies, ie, cardiac hypertrophy, fibrosis, and vascular diseases, can cause HF, all of the miRNAs discussed so far are also relevant to this disease entity.

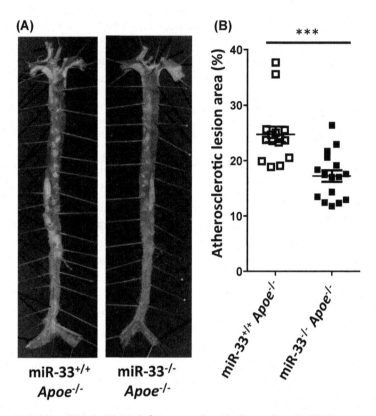

(A)

mir-33⁺/⁺   mir-33⁻/⁻
Apoe⁻/⁻   Apoe⁻/⁻

Figure 10.2 MicroRNA (miR)-33 deficiency-reduced atherosclerosis. (A) Representative images of the en face analysis of the total aorta in miR-33$^{+/+}$Apoe$^{-/-}$ and miR-33$^{-/-}$Apoe$^{-/-}$ male mice. (B) Quantification of the atherosclerotic lesion area in en face analysis of the total aorta in male mice. Values are mean±SE ($n=15$ to 16 each); ***$P<.001$. *From Horie T, Baba O, Kuwabara Y, Chujo Y, Watanabe S, Kinoshita M, et al. MicroRNA-33 deficiency reduces the progression of atherosclerotic plaque in ApoE(–/–) mice. J Am Heart Assoc 2012;1:e003376.*

miRNA deregulation is a common feature in cancer, central nervous system disorders, inflammation, cardiovascular diseases, and metabolic disorders, suggesting that miRNAs could serve as targets for therapeutic intervention. Many profiling studies have been conducted and revealed a large number of miRNAs that are differentially expressed in HF [18,22,72–75]. A complete description of all such molecules is beyond the scope of this chapter. Some of miRNAs whose functions are not mentioned previously are described here.

6.1 MiR-146

miRNAs are also related to a more specific cause of HF, such as chemotherapy-induced HF or obesity-related HF. It has been proposed that miRNAs can exert their roles in response to treatment with chemotherapeutic agents. It was reported that there is a significant reduction in ErbB4 expression in the hearts of mice after doxorubicin (Dox) treatment. It was suggested that upregulation of miR-146a after Dox treatment was involved in acute Dox-induced cardiotoxicity by targeting ErbB4 [76]. It is known that neuregulin-1-ErbB signaling is essential for maintaining adult cardiac function, and that ErbB2 and 4 are the major receptors for neuregulin-1 in the heart. Thus, inhibition of both ErbB2 and ErbB4 signaling was suggested to be one of the reasons why those patients who receive concurrent therapy with Dox and trastuzumab (a humanized anti-Her2; ErbB2 antibody) suffer from HF [77].

6.2 MiR-451

miRNA microarray analyses and real-time PCR have revealed that miR-451 levels were significantly increased in type 2 diabetes mellitus mouse hearts [78]. Calcium-binding protein 39 (Cab39) is a scaffold protein of liver kinase B1 (LKB1), an upstream kinase of AMP-activated protein kinase (AMPK). Cab39 is a direct target of miR-451 in neonatal rat cardiac myocytes, and Cab39 overexpression rescued lipotoxicity. Protein levels of Cab39 and phosphorylated AMPK were increased, and phosphorylated mammalian target of rapamycin was reduced in cardiomyocyte-specific miR-451 knockout mouse hearts compared with control mouse hearts. Thus, these results demonstrated that miR-451 is involved in diabetic cardiomyopathy through suppression of the LKB1/AMPK pathway.

7. CIRCULATING MICRORNAs

Numerous studies have demonstrated that miRNAs are not only found intracellularly, but also detectable outside cells, including various body fluids (ie, serum, plasma, saliva, urine, breast milk, and tears). Interestingly, ~90% of extracellular miRNAs are packaged with proteins (ie, Ago2, HDL, and other RNA-binding proteins) and ~10% are wrapped in small membranous particles (ie, exosomes, microvesicles, and apoptotic bodies) [79–82]. Recently, results obtained in studies of cancer suggest that the profiles of blood-circulating miRNAs might reflect the changes observed in cancerous tissue [83]. This concept has also proved valid in cardiovascular disease and it was reported that specific miRNAs were present in plasma/serum of patients with MI [84–86]. Serum miR-133 levels were upregulated rapidly even in patients without elevation of cardiac troponin T (Fig. 10.3). Moreover, plasma levels of endothelial cell-enriched miRNAs, such as

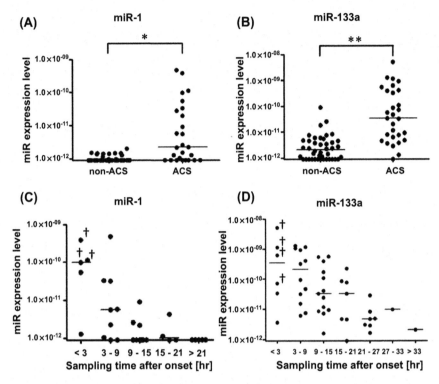

Figure 10.3 MicroRNA (miR)-1 and miR-133a levels increased in the serum of patients with acute coronary syndrome (ACS). Expression levels of miR-1 (A) and miR-133a (B) in the serum of patients with non-ACS versus ACS. *$P < .0005$, **$P < .0001$. The levels of circulating miRNAs decreased over time in the serum of patients with ACS. MiR-1 (C) and miR-133a (D) expression levels are shown according to the time of blood sampling after onset. *Dots with daggers* indicate samples without elevation of creatine phosphokinase or cardiac troponin T. *Horizontal lines* indicate the median. *From Kuwabara Y, Ono K, Horie T, Nishi H, Nagao K, Kinoshita M, et al. Increased MicroRNA-1 and MicroRNA-133a levels in serum of patients with cardiovascular disease indicate myocardial damage. Circ Cardiovasc Genet 2011;4:446–54.*

miR-126, -17, and -92a, inflammation-associated miR-155, and smooth-muscle-enriched miR-145 were reported to be significantly reduced in coronary artery disease (CAD) patients compared with healthy controls. These results also indicated that they can be used as biomarker candidates for CAD [87].

A diverse range of circulating miRNAs have been studied for the detection of HF. Tijsen et al. tried to determine whether miRNAs can distinguish clinical HF not only from healthy controls but also from non-HF forms of dyspnea [84]. They revealed that miR-423-5p was most strongly related to the clinical diagnosis of HF and receiver–operator characteristics curve analysis showed miR-423-5p to be a diagnostic predictor of HF, with an area under the curve of 0.91 ($P < .001$). From a diagnostic perspective, Goren et al. tried to evaluate a multimarker approach to HF diagnosis [88]. They measured the levels of 186 miRNAs in the sera of 30 stable chronic systolic HF patients and 30 controls. The differences in miRNA levels between the two groups were characterized, and a score, based on the levels of four specific miRNAs with the most significant increase in the HF group (miR-423-5p, -320a, -22, and -92b) was defined. Interestingly, the score was utilized to discriminate HF patients from controls with a sensitivity and specificity of 90%. Moreover, in the HF group, there was a significant association between the score and important clinical parameters such as elevated serum natriuretic peptide levels, a wide QRS, and dilatation of the left ventricle and left atrium. These results suggested that a multimarker approach is useful for the detection of not only HF but also left ventricular structure and function.

On the other hand, there remain concerns regarding the use of circulating miRNAs as biomarkers, because there remains a lack of consensus regarding optimal quantification methods, reference genes, and quality control of samples. Actually, hemolysis affects the accuracy of the results. It was shown that miR-21, -106a, -92a, -17, and -16 were hemolysis-susceptible miRNAs although they have been previously proposed as plasma/serum biomarkers of disease [89]. Because low level hemolysis is a frequent occurrence during sample collection, it is critical that this is taken into account in the measurement of any candidate miRNAs.

8. THERAPEUTIC STRATEGIES TO MODULATE THE FUNCTIONS OF MICRORNAs

Gain- or loss-of-function studies of specific miRNAs have revealed that miRNAs are viable targets for therapeutics. miRNAs have several advantages as a therapeutic modality [90]. They are easy to target because the mature sequence of miRNAs is short. Moreover, miRNAs often have many targets in specific pathways, which enable us to modulate the entire pathway by targeting several miRNAs for therapeutic strategies. Currently, there are two ways to modulate miRNA functions. One is to overexpress miRNAs by transducing double-stranded RNAs or transfecting viral-based vectors. The other is to inhibit the function of miRNAs by the use of chemically modified antisense miRNA oligonucleotides, such as antimiRs.

8.1 Augmentation or Restoration of MicroRNA Function

Transduction of double-stranded RNA with chemical modification is commonly used to therapeutically restore the function of miRNAs [91]. 2'-fluoro (2'-F) or phosphorothioate modification helps to protect against exonucleases, hence making the guide strand more stable [92]. It is known that double-stranded RNA can potentially induce a nonspecific interferon response through toll-like receptors [93]. Thus, longer oligonucleotides are commonly used for this purpose. Another way to express miRNAs is to use lenti-, adeno-, or adeno-associated viruses [94–96]. However, systemic delivery of miRNAs can result uptake in nontarget tissues that normally do not express such an miRNA of interest, resulting off-target effects. Therefore, targeted delivery of miRNAs is important to prevent undesirable side effects.

8.2 Inhibition of MicroRNA Function

miRNA function can be inhibited by so-called "miRNA sponges" or antisense oligonucleotides, such as antimiRs [97]. An miRNA sponges use overexpression of vector-harboring complementary binding sites to an miRNA of interest to inhibit the function of a specific miRNA. The larger the number of the binding sites in the vector, the stronger the inhibitory effects of the miRNA sponge. Whereas antimiRs have shown greater effects from a therapeutic perspective. To improve the antimiRs' binding affinity, biostability, and pharmacokinetic properties, their chemical modifications have been optimized. The commonly used modifications for increasing the duplex melting temperature (Tm) and improving nuclease resistance of antimiRs include 2'-O-methyl (2'-O-Me), 2'-O-Methoxyethyl (2'-MOE) 2'-fluoro, and bicyclic LNA modifications [98–100]. Increased nuclease resistance is also achieved by the use of phosphorothioate (PS) linkages in antimiR oligonucleotides or by using peptide nucleic acid [101,102]. The PS backbone also enhances binding to plasma proteins leading to reduced clearance by the kidney. AntimiR oligonucleotides were first designed as fully complementary nucleotides to the mature miRNA sequence [103]. However, recent studies have shown that an 8-mer LNA-modified antimiR was effective in inhibiting several miRNAs [104] (Fig. 10.4).

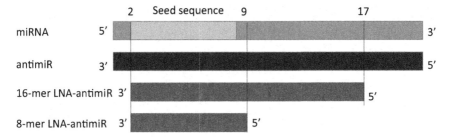

Figure 10.4 Structure of microRNAs (miRs) and locked nucleic acid-modified antimiRs.

9. SUMMARY AND CONCLUSIONS

miRNAs have emerged as powerful and dynamic modifiers of cardio-vascular diseases. The miRNA species discussed earlier are able to directly regulate the expression of transcription factors, signaling molecules, contractile proteins, and play critical roles in cardiovascular diseases. The existence of miRNAs in plasma/serum has also been demonstrated. The potential of circulating miRNAs as biomarkers for cardiovascular diseases has been examined. Their roles as prognostic biomarkers have yet to be elucidated, and larger studies with longer follow-up periods will be needed. Recently, two main strategies have been employed for the pharmacological modulation of miRNA activity in vivo: (1) restoring the function of an miRNA using either synthetic double-stranded RNAs or viral expression constructs, and (2) inhibition of miRNA function using chemically modified antimiR oligonucleotides. It is true that there may be side effects and potential off-target effects when long-term miRNA modulation is applied in vivo. Nevertheless, as described in this review, pharmacological modulation of disease-associated miRNAs has demonstrated promising therapeutic potential and appears to be well tolerated based on data from short-term studies in animal disease models and human patients. Thus, modulation of miRNAs may be novel and promising therapeutic approaches in the near future.

LIST OF ACRONYMS AND ABBREVIATIONS

AMPK AMP-activated protein kinase
CAD Coronary artery disease
DGCR8 DiGeorge syndrome chromosomal (or critical) region 8
DMD Duchenne muscular dystrophy

ERK-MAPK Extracellular signal-regulated kinase/mitogen-activated protein kinase
HF Heart failure
I/R Ischemia-reperfusion
IGF Insulin-like growth factor
LKB1 Liver kinase B1
LNA Locked nucleic acid
MHC Myosin heavy chain
MI Myocardial infarction
miRISC miRNA-induced silencing complex
miRNA MicroRNA
MMP Matrix metalloprotease-2
PTEN Phosphatase and tensin homologue
RasGAP Ras GTPase-activating protein
Rheb Ras homolog enriched in brain
SREBP Sterol regulatory element binding protein
TAC Transverse aortic constriction
TRBP TAR (HIV-1) RNA binding protein

REFERENCES

[1] Kairouz V, Lipskaia L, Hajjar RJ, Chemaly ER. Molecular targets in heart failure gene therapy: current controversies and translational perspectives. Ann N Y Acad Sci 2012;1254:42–50.
[2] Olson EN. Gene regulatory networks in the evolution and development of the heart. Science 2006;313:1922–7.
[3] Ambros V. The functions of animal microRNAs. Nature 2004;431:350–5.
[4] Lee RC, Feinbaum RL, Ambros V. The *C. elegans* heterochronic gene lin-4 encodes small RNAs with antisense complementarity to lin-14. Cell 1993;75:843–54.
[5] Wightman B, Ha I, Ruvkun G. Posttranscriptional regulation of the heterochronic gene lin-14 by lin-4 mediates temporal pattern formation in *C. elegans*. Cell 1993;75:855–62.
[6] Bohnsack MT, Czaplinski K, Gorlich D. Exportin 5 is a RanGTP-dependent dsRNA-binding protein that mediates nuclear export of pre-miRNAs. RNA 2004;10:185–91.
[7] Yi R, Qin Y, Macara IG, Cullen BR. Exportin-5 mediates the nuclear export of pre-microRNAs and short hairpin RNAs. Genes Dev 2003;17:3011–6.
[8] Lee Y, Ahn C, Han J, Choi H, Kim J, Yim J, et al. The nuclear RNase III Drosha initiates microRNA processing. Nature 2003;425:415–9.
[9] Yang JS, Lai EC. Alternative miRNA biogenesis pathways and the interpretation of core miRNA pathway mutants. Mol Cell 2011;43:892–903.
[10] Ladewig E, Okamura K, Flynt AS, Westholm JO, Lai EC. Discovery of hundreds of mirtrons in mouse and human small RNA data. Genome Res 2012;22:1634–45.
[11] Yang JS, Maurin T, Robine N, Rasmussen KD, Jeffrey KL, Chandwani R, et al. Conserved vertebrate mir-451 provides a platform for Dicer-independent, Ago2-mediated microRNA biogenesis. Proc Natl Acad Sci USA 2010;107:15163–8.
[12] Carthew RW, Sontheimer EJ. Origins and Mechanisms of miRNAs and siRNAs. Cell 2009;136:642–55.
[13] Rajabi M, Kassiotis C, Razeghi P, Taegtmeyer H. Return to the fetal gene program protects the stressed heart: a strong hypothesis. Heart Fail Rev 2007;12:331–43.

[14] Lee RC, Ambros V. An extensive class of small RNAs in *Caenorhabditis elegans*. Science 2001;294:862–4.

[15] Lagos-Quintana M, Rauhut R, Yalcin A, Meyer J, Lendeckel W, Tuschl T. Identification of tissue-specific microRNAs from mouse. Curr Biol 2002;12:735–9.

[16] Zhao Y, Ransom JF, Li A, Vedantham V, von Drehle M, Muth AN, et al. Dysregulation of cardiogenesis, cardiac conduction, and cell cycle in mice lacking miRNA-1-2. Cell 2007;129:303–17.

[17] Sayed D, Hong C, Chen IY, Lypowy J, Abdellatif M. MicroRNAs play an essential role in the development of cardiac hypertrophy. Circ Res 2007;100:416–24.

[18] Ikeda S, He A, Kong SW, Lu J, Bejar R, Bodyak N, et al. MicroRNA-1 negatively regulates expression of the hypertrophy-associated calmodulin and Mef2a genes. Mol Cell Biol 2009;29:2193–204.

[19] Elia L, Contu R, Quintavalle M, Varrone F, Chimenti C, Russo MA, et al. Reciprocal regulation of microRNA-1 and insulin-like growth factor-1 signal transduction cascade in cardiac and skeletal muscle in physiological and pathological conditions. Circulation 2009;120:2377–85.

[20] Li Q, Song XW, Zou J, Wang GK, Kremneva E, Li XQ, et al. Attenuation of microRNA-1 derepresses the cytoskeleton regulatory protein twinfilin-1 to provoke cardiac hypertrophy. J Cell Sci 2010;123:2444–52.

[21] Kumarswamy R, Lyon AR, Volkmann I, Mills AM, Bretthauer J, Pahuja A, et al. SERCA2a gene therapy restores microRNA-1 expression in heart failure via an Akt/FoxO3A-dependent pathway. Eur Heart J 2012;33:1067–75.

[22] Ikeda S, Kong SW, Lu J, Bisping E, Zhang H, Allen PD, et al. Altered microRNA expression in human heart disease. Physiol Genomics 2007;31:367–73.

[23] Karakikes I, Chaanine AH, Kang S, Mukete BN, Jeong D, Zhang S, et al. Therapeutic cardiac-targeted delivery of miR-1 reverses pressure overload-induced cardiac hypertrophy and attenuates pathological remodeling. J Am Heart Assoc 2013;2:e000078.

[24] Care A, Catalucci D, Felicetti F, Bonci D, Addario A, Gallo P, et al. MicroRNA-133 controls cardiac hypertrophy. Nat Med 2007;13:613–8.

[25] Matkovich SJ, Wang W, Tu Y, Eschenbacher WH, Dorn LE, Condorelli G, et al. MicroRNA-133a protects against myocardial fibrosis and modulates electrical repolarization without affecting hypertrophy in pressure-overloaded adult hearts. Circ Res 2010;106:166–75.

[26] Callis TE, Pandya K, Seok HY, Tang RH, Tatsuguchi M, Huang ZP, et al. MicroRNA-208a is a regulator of cardiac hypertrophy and conduction in mice. J Clin Invest 2009;119:2772–86.

[27] van Rooij E, Sutherland LB, Qi X, Richardson JA, Hill J, Olson EN. Control of stress-dependent cardiac growth and gene expression by a microRNA. Science 2007;316:575–9.

[28] Montgomery RL, Hullinger TG, Semus HM, Dickinson BA, Seto AG, Lynch JM, et al. Therapeutic inhibition of miR-208a improves cardiac function and survival during heart failure. Circulation 2011;124:1537–47.

[29] van Rooij E, Quiat D, Johnson BA, Sutherland LB, Qi X, Richardson JA, et al. A family of microRNAs encoded by myosin genes governs myosin expression and muscle performance. Dev Cell 2009;17:662–73.

[30] Bell ML, Buvoli M, Leinwand LA. Uncoupling of expression of an intronic microRNA and its myosin host gene by exon skipping. Mol Cell Biol 2010;30:1937–45.

[31] Rossi MA. Pathologic fibrosis and connective tissue matrix in left ventricular hypertrophy due to chronic arterial hypertension in humans. J Hypertens 1998;16:1031–41.

[32] Swynghedauw B. Molecular mechanisms of myocardial remodeling. Physiol Rev 1999;79:215–62.

[33] Manabe I, Shindo T, Nagai R. Gene expression in fibroblasts and fibrosis: involvement in cardiac hypertrophy. Circ Res 2002;91:1103–13.

[34] Brown RD, Ambler SK, Mitchell MD, Long CS. The cardiac fibroblast: therapeutic target in myocardial remodeling and failure. Annu Rev Pharmacol Toxicol 2005;45:657–87.

[35] Khan R, Sheppard R. Fibrosis in heart disease: understanding the role of transforming growth factor-beta in cardiomyopathy, valvular disease and arrhythmia. Immunology 2006;118:10–24.

[36] Martos R, Baugh J, Ledwidge M, O'Loughlin C, Conlon C, Patle A, et al. Diastolic heart failure: evidence of increased myocardial collagen turnover linked to diastolic dysfunction. Circulation 2007;115:888–95.

[37] Haudek SB, Xia Y, Huebener P, Lee JM, Carlson S, Crawford JR, et al. Bone marrow-derived fibroblast precursors mediate ischemic cardiomyopathy in mice. Proc Natl Acad Sci USA 2006;103:18284–9.

[38] van Rooij E, Sutherland LB, Thatcher JE, DiMaio JM, Naseem RH, Marshall WS, et al. Dysregulation of microRNAs after myocardial infarction reveals a role of miR-29 in cardiac fibrosis. Proc Natl Acad Sci USA 2008;105:13027–32.

[39] Wang L, Zhou L, Jiang P, Lu L, Chen X, Lan H, et al. Loss of miR-29 in myoblasts contributes to dystrophic muscle pathogenesis. Mol Ther 2012;20:1222–33.

[40] van Rooij E, Sutherland LB, Liu N, Williams AH, McAnally J, Gerard RD, et al. A signature pattern of stress-responsive microRNAs that can evoke cardiac hypertrophy and heart failure. Proc Natl Acad Sci USA 2006;103:18255–60.

[41] Ichimura A. miRNAs and regulation of cell signaling. FEBS J 2011;278.

[42] Thum T, Gross C, Fiedler J, Fischer T, Kissler S, Bussen M, et al. MicroRNA-21 contributes to myocardial disease by stimulating MAP kinase signalling in fibroblasts. Nature 2008;456:980–4.

[43] Bronnum H, Andersen DC, Schneider M, Sandberg MB, Eskildsen T, Nielsen SB, et al. miR-21 promotes fibrogenic epithelial-to-mesenchymal transition of epicardial mesothelial cells involving Programmed Cell Death 4 and Sprouty-1. PLoS ONE 2013;8:e56280.

[44] Roy S, Khanna S, Hussain SR, Biswas S, Azad A, Rink C, et al. MicroRNA expression in response to murine myocardial infarction: miR-21 regulates fibroblast metalloprotease-2 via phosphatase and tensin homologue. Cardiovasc Res 2009;82:21–9.

[45] Park MJ, Kim MS, Park IC, Kang HS, Yoo H, Park SH, et al. PTEN suppresses hyaluronic acid-induced matrix metalloproteinase-9 expression in U87MG glioblastoma cells through focal adhesion kinase dephosphorylation. Cancer Res 2002; 62:6318–22.

[46] Zheng H, Takahashi H, Murai Y, Cui Z, Nomoto K, Niwa H, et al. Expressions of MMP-2, MMP-9 and VEGF are closely linked to growth, invasion, metastasis and angiogenesis of gastric carcinoma. Anticancer Res 2006;26:3579–83.

[47] Wang J, Huang W, Xu R, Nie Y, Cao X, Meng J, et al. MicroRNA-24 regulates cardiac fibrosis after myocardial infarction. J Cell Mol Med 2012;16:2150–60.

[48] Lorenzen JM, Kaucsar T, Schauerte C, Schmitt R, Rong S, Hubner A, et al. MicroRNA-24 antagonism prevents renal ischemia reperfusion injury. J Am Soc Nephrol 2014;25:2717–29.

[49] Ren XP, Wu J, Wang X, Sartor MA, Qian J, Jones K, et al. MicroRNA-320 is involved in the regulation of cardiac ischemia/reperfusion injury by targeting heat-shock protein 20. Circulation 2009;119:2357–66.

[50] Hermeking H. p53 enters the microRNA world. Cancer Cell 2007;12:414–8.

[51] Bernardo BC, Gao XM, Winbanks CE, Boey EJ, Tham YK, Kiriazis H, et al. Therapeutic inhibition of the miR-34 family attenuates pathological cardiac remodeling and improves heart function. Proc Natl Acad Sci USA 2012;109:17615–20.

[52] Boon RA, Iekushi K, Lechner S, Seeger T, Fischer A, Heydt S, et al. MicroRNA-34a regulates cardiac ageing and function. Nature 2013;495:107–10.

[53] Aurora AB, Mahmoud AI, Luo X, Johnson BA, van Rooij E, Matsuzaki S, et al. MicroRNA-214 protects the mouse heart from ischemic injury by controlling Ca(2) (+) overload and cell death. J Clin Invest 2012;122:1222–32.

[54] Wang S, Olson EN. AngiomiRs–key regulators of angiogenesis. Curr Opin Genet Dev 2009;19:205–11.

[55] Mendell JT. miRiad roles for the miR-17-92 cluster in development and disease. Cell 2008;133:217–22.

[56] Dews M, Homayouni A, Yu D, Murphy D, Sevignani C, Wentzel E, et al. Augmentation of tumor angiogenesis by a Myc-activated microRNA cluster. Nat Genet 2006;38:1060–5.

[57] Bonauer A, Carmona G, Iwasaki M, Mione M, Koyanagi M, Fischer A, et al. MicroRNA-92a controls angiogenesis and functional recovery of ischemic tissues in mice. Science 2009;324:1710–3.

[58] Nicoli S, Standley C, Walker P, Hurlstone A, Fogarty KE, Lawson ND. MicroRNA-mediated integration of haemodynamics and Vegf signalling during angiogenesis. Nature 2010;464:1196–200.

[59] Harris TA, Yamakuchi M, Ferlito M, Mendell JT, Lowenstein CJ. MicroRNA-126 regulates endothelial expression of vascular cell adhesion molecule 1. Proc Natl Acad Sci USA 2008;105:1516–21.

[60] Endo-Takahashi Y, Negishi Y, Nakamura A, Ukai S, Ooaku K, Oda Y, et al. Systemic delivery of miR-126 by miRNA-loaded Bubble liposomes for the treatment of hindlimb ischemia. Sci Rep 2014;4:3883.

[61] Cordes KR, Sheehy NT, White MP, Berry EC, Morton SU, Muth AN, et al. miR-145 and miR-143 regulate smooth muscle cell fate and plasticity. Nature 2009;460:705–10.

[62] Cheng Y, Liu X, Yang J, Lin Y, Xu DZ, Lu Q, et al. MicroRNA-145, a novel smooth muscle cell phenotypic marker and modulator, controls vascular neointimal lesion formation. Circ Res 2009;105:158–66.

[63] Rayner KJ, Suarez Y, Davalos A, Parathath S, Fitzgerald ML, Tamehiro N, et al. MiR-33 contributes to the regulation of cholesterol homeostasis. Science 2010;328:1570–3.

[64] Najafi-Shoushtari SH, Kristo F, Li Y, Shioda T, Cohen DE, Gerszten RE, et al. MicroRNA-33 and the SREBP host genes cooperate to control cholesterol homeostasis. Science 2010;328:1566–9.

[65] Marquart TJ, Allen RM, Ory DS, Baldan A. miR-33 links SREBP-2 induction to repression of sterol transporters. Proc Natl Acad Sci USA 2010;107:12228–32.

[66] Horie T, Ono K, Horiguchi M, Nishi H, Nakamura T, Nagao K, et al. MicroRNA-33 encoded by an intron of sterol regulatory element-binding protein 2 (Srebp2) regulates HDL in vivo. Proc Natl Acad Sci USA 2010;107:17321–6.

[67] Gerin I, Clerbaux LA, Haumont O, Lanthier N, Das AK, Burant CF, et al. Expression of miR-33 from an SREBP2 intron inhibits cholesterol export and fatty acid oxidation. J Biol Chem 2010;285:33652 61.

[68] Brown MS, Ye J, Goldstein JL. Medicine. HDL miR-ed down by SREBP introns. Science 2010;328:1495–6.

[69] Horie T, Nishino T, Baba O, Kuwabara Y, Nakao T, Nishiga M, et al. MicroRNA-33b knock-in mice for an intron of sterol regulatory element-binding factor 1 (Srebf1) exhibit reduced HDL-C in vivo. Sci Rep 2014;4:5312.

[70] Rayner KJ, Sheedy FJ, Esau CC, Hussain FN, Temel RE, Parathath S, et al. Antagonism of miR-33 in mice promotes reverse cholesterol transport and regression of atherosclerosis. J Clin Invest 2011;121:2921–31.

[71] Horie T, Baba O, Kuwabara Y, Chujo Y, Watanabe S, Kinoshita M, et al. MicroRNA-33 deficiency reduces the progression of atherosclerotic plaque in ApoE($-/-$) mice. J Am Heart Assoc 2012;1:e003376.

[72] Bagga S, Bracht J, Hunter S, Massirer K, Holtz J, Eachus R, et al. Regulation by let-7 and lin-4 miRNAs results in target mRNA degradation. Cell 2005;122:553–63.

[73] Duisters RF, Tijsen AJ, Schroen B, Leenders JJ, Lentink V, van der Made I, et al. miR-133 and miR-30 regulate connective tissue growth factor: implications for a role of microR-NAs in myocardial matrix remodeling. Circ Res 2009;104:170–8. 6p following 8.

[74] Cheng Y, Ji R, Yue J, Yang J, Liu X, Chen H, et al. MicroRNAs are aberrantly expressed in hypertrophic heart: do they play a role in cardiac hypertrophy? Am J Pathol 2007;170:1831–40.

[75] Tatsuguchi M, Seok HY, Callis TE, Thomson JM, Chen JF, Newman M, et al. Expression of microRNAs is dynamically regulated during cardiomyocyte hypertrophy. J Mol Cell Cardiol 2007;42:1137–41.

[76] Horie T, Ono K, Nishi H, Nagao K, Kinoshita M, Watanabe S, et al. Acute doxorubi-cin cardiotoxicity is associated with miR-146a-induced inhibition of the neuregulin-ErbB pathway. Cardiovasc Res 2010;87:656–64.

[77] Slamon DJ, Leyland-Jones B, Shak S, Fuchs H, Paton V, Bajamonde A, et al. Use of chemotherapy plus a monoclonal antibody against HER2 for metastatic breast cancer that overexpresses HER2. N Engl J Med 2001;344:783–92.

[78] Kuwabara Y, Horie T, Baba O, Watanabe S, Nishiga M, Usami S, et al. MicroRNA-451 exacerbates lipotoxicity in cardiac myocytes and high-fat diet-induced cardiac hyper-trophy in mice through suppression of the LKB1/AMPK pathway. Circ Res 2015;116:279–88.

[79] Arroyo JD, Chevillet JR, Kroh EM, Ruf IK, Pritchard CC, Gibson DF, et al. Argo-naute2 complexes carry a population of circulating microRNAs independent of ves-icles in human plasma. Proc Natl Acad Sci USA 2011;108:5003–8.

[80] Vickers KC, Palmisano BT, Shoucri BM, Shamburek RD, Remaley AT. MicroRNAs are transported in plasma and delivered to recipient cells by high-density lipoproteins. Nat Cell Biol 2011;13:423–33.

[81] Valadi H, Ekstrom K, Bossios A, Sjostrand M, Lee JJ, Lotvall JO. Exosome-mediated transfer of mRNAs and microRNAs is a novel mechanism of genetic exchange between cells. Nat Cell Biol 2007;9:654–9.

[82] Hunter MP, Ismail N, Zhang X, Aguda BD, Lee EJ, Yu L, et al. Detection of microRNA expression in human peripheral blood microvesicles. PLoS ONE 2008;3:e3694.

[83] Mitchell PS, Parkin RK, Kroh EM, Fritz BR, Wyman SK, Pogosova-Agadjanyan EL, et al. Circulating microRNAs as stable blood-based markers for cancer detection. Proc Natl Acad Sci USA 2008;105:10513–8.

[84] Tijsen AJ, Creemers EE, Moerland PD, de Windt LJ, van der Wal AC, Kok WE, et al. MiR423-5p as a circulating biomarker for heart failure. Circ Res 2010;106:1035–9.

[85] Wang GK, Zhu JQ, Zhang JT, Li Q, Li Y, He J, et al. Circulating microRNA: a novel potential biomarker for early diagnosis of acute myocardial infarction in humans. Eur Heart J 2010;31:659–66.

[86] Kuwabara Y, Ono K, Horie T, Nishi H, Nagao K, Kinoshita M, et al. Increased MicroRNA-1 and MicroRNA-133a levels in serum of patients with cardiovascular disease indicate myocardial damage. Circ Cardiovasc Genet 2011;4:446–54.

[87] Fichtlscherer S, De Rosa S, Fox H, Schwietz T, Fischer A, Liebetrau C, et al. Cir-culating microRNAs in patients with coronary artery disease. Circ Res 2010;107:677–84.

[88] Goren Y, Kushnir M, Zafrir B, Tabak S, Lewis BS, Amir O. Serum levels of microR-NAs in patients with heart failure. Eur J Heart Fail 2012;14:147–54.

[89] Kirschner MB, Edelman JJ, Kao SC, Vallely MP, van Zandwijk N, Reid G. The impact of hemolysis on cell-free microRNA biomarkers. Front Genet 2013;4:94.

[90] van Rooij E. Introduction to the series on microRNAs in the cardiovascular system. Circ Res 2012;110:481–2.

[91] Thorsen SB, Obad S, Jensen NF, Stenvang J, Kauppinen S. The therapeutic potential of microRNAs in cancer. Cancer J 2012;18:275–84.

[92] Chiu YL, Rana TM. siRNA function in RNAi: a chemical modification analysis. RNA 2003;9:1034–48.

[93] Peacock H, Fucini RV, Jayalath P, Ibarra-Soza JM, Haringsma HJ, Flanagan WM, et al. Nucleobase and ribose modifications control immunostimulation by a microRNA-122-mimetic RNA. J Am Chem Soc 2011;133:9200–3.

[94] Kota J, Chivukula RR, O'Donnell KA, Wentzel EA, Montgomery CL, Hwang HW, et al. Therapeutic microRNA delivery suppresses tumorigenesis in a murine liver cancer model. Cell 2009;137:1005–17.

[95] Miyazaki Y, Adachi H, Katsuno M, Minamiyama M, Jiang YM, Huang Z, et al. Viral delivery of miR-196a ameliorates the SBMA phenotype via the silencing of CELF2. Nat Med 2012;18:1136–41.

[96] Trang P, Medina PP, Wiggins JF, Ruffino L, Kelnar K, Omotola M, et al. Regression of murine lung tumors by the let-7 microRNA. Oncogene 2010;29:1580–7.

[97] Ebert MS, Sharp PA. MicroRNA sponges: progress and possibilities. RNA 2010;16:2043–50.

[98] Davis S, Lollo B, Freier S, Esau C. Improved targeting of miRNA with antisense oligonucleotides. Nucleic Acids Res 2006;34:2294–304.

[99] Elmen J, Lindow M, Schutz S, Lawrence M, Petri A, Obad S, et al. LNA-mediated microRNA silencing in non-human primates. Nature 2008;452:896–9.

[100] van Rooij E, Olson EN. MicroRNA therapeutics for cardiovascular disease: opportunities and obstacles. Nat Rev Drug Discov 2012;11:860–72.

[101] Lennox KA, Behlke MA. A direct comparison of anti-microRNA oligonucleotide potency. Pharm Res 2010;27:1788–99.

[102] Fabani MM, Gait MJ. miR-122 targeting with LNA/2′-O-methyl oligonucleotide mixmers, peptide nucleic acids (PNA), and PNA-peptide conjugates. RNA 2008;14:336–46.

[103] Esau C, Davis S, Murray SF, Yu XX, Pandey SK, Pear M, et al. miR-122 regulation of lipid metabolism revealed by in vivo antisense targeting. Cell Metab 2006;3:87–98.

[104] Rottiers V, Obad S, Petri A, McGarrah R, Lindholm MW, Black JC, et al. Pharmacological inhibition of a MicroRNA family in nonhuman primates by a seed-targeting 8-Mer AntimiR. Sci Transl Med 2013;5:212ra162.

CHAPTER 11

The Clinical Potential of Heart Failure–Related miRNAs

A.J. Tijsen, Y.M. Pinto, E.E. Creemers

Key concepts

- Cardiac remodeling is an important underlying cause for the development of heart failure (HF).
- MicroRNAs (miRNAs) comprise a family of short noncoding RNAs that posttranscriptionally inhibit gene expression.
- Many miRNAs are deregulated in cardiac remodeling and are involved in different features of this remodeling process.
- Interference with miRNA expression might provide new therapeutics for HF.
- Different chemistries and delivery methods can be used to therapeutically interfere with miRNA expression levels.
- Circulating miRNAs show aberrant patterns in HF patients and might be used as biomarkers.

1. INTRODUCTION

1.1 Heart Failure

HF is defined as a complex clinical syndrome that can result from any structural or functional cardiac disorder that impairs the ability of the ventricle to fill with or eject blood and is characterized by symptoms such as dyspnea, fatigue, and edema [1]. A common cause of HF is cardiac remodeling due to coronary artery disease, hypertension, valvular heart disease, myocardial infarction (MI), diabetes mellitus, congenital heart disease, and arrhythmias [2]. HF is the most common diagnosis in patient >65 years admitted to the hospital and it is a global problem with an estimated prevalence of 38 million patients worldwide [2]. This number is currently increasing due to the aging population and paradoxically, due to improved treatment regimens for acute cardiac diseases such as MI. This has increased the survival of these patients, but since these patients experience some extent of cardiac damage

Translating MicroRNAs to the Clinic
ISBN 978-0-12-800553-8

they are more prone to develop HF [2]. The 5- and 10-year survival rates after the diagnosis of HF are estimated at 50% and 10%, respectively, and thereby worse than survival rates for most cancers [2].

Cardiac remodeling (Fig. 11.1), a process that underlies HF, is the response of the heart to injury or stress and is characterized by a number of structural alterations. The first aspect of this remodeling process is cardiac hypertrophy, which is an increase in cardiomyocyte cell size that results in an increase in cardiac mass [3]. Cardiac hypertrophy is accompanied by changes in expression of genes that regulate energy metabolism, calcium

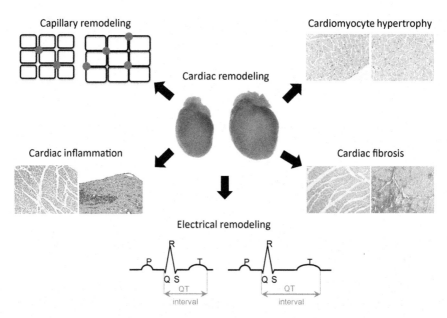

Figure 11.1 *Processes occurring during cardiac remodeling.* Cardiac remodeling is the response of the heart to injury or stress and is characterized by a number of structural alterations. The first aspect of this remodeling process is cardiac hypertrophy, which is an increase in cardiomyocyte cell size that results in an increase in cardiac mass. The second aspect is cardiac fibrosis, which is the excessive accumulation of extracellular matrix proteins in the interstitium and perivascular regions of the myocardium. The third aspect is capillary remodeling, where the growth of the microvasculature does not keep up with the cardiomyocyte hypertrophy. The fourth aspect is inflammation, of which the severity is correlated to severity of heart failure (HF). The last aspect is electrophysiological remodeling, where differences in expression of ion channels lead to prolongation of the action potential. Together these structural and cellular changes result in a dysfunctional myocardium, which contributes to the development of HF and the occurrence of arrhythmias in HF patients. For every aspect of remodeling, we depicted the healthy heart on the left and the remodeling heart on the right.

handling, and genes normally expressed in the embryonic heart [3]. The second aspect of cardiac remodeling is cardiac fibrosis, which is the excessive accumulation of extracellular matrix (ECM) proteins in the interstitium and perivascular regions of the myocardium [4]. Fibrosis increases mechanical stiffness of the heart, impairs myocyte contractility, disrupts electrical coupling, and worsens tissue hypoxia [4]. The third aspect of cardiac remodeling is capillary remodeling, where the growth of the microvasculature does not keep up with the cardiomyocyte hypertrophy. This leads to myocardial hypoxia, which may cause cardiomyocyte apoptosis [5]. Together these structural and cellular changes result in a dysfunctional myocardium, which contributes to the development of HF.

Current treatment of HF is merely based on relief of symptoms rather than reversal of the adverse remodeling process. Current therapies include angiotensin converting enzyme inhibitors (ACEi), diuretics, β-blockers, and/or combinations thereof [2]. These medicinal treatments intend either to decrease blood pressure (ACEi, diuretics) and reduce afterload to the heart, so that it will be easier for the heart to eject the blood, or they intend to decrease the heart rate (β-blockers), which allows the heart to fill with and eject blood more efficiently and thus to increase its ejection fraction (EF). Besides medicinal treatment, HF patients may also receive several devices [6]. HF patients, with a high risk of developing ventricular arrhythmias may receive implantable cardioverter defibrillators (ICDs). These ICDs prevent arrhythmias by delivering a shock to the heart when they detect an irregular beat. When a patient fails to respond to therapy, the final treatment option is heart transplantation [6]. Finally, left ventricular assist devices (LVADs) are pumps that will partially or completely take over the function of the heart and can be applied for short-term use in patients awaiting a transplantation or long-term in patients unsuitable for transplantation [6]. However, device treatment and transplantation are invasive treatment options with many risks, which require lifestyle modifications.

The increasing prevalence of HF combined with the fact that current treatment of HF is based on relief of symptoms emphasizes on the importance of developing novel therapies for HF. In the last two decades, it became clear that miRNAs constitute a completely new layer of gene regulation. And also in the heart miRNAs are identified as important regulators of gene expression which are involved in many biological processes in cardiomyocytes and other cardiac cells in both the healthy and the remodeling heart [7]. Preclinical studies using synthetic miRNA inhibitors (antimiRs) have shown that it is relatively easy to inhibit miRNA expression in the heart. Moreover, these studies revealed profound effects of inhibition of specific miRNAs on cardiovascular function [7]. This implies that miRNAs might develop into new therapeutic targets for the treatment of HF.

1.2 MicroRNAs

miRNAs are single-stranded, noncoding RNAs of ~22 nucleotides in length, which are highly conserved among species. There are estimated to be up to 1000 miRNAs in the human genome and together they are predicted to regulate as many as 30% of mRNA transcripts and thereby virtually every biological system [8]. Many of these miRNAs are classified in miRNA families based on a shared "seed" sequence, which are nucleotide 2–7 at the 5′-end of the miRNA. MiRNAs are located in intronic or intragenic regions of the genome either as individual miRNAs or in clusters of several miRNAs. During their biogenesis, hairpin-shaped miRNA precursors are processed into miRNA duplexes of which the mature miRNA strand is incorporated into the RNA-induced silencing complex (RISC) to become functional (Fig. 11.2) [9].

The mature miRNA which has been incorporated in the RISC-complex is able to bind the 3′-untranslated region (3′UTR) of specific mRNA sequences and to inhibit translation or promote degradation of these transcripts [9,10]. The primary determinant for binding of a miRNA to a 3′UTR is the "seed" region, which requires Watson–Crick pairing between the miRNA and the mRNAs 3′UTR. An important characteristic of miRNA function is that a specific miRNA is able to regulate multiple mRNA targets that are involved in the same biological pathway or cellular process, which makes them good candidates for novel therapeutics. This is because the modulation of one specific miRNA is probably more effective in interfering with a cellular process than the modulation of one single protein involved in this process. And indeed within 20 years after their discovery in 1993 [11], the therapeutic potential of miRNA inhibition is already investigated in phase 2a clinical trials, where the effectiveness and safety of inhibition of miR-122 in hepatitis C patients was shown [12]. In the last decade it has become increasingly clear that numerous miRNAs govern cellular processes underlying cardiac remodeling and therefore miRNAs also seem promising candidates for HF therapy.

Remarkably, miRNAs also appear to be highly stable in plasma and specific profiles of circulating miRNAs have been identified for a number of cardiac diseases. This has raised the possibility that miRNAs may be measured in the circulation to serve as diagnostic or prognostic biomarkers.

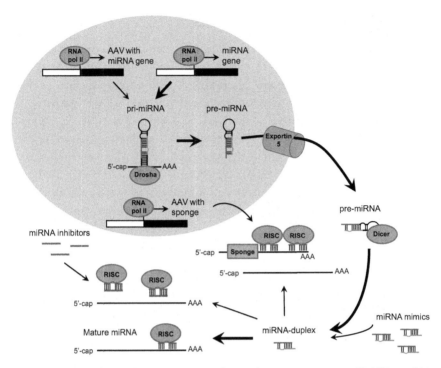

Figure 11.2 *Biogenesis of miRNAs and interference by overexpressors and inhibitors.* Biogenesis of miRNAs (*bold black arrows*) starts with transcription of the miRNA precursor by RNA polymerase II. This miRNA precursor, the pri-miRNA, contains a hairpin-shaped structure and can be several thousands of nucleotides long. Next, this pri-miRNA is recognized by Drosha, a ribonuclease which cleaves the pri-miRNA into 70–100 nucleotide long hairpins called pre-miRNAs. Pre-miRNAs are exported from the nucleus to the cytoplasm by exportin-5, and in the cytoplasm Dicer further processes the pre-miRNAs into a miRNA duplex. The two strands of this duplex are separated when the single-stranded mature miRNAs are incorporated into the RNA-induced silencing complex (RISC), which will bind to the 3′UTR of mRNA to repress the translation. MiRNA mimics resemble the miRNA duplex, which means that the antisense strand of the mimic will also be incorporated in the RISC complex. MiRNA inhibitors are single stranded and complementary to the mature miRNA. They will bind to the miRNA in the RISC complex, which prevents this miRNA from binding to the 3′UTR of its targets. Adeno-associated-virus-encoded precursors and sponges are transcribed in the nucleus. Precursors resemble the pri-miRNA and will be processed into functional mature miRNAs by the miRNA biogenesis pathway. Sponges contain tandem repeats of miRNA-binding sites and prevent the miRNA to bind the 3′UTR of its targets.

In this chapter, we will summarize the miRNAs that currently represent the most promising candidates for HF therapeutics. We also discuss the methods that are currently used to increase or decrease miRNA levels in the heart. And finally, we discuss the use of circulating miRNAs as biomarkers for HF.

2. MICRORNAs AS THERAPEUTICS

As discussed earlier, miRNAs could efficiently interfere with cardiac remodeling because of their ability to either target key genes or multiple genes within a pathway or cellular process. It will depend on the function of that pathway whether an increased or decreased function of this pathway is desired, and therefore whether an increased or decreased miRNA function is needed. For instance, for miR-29, increased levels would be desirable in HF. This miRNA is known to inhibit the expression of ECM-related genes and therefore counteracts the development of cardiac fibrosis. Several methods are developed to either systemically or cardiomyocyte-specifically manipulate miRNA levels (Fig. 11.2), which we will discuss in this section.

2.1 Systemic Delivery: Inhibition of MicroRNAs

Systemic miRNA inhibition can be induced by injections of chemically modified antisense oligonucleotides. These "antimiRs" harbor the full or partial reverse complementary sequence of the mature miRNA and therefore they hybridize to the mature miRNA incorporated into the RISC complex. This binding prevents the mature miRNA to bind and repress its mRNA target and thereby the antimiRs antagonize endogenous miRNA effects.

Several chemical modifications are applied to antimiRs to increase thermostability of the miRNA/antimiR duplex, enhance cellular uptake, and protect against degradation by nucleases. The increased thermostability, as indicated by an increase in duplex melting temperature (T_m), can be achieved by 2′-modifications of the antimiR's sugar backbone: 2′-O-methoxyethyl (2′-MOE), 2′-fluoro (2′-F), and locked nucleic acid (LNA). LNA is a 2′-sugar modification in which the ribose is locked in a C3′-endo conformation by the covalent linkage of its 2′-oxygen and the 4′-carbon and this modification induces the highest increase in T_m of 2–10°C per introduced modification [13]. Besides the increase in thermostability, these modifications also protect against degradation by nucleases. However, to further improve nuclease resistance, most antimiRs harbor phosphorothioate backbone linkages. These linkages also reduce clearance of antimiRs via urinary excretion by promotion of plasma protein binding [14]. Cholesterol linked to the 3′-side of the antimiR enhances cellular uptake and protection to nucleases [15].

The two most widely used chemistries for in vivo inhibition of miRNAs in the cardiovascular field are *antagomirs* and *LNA-based antimiRs* (Fig. 11.3).

Antagomirs are fully complementary to the mature miRNA sequence, contain a cholesterol and 2'-O-methyl sugar modifications and are partially conjugated via phosphorothioate linkage [15,16]. LNA-based antimiRs consist of LNA and DNA nucleotides with a phosphorothioate backbone and can be of several lengths. The LNA–DNA mixmers can be complementary to the full mature miRNA, but the increase in T_m by the LNA also allows shortening to 15–16 nucleotides targeting the 5' region of the miRNA (which includes the seed region). In addition, an antimiR with only LNA nucleotides

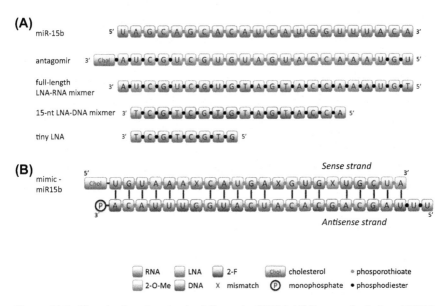

Figure 11.3 *Chemistries of systemic-delivered miRNA inhibitors and mimics.* MiRNA inhibitors and mimics use several modifications to increase their stability and function. These modifications are either 2' sugar modifications (2'-O-methyl, 2'-fluoro) or locked nucleic acids (LNAs). Furthermore the nucleotides within these synthetic molecules can be bound by a modified phosphorothioate linkage to further increase their stability. Here miR-15b is used as an example miRNA how several inhibitor chemistries are designed. (A) We show the full length complementary antagomir, which is bound to cholesterol; the full length RNA-LNA mixmer composed of LNA and 2'-O-methyl-modified RNA molecules; the 15 nucleotide long DNA–LNA mixmer composed of LNA and DNA molecules; and the tiny LNA, which are fully comprised of LNA nucleotides. (B) We also show the chemistry of the 'naked' in vivo mimic, a duplex of which the antisense strand will be incorporated in the RISC complex to become functional. This antisense strand is composed of unmodified and 2'-F modified RNA and contains a monophosphate at the 3' end. This antisense strand is bound to the sense strand, with inclusion of several mismatches to prevent function as antimiR by this sense strand. The sense strand is bound to cholesterol at its 5' end and composed of unmodified and 2'-O-methyl-modified RNA.

can even be shortened to eight nucleotides targeting the seed region of the miRNA [17]. These so-called tiny LNAs are thereby not able to silence one specific miRNA, but a full miRNA family of which all members share the same seed region.

Although the chemistries of antagomirs and LNA-based antimiRs show important differences, they share several characteristics in their functionality. For instance, their biodistribution after intravenous injection in mice is comparable. All chemistries fail to enter and inhibit miRNA levels in the brain and they all show high levels of the antimiRs and efficient inhibition in liver and kidney. Important for cardiovascular therapeutics, all chemistries enter the heart at high levels, which drop tremendously after 12 h, but result in substantial amounts remaining at a steady level of detection. These levels are sufficient to result in efficient inhibition of miRNA levels up to a month after the first injection [15,17,18]. Hullinger et al. [18] investigated whether the mode of administration of the antimiRs influenced the extent of miRNA knockdown in the heart and detected comparable levels of knock-down after single intravenous, intraperitoneal, and subcutaneous injections for both the LNA–DNA mixmer and the tiny LNA chemistry. Gavage administration of these antimiRs also resulted in cardiac inhibition, how-ever, with lesser efficiency, which might be because their size and charge result in limited intestinal absorption. A second shared characteristic between antagomirs and LNA-based antimiRs is that none of the chemis-tries showed signs of toxicity up to 4 weeks after injection as shown by no decrease in body weight; no increase in serum markers of liver toxicity; and no histopathological changes in cardiac, kidney, or liver tissue [15,17,18].

However, also some important differences between the chemistries and their way of function exist. Antagomirs reduce miRNA levels by degrada-tion of the mature miRNA, which results in the miRNA being undetect-able by northern blot [15,19]. On the other hand, LNA-based antimiRs (both LNA–DNA mixmers and tiny LNAs) inhibit miRNA activity by formation of a stable heteroduplex with the mature miRNA, visualized by a slower migrating complex on northern blot [17,18]. Furthermore, despite effective binding of the miRNA by both antagomirs and LNA-based anti-miRs, the antagomirs require higher doses of around 80 mg/kg while the LNA-based antimiRs can induce complete inhibition of the miRNA at doses as low as 0.33 mg/kg [15,18]. In this regard, the LNA–DNA mixmers show a higher efficiency of inhibition than the tiny LNAs, most probably because the latter also inhibits other family members of the same miRNA family [18].

Therapeutic miRNA inhibition using antimiRs is already tested in clinical trials. In this regard the inhibition of miR-122 by subcutaneous injection of an LNA-based antimiR called Miravirsen, in patients with hepatitis C results in decreased viral stability and increased sensitivity of the virus to the host immune response. A phase 2a clinical trial showed that administration of Miravirsen to hepatitis C patients resulted in dose-dependent viral suppression and during the 18 weeks follow-up in this trial no dose-limiting adverse events and no escape mutations in the hepatitis C genome occurred [12]. Further long-term follow-up of the same patients with a median follow-up of 24 months did not reveal any adverse side effects, but did show prolonged effectiveness of the single Miravirsen injection, where one patient was even viral RNA negative up to 7 months after injection [20]. Although these first in-human antimiR results show that miRNA inhibition is feasible, for other miRNAs, their therapeutic inhibition might be more challenging. An important underlying reason for the fact that miR-122 inhibition did not have adverse side effects probably relates to its specific expression pattern in the liver. Inhibition of other miRNAs with a broader expression pattern might therefore have unwanted side effects in other tissues. Furthermore, since all antimiR chemistries tend to accumulate in the kidney and liver, higher doses might be required for efficient inhibition in the cardiovascular system. This will not only increase the chance of off-target effects in unwanted tissue, but also increase the chance of renal and hepatotoxicity. In addition, for tiny LNAs, which inhibit a whole miRNA family, the risk of side effects further increases since the expression pattern of the miRNA family members combined is often broader than that of one specific miRNA. Another reason why antimiR-based therapeutics might be challenging relates to the timing of antimiR administration, whereby both the long half-life of the antimiRs and their delayed onset of activity might be important. The delayed onset of activity might be caused by regulation of specific targets in a cellular pathway, where time is needed to affect the downstream proteins in this pathway [21]. This indicates that some antimiRs are not useful for acute diseases. Furthermore, antimiRs are able to inhibit miRNA levels up to a month after the last injection, which indicates that they will influence cellular functions for a relative long time. In the cardiovascular field, inhibition of miR-15 is a good example why this might be problematic. Inhibition of this miRNA is shown to decrease apoptosis and increase proliferation of surviving cardiomyocytes directly after MI [18,22]. However, blockage of this miRNA has also been shown to increase tissue fibrosis during cardiac remodeling on the longer term [23].

This indicates that prolonged inhibition of miR-15 after MI might induce more severe cardiac remodeling and eventually adverse effects.

Another problem of inhibition of miRNAs is that not all miRNA targets are known, which indicates that we are unable to predict side effects in other tissues, but also in the tissue of interest by regulation of other important cellular processes. If a specific mRNA target is responsible for the beneficial effect of the inhibition of the miRNA and we are able to identify the binding site within the transcript, a solution might be to block the miRNA binding-site within the 3′UTR of this specific gene. This will allow the specific upregulation of a particular gene without affecting other cellular processes regulated by the same miRNA. The miRNA binding site might be blocked by LNA-based nucleotides completely antisense to the binding site in the mRNA. These so-called target site blockers bind with a higher affinity to the binding site as the miRNA and thereby hinder binding of the miRNA to this site. These target site blockers are not incorporated in a RISC complex and therefore they will not repress the translation of the mRNA target [24]. However, many miRNAs will exert their function via regulation of many targets involved in the same cellular process, which makes target blocking of all these binding sites unfeasible and inhibition by antimiRs a preferred therapeutic for these miRNAs.

2.2 Systemic Delivery: Increasing MicroRNA Levels

Increasing miRNA levels by systemic delivery can be achieved by injection of miRNA mimics. These are double-stranded RNA duplexes similar to the miRNA duplex during miRNA biogenesis and therefore they enter the biogenesis pathway to incorporate a functional strand in the RISC complex. The use of miRNA mimicry in vivo is even more challenging than the use of antimiRs and therefore efforts to restore miRNA levels have been lagging behind. The use of mimics is more challenging because cells have limited capacity to take up double-stranded RNA and due to the fact that the proper precursor strand has to be incorporated into the RISC complex [25]. This RISC incorporation increases the risk of side-effects as (1) incorporation of the other precursor strand might produce off-target effects; (2) miRNAs might be incorporated in cells and tissues where they are endogenously not expressed; (3) incorporation of these synthetic miRNAs might saturate the RISC complexes and thereby interfere with the endogenous miRNA processing and function.

Recently, Mongomery et al. [25] published the first in vivo data using a "naked" miRNA mimic for miR-29. They based the design and use of chemical modifications for stability and cellular uptake on data available for siRNAs, which are comprised of comparable duplexes. Therefore they placed a UU overhang on the 3'end of the "antisense" strand, which is the strand identical to the mature miRNA, and phosphorylated the 5'end. Furthermore they used 2'-F-modifications in this antisense strand to protect it against exonucleases without interfering with RISC loading. The nonfunctional "sense" strand allowed more modifications, this strand is at its 5'end linked to cholesterol for enhanced cellular uptake and contains 2'-O-Me modifications to increase stability of the duplex and prevent loading of this strand into RISC. Furthermore they induced mismatches between both strands of the duplex to prevent this "sense" strand to become functional as antimiR [25]. Injection of these mimics into mice showed a distribution completely different as for antimiRs. Where antimiRs tend to accumulate in kidney and liver, mimics were taken up by these organs but rapidly cleared within 1 day. The clearance in lung and spleen was slower, with increased levels of miR-29 up to 4 days after delivery. Results in the heart were variable, also in this organ mimics were detected after 1 day, and for some mice overexpression of miR-29 was still observed after 4 days, while other mice already returned to endogenous miR-29 levels [25]. They used a mouse model of bleomycin-induced pulmonary fibrosis. In this model, early injection of the mimics (3 days after bleomycin) blunted the fibrotic and inflammatory response and prevented the increase of ECM-related miR-29 target genes. Furthermore, injection of mimics 10 days after bleomycin treatment reversed the established fibrosis and also decreased the expression of the miR-29 target genes [25]. Interestingly, they only observed repression of miR-29 target genes after mimic injection in the "stressed" lungs and not in any organ under normal conditions indicating that the occurrence of off-target effects in nontargeted tissues might be lower as we would expect.

Besides the design and use of chemical modifications for miRNA mimics as described earlier, we might learn other lessons from the efforts toward siRNA-based therapeutics to improve miRNA mimic strategies [26]. One method developed in the siRNA field is the use of shielding agents such as polyethylene glycol, which shield the surface of delivery vehicles with hydrophilic polymers that minimize the nonspecific interactions with serum proteins and cells of the innate immune system and thereby prevent clearance by the immune system [26]. Another method is the exploitation of targeting ligands that improve uptake by target cells. These ligands enhance the uptake of the siRNAs or mimics by inducing receptor-mediated endocytosis. Besides the cholesterol modification used in the mimics described by Mongomery et al. [25] and used in antagomirs, other ligands that can be conjugated to the siRNAs are polymers, peptides, antibodies and small molecules [26]. The use of a cell type–specific ligand would result in a targeted delivery and thereby reduce the chance of off-target effects in untargeted tissues; however, this approach has not yet been developed for cardiac targeting. The shielding agents and the targeting ligands can be combined with the use of lipid nanoparticles. These are particles ranging in size between 20 and 200 nm in size, where the negatively charged siRNAs/mimics are entrapped in the positive liposomes. This entrapment protects the oligonucleotides from nuclease degradation, prevents their renal clearance because of the size of the particles, and the lipids in the nanoparticles promotes their cellular uptake [26]. Mirna Therapeutics Inc. successfully translated this last approach from the siRNA field toward the miRNA mimic field. This company packed a mimic for miR-34 into an ionizable liposome (SMARTICLES) that forms a particle with a diameter of ~120 nm and called this compound MRX34 [27]. MiR-34 is a known tumor suppressor miRNA, which is lost or downregulated in a variety of cancers. The efficiency and safety of MRX34 was tested in an orthotopic model of hepatocellular carcinoma, which showed efficient delivery to the tumor cells as indicated by a more than 100-fold increase in miR-34 levels and a reduction in the miR-34 target genes. Furthermore, MRX34-induced tumor regression (several mice became tumor free), prolonged the survival of the mice and inhibited the growth of other nonhepatic tumors [27]. Consequently, Mirna Therapeutics Inc. is now recruiting participants with primary liver cancer or solid cancers with liver involvement for a phase 1 clinical trial, which was completed by late 2015. Because the cellular uptake of these liposomes is based on endocytosis promoted by the lipids in the surface of these nanoparticles, they are expected to be taken up by all cell types including cardiomyocytes.

2.3 Targeted Delivery

As discussed earlier an important challenge for miRNA therapeutics is to prevent side effects due to inhibition or overexpression of the miRNA in non-diseased organs. Therefore the therapeutic potential of miRNA manipulation could be improved by the implementation of direct targeting strategies for specific cells and organs. A tool to provide this directed targeting are the numerous different serotypes of adeno-associated viruses (AAVs). The natural tropism of these serotypes and the cellular receptors with which they interact allow for specific tissue and cell targeting [28]. The use of cell-specific promoters in the AAV-encoding vectors can even further increase the specific targeting by these AAVs. In the cardiovascular field the first clinical trials using AAVs are already performed. In these trials AAV1-mediated overexpression of SERCA2 in the heart of patients with advanced HF appeared effective and revealed no major side effects [29,30]. These AAVs could be used for both targeted overexpression and knockdown of miRNAs depending on the type of insert they are expressing. When this insert is an miRNA precursor it will be converted into a RISC-bound mature miRNA via the endogenous miRNA biogenesis pathway. Compared to the systemic delivery via mimics, this approach overcomes the limited capacity to take up double-stranded RNA. However, because this approach relies on the endogenous miRNA biogenesis pathway caution is needed with regard to saturation of this pathway and thus function of all other endogenously expressed miRNAs. For targeted knockdown of miRNAs, these AAVs can be combined with so-called miRNA sponges. These sponges often contain a reporter gene (for instance GFP) combined with an artificial "3'-UTR," which consists of multiple, tandem binding sites to the miRNA of interest [31]. These sites are able to bind the miR-NAs and compete with its interaction with endogenous targets, which relieves endogenous target repression by the miRNAs. This competition of miRNA binding between the sponge and the endogenous targets indicates that a high level of sponge expression would be the most effective [32]. However, because the same rules for miRNA binding apply to endogenous targets and sponges, binding of only the seed sequence of the miRNA to the sponge would be enough to compete with the endogenous levels. So especially at high expression levels of the sponge, other family members would also bind to the sponge, which indicates that these sponges can be suitable to inhibit a complete miRNA family and less suitable to inhibit a specific miRNA [31,32].

Another future tool for targeted delivery of miRNA therapeutics might be the use of microbubbles in combination with ultrasound [33]. Microbubbles are several microns in size and are originally used as contrast agents in diagnostic ultrasounds. They are pushed away from the transducer face by radiation force, which allows for concentration of microbubbles at a specific target zone. Combination of this concentration strategy with a high acoustic intensity, which makes the microbubbles oscillate and can force them to collapse, results in ultrasound-targeted microbubble destruction [33]. When these microbubbles are loaded with antimiRs or mimics this ultrasound-targeted microbubble destruction forces them to release their content locally.

3. MICRORNAs IN CARDIAC REMODELING

The crucial role of miRNAs in postnatal cardiac biology was demonstrated by cardiomyocyte-specific deletion of dicer and DGCR8 (a required component of the Drosha enzyme complex), both essential components of the miRNA biogenesis pathway [34–36]. Mice in which either Dicer or DGCR8 is cardiomyocyte specifically deleted directly after birth (via an αMHC-cre or MCK-cre system, respectively) die prematurely. Ablation of Dicer resulted in early postnatal death due to impaired cardiac function, as evidenced by dilated ventricles and a decrease in fractional shortening [34]. Ablation of DGCR8 resulted in death within 2 months after birth also due to impaired cardiac function, which was already detected by a decrease in fractional shortening at 3 weeks of age [35]. Da Costa Martins et al. [36] used a tamoxifen-inducible system to delete Dicer specifically in cardiomyocytes in juvenile (3 weeks) and adult mice. In juvenile mice, this resulted in increased cardiac size and impaired cardiac function combined with sudden death within 3 weeks. In adult mice, there were no signs of reduced survival, but these mice developed severe HF as evidenced by fibrosis, dilation of the left ventricle and a rapid decline in cardiac function. Combined these studies indicated the crucial role of miRNAs in cardiac remodeling and development of HF. Interestingly, these studies also show that prevention of miRNA maturation also results in electrophysiological remodeling and dysfunction as evidenced by a decrease in heart rate and increase in PR interval and QRS width on the ECG in both the postnatal Dicer and DGCR8 knockout mice, which indicates conduction defects [34,35]. Furthermore, this is also evidenced by the sudden death of the juvenile dicer-ablated mice, which is most probably due to fatal arrhythmias [36].

Several groups performed miRNA profiling to investigate which specific miRNAs were deregulated in cardiac remodeling (Table 11.1) [37–41]. Van Rooij et al. [37] revealed a signature miRNA pattern in response to cardiac stress by performing miRNA microarrays in two mouse models of cardiac hypertrophy. The first model, the thoracic aorta constriction (TAC) model, is a model where increased afterload on the heart induces hypertrophy. In these TAC hearts, van Rooij et al. [37] found 27 up- and 15 downregulated miRNAs. In the second model a severe form of hypertrophy is induced by overexpression of activated calcineurin A (CnA tg), which

Continued

Table 11.1 Overview of microRNA arrays in cardiac remodeling

Species	Model/disease	Upregulated miRNAs	Downregulated miRNAs	References
Mouse	TAC + CnA tg	miR–10b, miR–19a, miR–21, miR–23a/b, miR–24, miR–25, miR–27a/b, miR–125, miR–126, miR–154, miR–195, miR–199a, miR–199a★, miR–210, miR–214, miR–217, miR–218, miR–330, miR–351	miR–29c, miR–30e, miR–93, miR–133a/b, miR–150, miR–181b	[37]
Human	End-stage DCM	Let–7b/c/d/e, miR–1, miR–17–3p, miR–21, miR–29a, miR–34b, miR–125a, miR–126_AS, miR–130a, miR–132, miR–212, miR–215, miR–292-3p, miR–295, miR–297, miR–320, miR–322, miR–330, miR–333, miR–372, miR–423, miR–213, miR–302a, miR–32, miR–365, miR–373, miR–424, miR–429, miR–525_AS	miR–30a/b, miR–182, miR–452_AS, miR–494, miR–515-5p, miR–520d_AS	[38]
Human	Ischemic CM	Let–7b/c, miR–23a, miR–24, miR–27a, miR–100, miR–103, miR–140★, miR–191, miR–195, miR–199a★, miR–214, miR–320	miR–126★, miR–222, miR–422b	[39]
Human	DCM	Let–7b/c, miR–15b, miR–23, miR–99, miR–100, miR–103, miR–125b, miR–140★, miR–195, miR–199a★, miR–214, miR–342	miR–1, miR–17-5p, miR–19a/b, miR–20a/b, miR–28, miR–30e-5p, miR–101, miR–106a, miR–126★, miR–222, miR–422b	[39]

Table 11.1 Overview of microRNA arrays in cardiac remodeling—cont'd

Species	Model/disease	Upregulated miRNAs	Downregulated miRNAs	References
Human	Aorta stenosis	Let-7c/e, miR-15b, miR-23a/b, miR-24, miR-27a/b, miR-93, miR-99a/b, miR-103, miR-125b, miR-140★, miR-145, miR-181a, miR-191, miR-214, miR-320, miR-342, miR-423★	miR-1, miR-10a, miR-19a/b, miR-20a/b, miR-26b, miR-30e-5p, miR-101, miR-126, miR-126★, miR-374, miR-451, miR-499	[39]
Rat	PE-treated NRCM	miR-18b, miR-20b, miR-21, miR-23a, miR-106b, miR-125b, miR-133a	miR-187, miR-292-5p, miR-373★, miR-466	[40]
Mouse	TAC	miR-17-5p, miR-18b, miR-19b, miR-20b, miR-21, miR-106a, miR-125b, miR-140, miR-142-3p, miR-153, miR-184, miR-200a, miR-208, miR-210, miR-211, miR-221, miR-222	miR-30b/c, miR-150	[40]
Mouse	TAC	miR-21, miR-27a/b, miR-146, miR-214, miR-341, miR-424	miR-29a/b/c, miR-30e, miR-126-5p, miR-133a/b, miR-149, miR-150, miR-185, miR-451, miR-486	[41]

TAC, thoracic aorta constriction; *CnA tg*, calcineurin A transgene; *DCM*, dilated cardiomyopathy; *CM*, cardiomyopathy; *PE*, phenylephrine; *NRCM*, neonatal rat cardiac myocytes.

activates a pathway causal for hypertrophy. In this second model, they found 33 up- and 14 downregulated miRNAs. They concluded that the 21 up- and 7 downregulated miRNAs in common between these two models probably provided a miRNA signature of the cardiac stress response [37]. Thum et al. [38] profiled miRNAs in human end-stage HF and in human fetal cardiac tissue and revealed similar patterns compared to healthy adult hearts, which indicates that in cardiac remodeling reexpression of fetal genes not only occurs on the mRNA level, but also on the miRNA level. Interestingly, only a small overlap (miR-21, miR-29, miR-30, miR-125, miR-126, miR-330) between the patterns in the two studies existed, which might be explained by important differences between the studies. The most plausible explanation is the difference in stage of the disease and the underlying etiology between the different models. In the study by van Rooij et al. [37] two mouse models leading to hypertrophy are used, while in the study by Thum et al. [38] human end-stage HF samples are studied. On the mRNA level, different signaling pathways are activated during initiation and progression of cardiac hypertrophy compared to the failing heart [38], therefore it is not surprising that also on the miRNA level differences exist. The effect of the underlying etiology on the detected miRNA patterns is further exemplified by the study by Ikeda et al. [39], who performed miRNA profiling in patients with ischemic cardiomyopathy (ICM), dilated cardiomyopathy (DCM), and aorta stenosis (AS). They detected miRNA expression profiles that could distinguish between these different etiologies of HF [39].

In the last years, tremendous efforts have been undertaken to investigate the function of the deregulated miRNAs in the heart at baseline and during cardiac remodeling. We refer the interested reader to recent reviews to get an overview of these miRNAs and their function [7,42,43]. Here, we will provide examples of miRNAs in all aspects of cardiac remodeling that are good candidates to become therapeutics for the diseased heart.

3.1 MicroRNAs in Cardiomyocyte Hypertrophy

Cardiomyocyte hypertrophy is one of the main underlying processes of cardiac remodeling in response to stress. Hypertrophy is defined as an increase in cell size, and in the heart, this results in an increase in left ventricular wall thickness and cardiac mass [3]. The increase in cell size is accompanied by changes in expression of genes that regulate energy metabolism, calcium handling and contractility, and genes normally expressed in the embryonic heart. And although the hypertrophic response is initially a compensatory response to stress on the heart, chronic hypertrophy increases the risk for development of HF [3].

3.1.1 MiR-208

MiR-208a is encoded within intron 27 of the αMHC gene and is thus far the only identified miRNA with a cardiomyocyte-specific expression pattern [44]. It is part of a family of miRNAs referred to as myomiRs, which also includes miR-208b and miR-499. Both other family members are also encoded within myosin genes; miR-208b in the βMHC gene and miR-499 in the Myh7b gene [45].

The miR-208a knockout (KO) mouse provided the first indication that inhibition of miR-208 could provide a new therapy for hypertrophy and HF development. Normally during hypertrophy, a switch from the fast ATPase αMHC to the slow ATPase βMHC occurs, which results in a decrease in cardiac contractility and performance. The miR-208a KO animals failed to upregulate βMHC expression in response to cardiac stress, while other stress-responsive genes such as ANF and BNP were readily upregulated. Furthermore, these mice showed virtually no sign of cardiomyocyte hypertrophy or fibrosis, which suggests that inhibition of miR-208 may be an interesting therapeutic target for HF [44]. However, caution is warranted as another group detected cardiac arrhythmias in both their cardiomyocyte-specific miR-208a transgenic mouse model (first and second degree AV-blocks) and their miR-208a KO mice (atrial fibrillation) [46].

Interestingly, further investigation into this myomiR family using the miR-208a KO mouse revealed the regulatory function of this miRNA on the expression of not only the myosin genes but also the myomiRs encoded within [45]. In these mice, which failed to upregulate βMHC in response to stress, also no expression of Myh7b, miR-208b, and miR-499 was detected. Interestingly, the cardiac phenotype of the miR-208a KO mouse could be rescued by transgenic overexpression of miR-499 and in these double mutant mice, both miR-208b and βMHC were restored to normal levels. This indicates that miR-499 is a downstream mediator of miR-208a, which would suggest that inhibition of miR-499 could also be used therapeutically. However, in contrast to the miR-208a KO mice, miR-499 KO mice were able to induce βMHC expression, which indicates that loss of both miR-208a and miR-499 is required for the therapeutic effect [45]. Importantly, the situation in human is slightly different than in mice. In the human heart, the predominant myosin is βMHC instead of αMHC and thus the predominant miRNA is miR-208b instead of miR-208a. As miR-208a, miR-208b is also able to induce expression of miR-499, which suggest that in human miR-208b might be the miRNA required for pathological cardiac remodeling and that in human miR-208b is the actual miRNA that should be inhibited therapeutically [45].

In the next step toward the development of miR-208 inhibition as therapy, Montgomery et al. [47] tested the effectiveness of LNA-based antimiR-208a in Dahl salt-sensitive rats on a high salt diet. These rats develop HF in response to hypertension induced by the high salt diet. In these rats inhibition of miR-208a by subcutaneous antimiR injections dose dependently reduced the expression of βMHC, Myh7b and their myomiRs miR-208b, and miR-499. Inhibition of these three myomiRs by antimiR-208a injections prevented pathological myosin switching, cardiomyocyte hypertrophy, and fibrosis, while improving cardiac function, overall health, and survival [47]. In contrast to the miR-208a KO mice, mice and rats injected with these LNA-based antimiR-208a showed proper cardiac electrophysiology on ECG up to 8 weeks after injection [47]. Interestingly, the mice treated with antimiR-208a turned out to be leaner than control littermates and they failed to develop obesity on a high fat diet [48]. As an underlying mechanism, antimiR-208a improved systemic insulin sensitivity and glucose tolerance, which indicates that the heart influences systemic metabolism and that inhibition of miR-208 might also be a potential therapeutic for metabolic disorders. The target gene responsible for the effect on metabolism appeared to be MED13, a regulator of metabolic genes in the heart [48]. Whether the effect on systemic metabolism is the result of a compensatory metabolic activity in noncardiac tissues or caused by the release of paracrine factors acting on distal tissues to increase systemic metabolic activity remains to be seen.

In summary, therapeutic inhibition of miR-208a by antimiR injections might be effective and safe for both HF and metabolic disorders. However, large animal models are necessary to determine which miR-208 family member should be inhibited in animals were βMHC is the predominant myosin.

3.1.2 Mir-25

Reduced expression and activity of SERCA2a is a hallmark of HF, and this has been shown to contribute to impaired Ca^{2+} uptake by the sarcoplasmatic reticulum (SR) during excitation–contraction coupling in cardiomyocytes, which results in a reduced contractility of the failing heart [29,30]. AAV-mediated restoration of SERCA2a expression has been proven effective as therapy in clinical trials with HF patients [29,30]. Therefore, inhibition of a miRNA that targets SERCA2a and is upregulated in HF might be a good candidate for miRNA therapeutics to restore SERCA2a levels. Wahlquist et al. [49] performed a GFP-based screening of 875 miRNAs to identify which miRNAs are able to target SERCA2a; and they identified 15 miRNAs that were both able to target SERCA2a and upregulated in human HF. MiR-25 was identified as the most potent regulator of SERCA2a, and its inhibition by intravenous injection of antimiRs revealed its therapeutic potential. In a model of established HF (3 months after TAC), these antimiR-25-injected mice showed restoration of SERCA2a expression and activity, reduced fibrosis and cardiomyocyte hypertrophy, increased survival, and most importantly cardiac function was restored to normal levels [49]. These results indicate that miR-25 upregulation in HF is (partly) responsible for the loss of SERCA2a and that inhibition of miR-25 might provide a new therapy for established HF. In contrast to these findings, Dirkx et al. [50] reported that miR-25 is downregulated in two mouse models of cardiac hypertrophy (TAC and CnA tg). Furthermore, they showed that inhibition of miR-25 by intraperitoneal antagomir injections resulted in mild cardiac disease at baseline and exacerbated TAC-induced cardiac remodeling [50]. Possible explanations for these contrasting findings are the use of different chemistries and different timing of antimiR injections after TAC surgery. Wahlquist used 2'-O-methyl-modified antimiRs, injected them 3 months after TAC, a time-point that represents established HF, and performed their functional assessments up to 2.5 months after antimiR injection [49]. On the other hand, Dirkx used antagomirs, injected them 3 days after TAC surgery, a time-point in which compensatory hypertrophy takes place, and performed their functional assessments up to 4 weeks after TAC [50]. This indicates that miR-25 could have a beneficial role during hypertrophy development and maladaptive effects in chronic HF. Further studies with the same chemistry and dosing regimen at different time-points after TAC are needed to further investigate these apparently dual effects.

3.2 MicroRNAs in Fibrosis

Cardiac fibrosis is the excessive accumulation of ECM proteins in interstitium and perivascular regions within the myocardium. Fibrosis is an important feature of cardiac remodeling and is a hallmark of maladaptive hypertrophy and HF. It leads to disruption of normal myocardial structures and increases tissue stiffness, which together contribute to contractile dysfunction of the heart [51]. Fibroblasts are the cells responsible for ECM synthesis and fibrosis. In the stressed heart fibroblasts differentiate and become activated in response to cytokines and growth factors, such as TGFβ and CTGF. These activated fibroblasts, called myofibroblasts, proliferate, migrate, and remodel the cardiac interstitium to develop fibrosis [52].

MiR-29 is a miRNA family composed of three members (miR-29a, b, and c), which are mainly expressed in fibroblasts. In the heart, this family is downregulated in response to TAC, chronic calcineurin activation, and in the viable myocardium after MI [37,53]. Furthermore, miR-29 is downregulated in cultured fibroblasts stimulated by TGFβ, which suggests that the loss of miR-29 in cardiac remodeling is mediated by this growth factor [53]. The miR-29 family targets a multitude of ECM genes involved in development of fibrosis, including collagens, fibrillins, and elastin. Inhibition of miR-29 using antagomirs induced the expression of collagens in the healthy mouse heart and in the heart subjected to TAC surgery, this resulted in excess perivascular fibrosis without affecting cardiac function [53,54]. This suggests that the loss of miR-29 in cardiac remodeling derepresses the expression of its ECM target genes and contributes to the development of fibrosis.

Overexpression of miR-29 by mimics or by vectors encoding an miR-29 precursor in cultured cardiac fibroblasts reduced collagen expression and secretion, which suggest that overexpression of miR-29 might be used therapeutically to prevent or regress fibrosis [53–55]. This was further supported in vivo by the overexpression of miR-29 using an overexpression plasmid delivered to the heart by microbubble-mediated ultrasound in a mouse model of Angiotensin II-induced cardiac remodeling [55]. In this model, overexpression of miR-29b prevented the Angiotensin II-induced decline in cardiac function and development of fibrosis. Furthermore, delivery of miR-29 using the same method 14 days after start of the Angiotensin II infusion blocked further deterioration of cardiac function and development of fibrosis [55]. As discussed earlier (under the paragraph miRNAs as therapeutics), the first study towards in vivo mimicry was also directed to overexpress miR-29b to treat fibrosis [25]. In this study, injection of mimic-29b both blunted and reversed the fibrotic and inflammatory response in the lung in response to bleomycin. Furthermore, this study revealed that these mimics could target the heart, however, they seemed to be cleared rather fast from this organ [25]. Together these studies indicate that overexpression of miR-29b by microbubble-mediated ultrasound delivery of mimics might be used therapeutically to prevent and maybe even reverse cardiac fibrosis.

3.3 MicroRNAs in Capillary Remodeling

Pathological cardiac remodeling is characterized by a reduction in capillary density. This means that when the growth of the microvasculature does not keep pace with the extent of cardiomyocyte hypertrophy, myocardial hypoxia will occur and eventually cause myocardial dysfunction [5]. This was illustrated in a model of pressure overload-induced hypertrophy, where blockade of VEGF reduced capillary density and accelerated the development of HF [56]. MI also leads to cardiac remodeling, and in this situation neovascularization is of crucial importance to ensure restoration of blood flow to the injured myocardium. Therefore miRNA interference enhancing neovascularization could improve cardiac function after MI and thus slow down or reverse the progression to HF.

MiR-92a is part of the miR-17–92 cluster, which also includes miR-17, miR-18a, miR-19a, miR-20a, and miR-19b-1. All members of this cluster are expressed in endothelial cells, where miR-92a has the highest expression level of this family [57]. In endothelial cells, miR-92a turned out to have antiangiogenic effects as shown by the study of Bonauer et al. [57] where forced overexpression of miR-92a in cultured endothelial cells blocked sprout formation in a three-dimensional model of angiogenesis, inhibited vascular network formation in matrigel assays and reduced endothelial cell migration. Furthermore, inhibition of miR-92a in vivo by antagomirs promoted neovascularization in both a model of hind limb ischemia and a model of MI. In the MI model, inhibition of miR-92a increased the number of vessels, decreased the number of apoptotic cells, reduced the infarct size, and most importantly improved cardiac function 2 weeks after the infarction [57]. In a subsequent study, Hinkel et al. [58] tested the inhibition of miR-92a by local or systemic delivery of LNA-based antimiRs in a pig model of ischemia reperfusion. They found that both local and systemic delivery were able to reduce cardiac miR-92a levels to an almost similar extent. Interestingly, only local delivery was capable to decrease the infarct size, reduce cell death, enhance neovascularization, reduce inflammation, and improve cardiac function at 3 or 7 days after ischemia/reperfusion [58]. The observation that the outcome strongly depends on the delivery method in this large animal model emphasizes the importance of determining the right dose and delivery method before miRNA therapeutics for the heart can be applied in humans. Besides these promising short-term effects, miR-92a inhibition by intracoronary delivery of microparticles containing LNA-based antimir-92a in pigs with ischemia followed by reperfusion for 1 month revealed that also on the long-term miR-92a inhibition has a beneficial effect on cardiac remodeling after ischemia [59]. This delivery method turned out to be infarct zone specific as the microparticles were retained in the capillaries of the damaged area. After 1 month of reperfusion pigs with inhibition of miR-92a had a higher vascular density in the infarcted area and a better cardiac function [59]. Together these data indicate that inhibition of miR-92a after a MI might be used therapeutically to enhance neovascularization of the infarcted area and reduce maladaptive cardiac remodeling, which might preserve cardiac function and prevent the development of HF after MI.

3.4 MicroRNAs in the Inflammatory Response During Cardiac Remodeling

In patients with congestive HF a chronic activation of the innate immune system is often observed, as evidenced by increased numbers of macrophages and lymphocytes and increased levels of proinflammatory cytokines in the myocardium [60]. Furthermore, the levels of these cytokines have been shown to correlate with the severity of symptoms in HF [61]. Angiotensin II, a hormone that induces hypertension and subsequently cardiac remodeling, promotes the underlying inflammatory reaction in cardiac remodeling [62]. In response to Angiotensin II infusion in the heart, patches of infiltrating leukocytes expressing miR-155 were detected [62]. MiR-155 turns out to be a key inflammatory miRNA, as shown by the fact that lipopolysaccharides failed to activate macrophages of miR-155 KO mice. This observation would predict that inhibition of miR-155 during Angiotensin II infusion protects against the inflammatory response. Indeed, genetic loss or pharmacological inhibition of miR-155 by LNA-based antimiRs reduced cardiac inflammation, hypertrophy and preserved cardiac function in response to angiotensin II infusion or TAC. To further prove that this was caused by the loss of miR-155 in macrophages and not in cardiomyocytes, Heymans et al. [62] inhibited miR-155 cardiomyocyte-specifically using AAV9-expressing miR-155 sponges, which did not affect the cardiac remodeling after Angiotensin II infusion. On the other hand, bone marrow transplantation from miR-155 KO to wild-type mice showed a blunted leukocyte infiltration into the myocardium and protected the heart to hypertrophy and dysfunction [62]. Seok et al. [63] confirmed the beneficial effect of inhibition of miR-155 in cardiac remodeling. They subjected the miR-155 KO mouse to 4 weeks of TAC and crossed them to CnA transgenic mice; in both hypertrophy models loss of miR-155 prevented cardiac remodeling and preserved cardiac function. Interestingly, they also detected a suppressed fibrotic response, which is in contrast to the results described by Heymans et al. [62,63]. Furthermore, results of Seok et al. questioned the finding by Heymans et al. that the in vivo effects were only due to loss of miR-155 in macrophages and not in cardiomyocytes [62,63]. Seok et al. stimulated both mouse neonatal cardiomyocytes of the miR-155 KO mouse and rat neonatal cardiomyocytes transfected with miR-155 inhibitors with phenylephrine, which resulted in a repressed hypertrophic response in the myocytes without miR-155 [63].

So whether the loss of miR-155 in cardiomyocytes adds to the beneficial effects of in vivo inhibition of miR-155 in models of hypertrophy remains unclear. Nevertheless, inhibition of miR-155 resulted in reduced hypertrophy and preserved cardiac function in three different models of hypertrophy, strongly suggesting that inhibition of miR-155 might be used therapeutically to reduce cardiac remodeling.

3.5 MicroRNAs in Electrical Remodeling

Ventricular and atrial arrhythmias are important contributors to mortality among HF patients. On the one hand, these arrhythmias are caused by fibrosis which disturbs the electrical continuity between cardiomyocytes, leading to conduction slowing and facilitating the occurrence of arrhythmias [64]. On the other hand, cardiac remodeling is also associated with electrical remodeling, which means that differences in expression of ion channels responsible for the cardiac action potential occur. This electrical remodeling process results in prolongation of the action potential in HF, which can be seen as a prolongation of the QT-time on the ECG, a phenomenon known to be associated with the occurrence of fatal arrhythmias [65,66].

The miRNA most clearly demonstrated to have a role in arrhythmogenesis is miR-1. MiR-1 is encoded by two regions in the genome, miR-1-1 on mouse chromosome 2 and miR1-2 on chromosome 18 and in both regions it is clustered with miR-133a. Zhao et al. [67] specifically deleted miR-1-2 from the mouse genome without affecting expression of miR-133a or miR-1-1. Half of the miR-1-2 KO mice displayed cardiac developmental defects and died before or within hours after birth. Many of the mice surviving until adulthood died suddenly, which indicates that fatal arrhythmias might have occurred. Indeed, surface ECGs showed a lower heart rate, shortened PR intervals, and prolonged ventricular depolarization, which indicates abnormal cardiac conduction [67]. Targets of miR-1 that may explain these effects on conduction include the transcription factor Irx5, which controls expression of several ion channels and gap junction proteins (KCNJ2, GJA1, KCNE1) [67,68]. Strikingly, in another KO mouse model where the entire miR-1-2/miR-133a-1 cluster was removed, mice were not showing any embryonic or postnatal lethality, neither did these mice show any sign of arrhythmias or conduction defects on the ECG and no sudden death was observed [68]. This difference between these two KO models might be explained by differences in genetic background of the mice, however, both mouse lines were on an Sv129/C57bl6 mixed background but it is unclear how far they were crossed to one specific background. Another possible explanation is that miR-1 and miR-133 have opposing effects on cardiac conduction and therefore they counterbalance each other's effects.

Interestingly, the reduction in miR-1/133a dosage in the KO mice of Besser et al. [68] induced QT prolongation. MiR-1 is downregulated in several forms of HF [39]. This indicates that the loss of miR-1 during HF might have a role in the observed QT prolongation in HF. Especially because it was revealed that the QT prolongation in the miR-1/133a KO mice was due to an increased L-type calcium channel activity and changes in calcium handling are known to occur during cardiac remodeling. Furthermore, miR-1 might play a role in these changes in calcium handling, as was revealed by the study of Terentyev et al. [69] in which overexpression of miR-1 resulted in changes in calcium handling comparable to changes in the failing heart. They overexpressed miR-1 in rat ventricular myocytes and showed an increase in spontaneous calcium release from the SR and a decrease in the calcium content in the SR due to hyperphosphorylation of RYR2 by CAMKII and enhanced activity of RYR2. These changes in calcium handling, sensitize cardiomyocytes to arrhythmogenic calcium cycling [69].

With so many effects on electrophysiology and calcium handling, the question still remains what will happen if miR-1 is therapeutically targeted in cardiac remodeling. Karakikis et al. [70] investigated this question in a rat model of pressure overload where they restored miR-1 levels using AAV9. They started the AAV9-mediated treatment 2 weeks after induction of pressure overload and investigated the effects after 7 weeks. This restoration of miR-1 levels regressed cardiac hypertrophy, preserved cardiac function and halted the disease progression towards HF by the persisting pressure overload. Furthermore, AAV9-miR-1 treatment prevented upregulation of hypertrophic markers, improved calcium handling, and decreased fibrosis and cardiomyocyte apoptosis [70]. Unfortunately, these authors did not specifically study the occurrence of arrhythmias, however, if these arrhythmias had resulted in sudden death this would have been noticed. These are promising results toward the therapeutic use of miR-1. However, long-term overexpression of miR-1 might have detrimental effects, as evidenced by the cardiomyocyte-specific transgenic miR-1 mouse. This mouse model shows disrupted sarcomere assembly, which results in a reduced contractility and cardiac function at 2, 4, and 6 months of age [71]. These data indicate that studies toward the long-term effects of restoring miR-1 levels in cardiac remodeling are needed and a thorough investigation of the electrophysiological effects is required before miR-1 can be considered as a therapeutic option for HF.

3.6 MicroRNAs in Postmyocardial Infarction Remodeling

In the last decades, spectacular advances have occurred in cardiovascular medicine and surgery, with the consequence that many patients now survive an acute MI [2]. However, the insult still leads to a shortage of oxygen supply to the myocardium, which often does results in loss of viable tissue and contractility. In the long term, in the surviving myocardium secondary cardiac remodeling occurs, which can result in HF [2]. Therefore, many research efforts are now focused on decreasing the damage to the myocardium after MI and thereby preventing the development of HF. One miRNA family that appears highly relevant in this regard is the miR-15 family, which consists of six highly conserved miRNAs that are all abundantly expressed in cardiomyocytes and cardiac fibroblasts [18,23]. Inhibition of the miR-15 family has been shown to decrease infarct size by both induction of proliferation of cardiomyocytes and by protection of cardiomyocytes against apoptosis [18,22]. The first evidence that the miR-15 family regulates cardiomyocyte proliferation was provided by overexpression and knockdown experiments in neonatal mice. Mice overexpressing miR-195 during development under the βMHC-promoter were born with ventricular septal defects and a reduced ventricular weight due to a reduction in proliferation [72]. When these mice were subjected to MI on day 1 after birth they failed to regenerate the infarcted myocardium while wild-type mice were able to regenerate the myocardium via cardiomyocyte proliferation [22]. On the other hand, inhibition of the full miR-15 family by LNA-based antimiRs in neonatal and adult mice subjected to ischemia/reperfusion resulted in an increased regeneration of the infarcted myocardium due to increased survival and proliferation of cardiomyocytes [18,22,72]. Besides these effects in mice, Hullinger et al. also showed the upregulation of the miR-15 family in the hypoxic myocardium of pigs and the ability to repress the miR-15 family in the pig myocardium [18]. Unfortunately they did not combine both and investigate the effect of inhibition of the miR-15 family in pigs subjected to ischemia/reperfusion.

Earlier described studies suggested to inhibit the miR-15 family therapeutically after MI to reduce infarct size and increase the amount of viable myocardium. However, we recently showed that inhibition of the miR-15 family in the TAC-model in mice resulted in an increased fibrotic and hypertrophic response 3 weeks after surgery by activation of the TGFβ-pathway [23]. This indicates that inhibition of the miR-15 family immediately after MI might be used therapeutically to decrease infarct size and improve cardiac function, but on the long term adverse cardiac remodeling might occur due to unrestrained activation of the TGFβ-pathway. Therefore, future studies are needed to study the long-term effects of miR-15 family inhibition after MI and to determine the exact time period in which inhibition of the miR-15 family is beneficial.

4. CIRCULATING MICRORNAs AS BIOMARKER

One of the major challenges in cardiovascular medicine is the identification of reliable diagnostic and prognostic biomarkers that can be measured routinely in plasma. Currently, circulating biomarkers such as brain natriuretic peptide (BNP) facilitate the clinical measurement of HF. In general levels of BNP are directly related to the severity of HF and they respond to improvement due to therapy [73]. However, BNP levels do have a so-called "gray zone," where measurements are not informative as diagnostic biomarker. In this gray zone the levels are above the level to exclude HF diagnosis and below the age-related cut-off points to diagnose a patient with HF [73]. This indicates that other biomarkers are needed to reliably diagnose patients especially those with BNP levels within this gray zone. In 2008, miRNAs were discovered to be present in the circulation in a remarkably stable form, they are stable even in harsh conditions of boiling, low or high pH, long-term storage at room temperature and multiple freeze-thaw cycles [74,75]. Mainly because of this stability, miRNAs are currently explored for their potential as biomarkers in a wide range of diseases, including HF and other cardiovascular diseases.

Besides their stability, circulating miRNAs also fulfill some other criteria linked to the ideal biomarker: (1) they are easily accessible in the circulation using noninvasive methods; (2) they are often regulated in a tissue- and pathology-specific manner and therefore they show a high degree of sensitivity and specificity to the disease; (3) they might already be detected during the cardiac remodeling stages and thus allow early detection; (4) their levels in the circulation have been shown to change in response to therapy, which means they are sensitive to changes in the disease state; (5) they can be detected with sequence-specific amplification, which allows highly sensitive and specific detection [76]. These qualities suggest that the discovery-validation pipeline for miRNA biomarkers might be very efficient [74].

miRNAs are proposed as diagnostic or prognostic biomarkers for several forms of cardiac diseases including HF [77]. Interestingly, the cardiomyocyte-specific miR-208 and the cardiomyocyte-enriched miRNAs (miR-133, miR-1, miR-499) are only shown to be elevated in the circulation after MI, where massive death of cardiomyocytes occurs [77]. While levels of these miRNAs in the circulation do not change in other cardiac disease were death of cardiomyocytes is less massive [77]. This indicates that in these diseases a pathology-specific release of certain miRNAs occurs, which allows for the identification of pathology-specific biomarkers. Here we will discuss the circulating miRNAs that are proposed as diagnostic or prognostic biomarkers for HF (Table 11.2). We will not discuss the suggested role of these miRNAs in cell–cell communication, but refer the reader to other reviews [77–79].

4.1 Circulating MicroRNAs as Biomarker for Heart Failure

Our laboratory was the first to show aberrant profiles of circulating miRNAs in HF. We determined the levels of 16 differentially expressed miRNAs selected from a microarray in a cohort of 39 healthy controls and 50 dyspnea patients, 30 of whom were diagnosed with dyspnea due to HF and 20 due to other causes [80]. We validated 7 of these 16 miRNAs to be enriched in plasma of HF patients in this cohort (miR-423-5p, miR-18b*, miR-129-5p, miR-622, HS_202.1, miR654-3p, and miR-1254), of which miR-423-5p was the most promising candidate as a biomarker. This miRNA distinguished HF patients from healthy controls with an area under the receiver-operating-characteristic curve (AUC) of 0.91 and from dyspnea patients without HF with an AUC of 0.83. Furthermore, miR-423-5p levels were associated with disease severity as evidenced by the inverse correlation with EF and higher levels of miR-423-5p in patients with a higher New York Heart Association (NYHA) classification. The levels of miR-423-5p in plasma were also correlated to BNP [80]. The elevation of circulating miR-423-5p in HF

Table 11.2 Circulating miRNAs as biomarker for heart failure

Groups	miRNAs	Source	RNA isolation	miRNA detection	Normalization	Age/sex differences	Multivariate analysis	Refs.
39 healthy control; 20 HF cases; 30 non–HF cases with dyspnea	miR–423–5p, miR–18b*, miR–129–5p, miR–622, HS_202.1, miR–654–3p, miR–1254	Citrate plasma	mirVana PARIS kit	Sybr–green based	miR–1249	Controls >10 years younger	Age and sex	[80]
30 chronic HF; 30 healthy control	miR–423–5p, miR–320a, miR–22, miR–92b	Serum	Phenol: chloroform purification	Taqman probes	Based on mean Ct of all miR–NAs	Age, sex, ethnically matched	No	[81]
45 DCM; 39 control	miR–423–5p	EDTA plasma	mirVana PARIS kit	Taqman probes	Cel-miR–39 spike in	Age, sex matched	No	[82]
44 HF; 32 COPD; 15 healthy control; 59 other breathless	miR–103, miR–142–3p, miR–199a–3p, miR–23a, miR–27b, miR–324–5p, miR–342–3p	Plasma	miRNeasy kit	Sybr green	4 reference miRs	HF 5–10 years older	Age and sex	[83]
41 right ventricular HF; 10 healthy controls	miR–423–5p	Plasma	MasterPure kit	Taqman probes	Cel-miR–54 spike in	Age and sex matched	No	[84]

Continued

Table 11.2 Circulating miRNAs as biomarker for heart failure—cont'd

Groups	miRNAs	Source	RNA isolation	miRNA detection	Normalization	Age/sex differences	Multivariate analysis	Refs.
246 MI patients	miR-423-5p, miR-133a	Plasma	miRNeasy kit	Taqman probes	Cel-miR-39 spike in	NA	Age, sex, hypertension, DM, reperfusion therapy, peak creatine kinase, LVEF, HF during hospitalization	[85]
17 HF; 10 control	miR-29b, miR-133a, miR-423-5p	EDTA plasma	mirVana PARIS kit	Taqman probes	Cel-miR-39 spike in	HF >10 years older	No	[87]
75 HFpEF; 75 HFrEF; 75 no-HF	miR-30c, miR-146a, miR-221, miR-328, miR-375	Serum	miRNeasy kit	Taqman probes	None	No	Age, sex, BNP	[88]
30 HFpEF; 30 HFrEF; 30 control	miR-1233, miR-125a-5p, miR-183-3p, miR-190a, miR-193b-3p, miR-193b-5p, miR-211-5p, miR-494, miR-545-5p, miR-550a-5p, miR-638, miR-671-5p	EDTA plasma	Exiqon biofluids kit	Taqman probes	Cel-miR-39 spike in	No	No	[89]

20 stable HF; 22 decompensated HF; 15 control	miR-182	EDTA plasma	miRNeasy kit	Microarray	Array based	HF 5 years older	Age, gender, functional class, treatment	[93]
34 pediatric DCM	miR-646, miR-639, miR-155, miR-636	Plasma	Trizol LS and miRNeasy kit	Taqman probes	miR-16 and miR-573	No	No	[94]
53 HF; 39 control	miR-558, miR-122★, miR-520d-5p, miR-200b★, miR-622, miR-519e★, miR-1231, miR-1228★	Whole blood	miRNeasy kit	Sybrgreen based	RNA-U6	More male in HF	No	[90]
37 HF; 41 control	miR-548 family	PBMC	Trizol	Taqman probes	RNA-U6	No	No	[91]
19 healthy controls; 19 NIDCM; 15 ICM	miR-107, miR-139, miR-142-5p, miR-142-3p, miR-19b, miR-125b, miR-497	PBMC	Trizol	Taqman probes	miR-16	Unknown	No	[92]

HF, heart failure; DCM, dilated cardiomyopathy; COPD, chronic obstructive pulmonary disease; HFpEF, HF with preserved ejection fraction; HFrEF, HF with reduced ejection fraction; NIDCM, nonischemic DCM; ICM, ischemic cardiomyopathy; PBMC, peripheral blood mononuclear cell; NA, non applicable; DM, diabetes mellitus; LVEF, left ventricular ejection fraction.

was confirmed in several other patient cohorts. Goren et al. [81] determined the level of 186 serum miRNAs in 30 HF patients and 30 age, gender, and ethnically matched controls and found 26 miRNA with different levels, of which miR-423-5p showed the highest increase in HF patients. In this cohort, miR-423-5p was able to distinguish HF patients from controls with an AUC of 0.88, which could be increased to 0.90 by a multi-miRNA approach that also included three other miRNAs enriched in HF patients (miR-320a, miR-22, and miR-92b). Also in this study miR-423-5p levels were correlated to BNP, but no correlation to disease severity was detected [81]. Fan et al. [82] also detected elevated levels of miR-423-5p and a positive correlation with BNP in their cohort of 45 DCM patients and 39 age- and sex-matched controls. Ellis et al. [83] showed a trend toward higher levels in 44 HF patients compared to 15 healthy controls; however, they found similar increases of miR-423-5p in 91 non-HF-related dyspnea cases. By itself miR-423-5p was not a predictor of HF diagnosis, but it did significantly improve the diagnostic accuracy of BNP by 3.2% to 0.93. They also found seven other miRNAs that could predict HF diagnosis of which the AUC ranged from 0.62 to 0.67 [83].

MiR-423-5p seems to be specifically useful for left ventricular end-stage HF diagnosis and to be able to distinguish this end-stage HF from other forms of cardiac remodeling. The group of Thum, who investigated miR-423-5p levels in two other settings of cardiac remodeling showed that miR-423-5p was not useful in these forms of cardiac remodeling [84,85]. In the first setting they compared 41 patients with congenital transposition of the great arteries (aorta and pulmonary artery) where the systemic circulation is supported by the right ventricle to 10 age- and sex-matched healthy controls and showed no difference in miR-423-5p levels [84]. In a second report, Bauters et al. [85] measured miR-423-5p and miR-133a levels in plasma of 246 MI patients 1, 3, and 12 months after MI. In these patients they found an upregulation of miR-423-5p at all three time-points compared to baseline, however, no relation to cardiac function was detected indicating that miR-423-5p cannot be used as a prognostic biomarker for cardiac remodeling after MI [85]. The lack of a relation between miR-423-5p and cardiac remodeling in these two settings might be explained by the different nature of cardiac remodeling in the right and left ventricle and in cardiac remodeling within 1 year after MI versus end-stage HF as was used in the other studies. This might indicate that elevated miR-423-5p levels can be used as specific diagnostic biomarkers for end-stage HF.

One of the characteristics of an ideal biomarker is changing levels as disease severity changes in response to therapy. Dickinson et al. [86] showed that miR-423-5p is also able to fulfill this criterion. In this study they made use of the Dahl salt-sensitive rat model on a high salt diet and they treated these rats both with antimiR-208 and with ACEi. They have shown the efficacy of antimiR-208 in this model before as discussed earlier [47,86]. They performed a miRNA array on plasma of the antimiR-treated animals in comparison with the control animals on a high salt diet and detected 27 miRNAs decreased and 37 miRNAs increased in response to treatment, of which miR-423-5p was one of the most decreased miRNAs. Further validation by qPCR revealed six miRNAs (miR-423-5p, miR-16, miR-20b, miR-93, miR-106b, and miR-223) that were increased with disease progression as shown by a gradual increase in their levels in biweekly drawn samples of the rats on high salt diet up to 8 weeks after the start of the diet. Both antimiR-208a treatment and ACEi prevented the increase in plasma levels of all six miRNAs. Interestingly, only three miRNAs (miR-423-5p, miR-16, and miR-223) were significantly correlated to BNP levels and upregulation of MYH7 [86]. These data further indicate the potential of circulating miRNAs as diagnostic biomarkers for HF that represent disease progression and response to therapy.

Besides the elevation of circulating miR-423-5p levels in plasma of HF patients, this miRNA was also found to be upregulated in the human failing myocardium [38,80], which suggests that circulating miR-423-5p in HF is derived from the failing myocardium. Goldraich et al. [87] confirmed this suggestion by calculating transcoronary gradients. They sampled miRNA levels in the coronary sinus, which is the end of the coronary circulation and in the femoral artery and calculated the difference. They revealed a negative gradient of miR-423-5p in control subjects and a positive gradient in HF patients, which indicates that in HF patients miR-423-5p is released by the myocardium [87].

The earlier studies were aimed at distinguishing HF patients from non-HF patients. However, another challenge is to differentiate between HF with and without reduced EF, which is important as both patient might require different treatment. Watson et al. [88] studied five miRNAs (miR-375, miR-146a, miR-30c, miR-328, and miR-221), selected from an array, that had reduced levels in 150 HF patients compared to 75 no-HF samples, but with risk factors for HF. These miRNAs were outperformed by BNP in their ability to distinguish HF from no-HF, but addition of these miRNAs to BNP increased the AUC with 3%. However, in their ability to distinguish

75 HF patients with reduced EF from 75 patients with preserved EF, their performance was comparable to BNP and a combination of these miRNA with BNP resulted in an AUC of 0.86, which was a 20% increase compared to BNP alone [88]. Wong et al. [89] studied 12 miRNAs (Table 11.2) selected from an array in a cohort of 30 controls, 30 patients with and 30 patients without preserved EF. They found that several combinations of these miRNAs were able to distinguish these groups from each other in a comparable AUC as BNP alone, with some combinations with or without BNP even reaching perfect diagnostic performance [89].

A couple of groups decided not to study circulating miRNAs in plasma, but in other components of the blood. Vogel et al. [90] investigated miRNA profiles in 53 HF patients and 39 controls in whole blood, where the miRNA signature can be derived from the cellular and noncellular fraction. They performed an array and identified three miRNAs (miR-558, miR-122*, and miR-520d-5p) that could distinguish HF patients form non-HF patients with an AUC of 0.7, combination of these three miRNAs with five others increased this AUC to 0.81, which indicated the possible superior performance of a multimarker approach. Further analysis revealed that these miRNAs were correlated to EF and NYHA functional class. Interestingly, separation of groups on the basis of miR-519e* levels showed a clear difference in occurrence of events (hospitalization, heart transplantation, cardiovascular death) in these groups, which indicates a possible prognostic value of this miRNA [90]. Two other groups studied miRNA profiles in peripheral blood mononuclear cells (PBMCs). Gupta et al. [91] identified a downregulation of the miR-548 family in PBMCs of 37 HF patients compared to 41 controls and the least downregulated family member, miR-548c, was able to distinguish these groups with an AUC of 0.85. Voellenkle et al. [92] showed in a cohort of 19 control subjects, 19 subjects with nonischemic DCM, and 15 subjects with ICM that miR-107, miR-142-5p, and miR-139 were downregulated in both classes of HF, miR-125b and miR-497 in ICM only, and miR-142 and miR-29b were upregulated in nonischemic DCM only. That no clear overlap is detected between the whole blood, PBMC and plasma/serum samples suggest that the miRNAs detected in plasma are not released by the cellular fraction of the blood and that the results of the studies in plasma are not influenced by lysis of the cellular fraction during sample preparation.

Besides the possible use as diagnostic biomarkers, circulating miRNAs might also be used as prognostic biomarkers to predict which patients will be hospitalized or respond to a certain type of therapy. Cakmak et al. [93]

performed microarrays on plasma of 42 HF patients and analyzed the prognostic value of the miRNAs measured in this array after a follow-up of 6 months. They revealed that miR-182 was associated with cardiovascular mortality after 6 months and outperformed BNP in its prognostic value [93]. Unfortunately these results were not validated in another cohort by qRT-PCR. Miyamoto et al. [94] looked specifically in 34 pediatric patients (<18 years) with DCM, a population of which 15–35% show recovery of ventricular function while the others are in need for a cardiac transplantation. They identified four miRNAs (miR-155, miR-636, miR-646, and miR-639) in blood taken at the moment of transplant evaluation, that were able to distinguish children that recovered from those that received a transplant or died with an AUC of 0.875 [94]. That there is no clear overlap between the miRNAs identified in the diagnostic and prognostic studies suggests that eventually a multimarker approach is needed to cover all facets of HF.

5. CONCLUSIONS AND FUTURE DIRECTIONS

5.1 Challenges for MicroRNAs as Therapeutics

As discussed earlier, miRNAs might be used therapeutically to interfere with all possible aspects of cardiac remodeling to prevent or even reverse the development of HF. The field to develop miRNA-based therapeutics moves fast to overcome all possible hurdles to be used clinically. One of the main challenges will be to prevent the occurrence of side effects both in the targeted tissue and in other tissue. With regard to the targeted tissue, this problem might occur due to the fact that miRNAs target many mRNAs within a cell and although this might increase the effectiveness of the therapy as miRNAs tend to target multiple genes within one (deleterious) pathway, this might also lead to interference with other "beneficial" pathways. Currently, clinical trials are performed with systemically delivered miRNA inhibitors and mimics [12,27], and side effects do not seem to be a major problem, despite the fact that these substances interfere with miRNA levels in most organs where they have been taken up. In case of miRNA inhibitors, this might be explained by tissue-specific expression of the miRNA to be inhibited. However, many miRNAs are ubiquitously expressed, which might make the systemic delivery method less suitable. This indicates that further future research into chemistries of the systemic-delivered miRNA therapeutics is crucial. Future developments might enable us to deliver these chemistries to a specific tissue or cell-type of interest and thus reduce the chances of side effects in other tissues.

Minimalization of side effects by miRNA uptake in unrelated tissues can also be accomplished by the use of targeted delivery systems. However, in case of AAV9, this system relies on the delivery of genetically encoded overexpressors or inhibitors that will be present and active in nondividing cells such as cardiomyocytes for the rest of recipient's life. This might induce problems with regard to timing of the therapy. As for instance discussed earlier for inhibition of the miR-15 family in cardiac remodeling, this inhibition only seems beneficial immediately after the MI to inhibit cardiomyocyte apoptosis and induce their proliferation, but long-term inhibition might induce adverse cardiac remodeling leading to fibrosis [18,22,23]. This indicates that inhibition of miR-15 by targeted inhibition with genetically encoded inhibitors, could eventually result in a worsening of the cardiac remodeling because of an inability to withdraw the therapeutics. The chemistries used for systemic inhibition of miRNAs are very stable and also induce inhibition of miRNAs for a long time (up to 6 weeks) [18,47], however, they are eventually broken down, so methods to reduce their stability and half-life to shorten their time span of activity might be relatively easily to develop.

At the moment the role of miRNAs in cardiac remodeling is mainly studied in small animal models. Only for two miRNAs (miR-15 and miR-92a) the ability of LNA-based antimiRs to inhibit miRNA levels in the pig heart is shown and only for miR-92a the beneficial effect of this inhibition on ischemia/reperfusion is shown [18,58,59]. To be able to progress to testing miRNA therapeutics for cardiac remodeling in clinical trials, studies in large animal models are required to show both efficacy and safety.

5.2 Challenges for Circulating MicroRNA as Biomarker

Although earlier described results indicate that circulating miRNAs offer many attractive features of biomarkers, still several challenges remain before they can be clinically used as biomarkers.

First, most studies to date are performed in populations with less than 100 subjects. Although some candidate miRNAs are confirmed in more than one study, other studies questioned the use of these miRNAs in other settings of cardiac remodeling. This indicates that validation in larger patient cohorts is needed. In these large studies all aspects for a clinically useful biomarker should be investigated, for instance their diagnostic performance, prognostic performance, specificity for underlying etiology, disease state, and response to therapy. Furthermore, in larger cohorts the influence of confounding variables such as age, gender, ethnicity, and medication should be investigated.

Second, at the moment miRNAs are detected after RNA isolation and by sequence-specific amplification, which is a rather time-consuming method. With regard to HF, where results are usually not urgently needed this is not a problem. However, for instance after MI, where treatment has to be started as soon as possible this is a major disadvantage. Therefore, optimized methods to detect miRNAs for instance by direct amplification in the plasma without RNA isolation would improve their clinical utility. Furthermore, it is expected that a combination of multiple miRNAs may provide a better accuracy than measurement of a single miRNA, because it could cover many aspects of the disease. For instance miRNAs with a good diagnostic ability could be combined with miRNAs with a good prognostic ability. A combination of miRNAs could also detect all forms of cardiac remodeling independent of the etiology of the disease as for instance the ischemic and nonischemic forms of HF. Furthermore, a combination of miRNAs could ideally reflect the severity of the disease, which means that it could also allow us to detect cardiac hypertrophy before the heart is failing or the improvement in response to therapy. The ability of miRNAs to also detect cardiac hypertrophy is recently suggested by Derda et al. [95], who showed that the miR-29 family is both able to distinguish between obstructive and nonobstructive forms of hypertrophic cardiomyopathy and correlated to ventricular septum thickness. In summary, this indicates that we need a method of miRNA detection, which can be performed directly on plasma without the need to isolate RNA, and which allows simultaneous measurement of several miRNAs.

Third, a good normalization method is required. Because of the low amount of RNA in plasma and serum, concentration and quality measurements are virtually impossible, which means that variance in the amount of starting concentration and miRNA extraction might occur. At the moment no good "housekeeping" circulating miRNA has been identified. Many groups used different miRNAs as internal control, usually identified in their microarray as a stable miRNA. While the particular miRNA may have been stable in their study, this miRNA may change in other pathological conditions. Therefore this miRNA may not be suitable as an internal control in all studies or it might even have influenced the results in a validation cohort due to underlying comorbidities. Another widely used method is the addition of synthetic spike-in miRNAs, mainly *Caenorhabditis elegans* miRNAs without homology to any human miRNAs. Interestingly, Mitchell et al. [74] showed the importance of fully inactivated plasma RNA activity, as evidenced by the instability of spike-in miRNAs in crude plasma. Therefore, the moment of addition of these spike-ins is of vital importance to their usefulness. Another possible method for normalization is correcting for plasma volume, a method that is currently standardly used in the clinic for all biomarkers. All these possible solutions should be compared in future studies to identify the most reliable method of normalization.

Fourth, caution is needed with regard to the addition of the anticoagulant heparin to the blood. The addition of heparin is commonly used in cardiovascular diagnostics and interventions and thus many blood samples of cardiovascular patients contain heparin or traces thereof. However, heparin is shown to inhibit reverse transcriptase and DNA polymerase enzymes and thus to interfere with miRNA measurement by real-time PCR [96–98]. Heparinase treatment of heparin-contaminated samples could rescue the detection of miRNAs by real-time PCR [96,97]. Strikingly, some specific miRNAs seemed to be more sensitive to heparin contamination than others and also for these specific miRNAs, heparinase treatment showed differences in effectivity [98]. In general, samples contaminated with heparin can still be used to detect miRNAs after heparinase treatment; however, some miRNAs that are very sensitive to heparin might not be reliably detected in these samples. Therefore, one should test whether the proposed miRNA is sensitive to heparin inhibition before it can be used as biomarker, or other miRNA detection methods insensitive to heparin should be developed.

In summary, the identification of stable circulating miRNAs launched a new generation of potential biomarkers, which are also promising to be useful for diagnosis and prognosis of HF. The detection of these miRNAs still has to overcome a couple of hurdles before they can be used clinically, however, they are expected to show unrivaled specificity and sensitivity.

5.3 Other Noncoding RNAs

By the discovery of miRNAs in 1993, a completely new layer of gene regulation was identified and a new research field focusing on noncoding RNAs in disease was uncovered [11]. Since this discovery, it became clear that only ~2% of the genome is transcribed into protein and that the nonprotein coding DNA formerly regarded as "junk DNA" actually encodes numerous species of noncoding RNAs with several functions [99].

LncRNAs are noncoding RNA transcripts of >200 nucleotides that regulate gene expression and they are dynamically expressed in development and disease [99]. Interestingly, lncRNAs are poorly conserved between species, which suggests a role in changing complexity of species [99]. They are mostly localized in the nucleus and here they regulate gene expression at the epigenetic level, which means they induce, for example, chromatin modifications. By doing so, LncRNAs are involved in several biological functions, including X chromosome inactivation, imprinting, transcription initiation, splicing, and decoy of transcription factors and miRNAs [99]. An interesting lncRNA with regard to cardiac remodeling is Mhrt. This lncRNA is cardiomyocyte-specifically expressed from the antisense strand of the βMHC locus and downregulated in the remodeling heart [100]. Transgenic restoration of Mhrt in mice subjected to TAC surgery prevented hypertrophy and HF. Mhrt antagonizes the function of Brg1, a chromatin remodeling factor activated by stress that is responsible for chromatin remodeling of the MHC promoters and the induction of the α- to β-MHC switch [100].

A subgroup of lncRNAs is involved in the regulation of miRNA levels, these are termed competing endogenous (ce)RNAs and they function as decoy for miRNAs. These ceRNAs might be derived from pseudogenes that probably function as sponge for miRNAs to relieve the translational suppression of mRNA targets [101]. CircRNAs, RNA molecules with a circular shape, are a specialized type of ceRNAs. Some of these circRNAs are shown to be able to bind multiple copies of miRNAs and other circRNAs with binding sites for one specific miRNA or a couple of (functionally related) miRNAs are predicted to exist; however, their functionality still needs to be uncovered [102,103]. Both ceRNAs and circRNAs are dynamically expressed in development and disease [101,103], which implies the existence of an additional layer of RNA-based regulation of gene expression via the regulation of miRNA expression.

The lncRNAs are a relatively new field of research and we are only starting to unravel their role in general and in cardiac remodeling specifically. Therefore their therapeutic or biomarker abilities have not been studied yet. However, because of their broad range of functions and their ability to control gene expression, they provide an interesting source of therapeutic targets to be explored. This possible therapeutic use of lncRNAs would be a great achievement for this "junk DNA."

GLOSSARY

Heart failure A complex clinical syndrome that can result from any structural or functional cardiac disorder that impairs the ability of the ventricle to fill with or eject blood.

Cardiac remodeling Structural alterations of the myocardium underlying the development of heart failure.

Cardiomyocyte hypertrophy An increase in cardiomyocyte cell size and cardiac mass.

Cardiac fibrosis The excessive accumulation of extracellular matrix proteins in the interstitium and perivascular regions of the myocardium.

MicroRNA 22-nt long noncoding RNA that posttranscriptionally represses gene expression.

MicroRNA seed Nucleotide 2–7 of the 5′-side of the miRNA, which is the most important part of the miRNA for target recognition.

MicroRNA family Group of miRNAs that share the same seed sequence.

AntimiRs Chemically modified RNA molecules complementary to miRNAs that bind the miRNAs and inhibit their function.

Tiny LNA Specific form of antimiR that is complementary to the seed region only and therefore able to inhibit a full microRNA family.

MicroRNA mimic Synthetic duplex that resembles the miRNA duplex and can be converted toward a functional miRNA.

LNA RNA backbone modification used in antimiRs to increase the thermostability of the miRNA/antimiR heteroduplex.

Antagomir A microRNA inhibitor linked to cholesterol.

miRNA target Endogenous mRNA regulated by a specific miRNA by binding to its 3′-UTR.

Target site blockers Synthetic molecules that bind the miRNA binding site within a 3′-UTR and prevent binding and function of the endogenous miRNA.

Ejection fraction The percentage of blood that is pumped from the heart with each heartbeat.

Nanoparticles Particles ranging in size between 20 and 200 nm, where the negatively charged siRNAs/mimics can be entrapped in the positive liposomes.

Adeno-associated virus Virus particles that can be used to deliver vectors for gene therapy.

miRNA sponge miRNA inhibitor that contains tandem repeats of miRNA binding sites.

Microbubbles Bubbles originally used as contrast agents in diagnostic ultrasound, which can be loaded with antimiRs/mimics and forced to release their content by ultrasound.

TAC Binding of the thoracic aorta to increase the afterload on the heart, which induces cardiac remodeling.

Myocardial infarction Arrest of blood flow to the myocardium leading to hypoxia and cardiomyocyte death.

Ischemia reperfusion Infarction model, where blood flow to a part of the heart is arrested for a short time frame.

Biomarker A measurable indicator of some biological state or condition.

Spike-in Synthetic RNA molecules added to plasma during RNA isolation for normalization purposes.

LncRNA Noncoding RNA transcripts of >200 nucleotides that regulate gene expression.

LIST OF ACRONYMS AND ABBREVIATIONS

3′-UTR 3′-Untranslated region
AAV Adeno-associated virus
ACEi Angiotensin-converting enzyme inhibitor
AS Aorta stenosis
AUC Area under the receiver-operating-characteristic curve

AV-block Atrioventricular block
BNP Brain natriuretic peptide
ceRNA Competing endogenous RNA
DCM Dilated cardiomyopathy
ECG Electrocardiogram
ECM Extracellular matrix
EF Ejection Fraction
HF Heart Failure
ICD Implantable cardiac defibrillator
ICM Ischemic cardiomyopathy
KO Knockout
LNA Locked nucleic acid
LVAD Left ventricular assist device
MI Myocardial Infarction
miRNA MicroRNA
NYHA New York Heart Association
PBMC Peripheral blood mononuclear cells
PEG Polyethylene glycol
RISC RNA-induced silencing complex
SR Sarcoplasmatic reticulum
T_m Melting temperature
TAC Thoracic aorta constriction

REFERENCES

[1] Yancy CW, Jessup M, Bozkurt B, Butler J, Casey DE, Drazner MH, et al. ACCF/AHA guideline for the management of heart failure: a report of the American College of Cardiology Foundation/American heart Association Task Force on Practice Guidelines. J Am Coll Cardiol 2013;62(16):e147–239.

[2] Braunwald E. The war against heart failure: the Lancet lecture. Lancet 2015;385(9970):812–24.

[3] Lorell BH, Carabello BA. Left ventricular hypertrophy: pathogenesis, detection, and prognosis. Circulation 2000;102(4):470–9.

[4] Swynghedauw B. Molecular mechanisms of myocardial remodeling. Physiol Rev January 1999;79(1):215–62.

[5] De Boer RA, Pinto YM, Van Veldhuisen DJ. The imbalance between oxygen demand and supply as a potential mechanism in the pathophysiology of heart failure: the role of microvascular growth and abnormalities. Microcirculation 2003;10(2):113–26.

[6] Mancini D, Colombo PC. Left ventricular assist devices: a rapidly evolving alternative to transplant. J Am Coll Cardiol 2015;65(23):2542–55.

[7] Tijsen AJ, Pinto YM, Creemers EE. Non-cardiomyocyte microRNAs in heart failure. Cardiovasc Res 2012;93(4):573–82.

[8] Berezikov E, Guryev V, van de Belt J, Wienholds E, Plasterk RHA, Cuppen E. Phylogenetic shadowing and computational identification of human microRNA genes. Cell 2005;120(1):21–4.

[9] Bartel DP. MicroRNAs: genomics, biogenesis, mechanism, and function. Cell 2004;116(2):281–97.

[10] Pasquinelli AE. MicroRNAs and their targets: recognition, regulation and an emerging reciprocal relationship. Nat Rev Genet 2012;13(4):271–82.

[11] Lee RC, Feinbaum RL, Ambros V. The *C. elegans* heterochronic gene lin-4 encodes small RNAs with antisense complementarity to lin-14. Cell 1993;75(5):843–54.

[12] Janssen HLA, Reesink HW, Lawitz EJ, Zeuzem S, Rodriguez-Torres M, Patel K, et al. Treatment of HCV infection by targeting microRNA. N Engl J Med 2013;368(18):1685–94.

[13] Kumar R, Singh SK, Koshkin AA, Rajwanshi VK, Meldgaard M, Wengel J. The first analogues of LNA (locked nucleic acids): phosphorothioate-LNA and 2′-thio-LNA. Bioorg Med Chem Lett 1998;8(16):2219–22.

[14] Geary RS, Yu RZ, Levin AA. Pharmacokinetics of phosphorothioate antisense oligo-deoxynucleotides. Curr Opin Investig Drugs 2001;2(4):562–73.

[15] Krützfeldt J, Rajewsky N, Braich R, Rajeev KG, Tuschl T, Manoharan M, et al. Silencing of microRNAs *in vivo* with "antagomirs". Nature 2005;438(7068):685–9.

[16] Krützfeldt J, Kuwajima S, Braich R, Rajeev KG, Pena J, Tuschl T, et al. Specificity, duplex degradation and subcellular localization of antagomirs. Nucleic Acids Res 2007;35(9):2885–92.

[17] Obad S, dos Santos CO, Petri A, Heidenblad M, Broom O, Ruse C, et al. Silencing of microRNA families by seed-targeting tiny LNAs. Nat Genet 2011;43(4):371–8.

[18] Hullinger TG, Montgomery RL, Seto AG, Dickinson BA, Semus HM, Lynch JM, et al. Inhibition of miR-15 protects against cardiac ischemic injury. Circ Res 2012;110(1):71–81.

[19] Thum T, Gross C, Fiedler J, Fischer T, Kissler S, Bussen M, et al. MicroRNA-21 contributes to myocardial disease by stimulating MAP kinase signalling in fibroblasts. Nature 2008;456(7224):980–4.

[20] van der Ree MH, van der Meer AJ, de Bruijne J, Maan R, van Vliet A, Welzel TM, et al. Long-term safety and efficacy of microRNA-targeted therapy in chronic hepatitis C patients. Antiviral Res 2014;111:53–9.

[21] Van Rooij E, Olson EN. MicroRNA therapeutics for cardiovascular disease: opportunities and obstacles. Nat Rev Drug Discov 2012;11(11):860–72.

[22] Porrello ER, Mahmoud AI, Simpson E, Johnson BA, Grinsfelder D, Canseco D, et al. Regulation of neonatal and adult mammalian heart regeneration by the miR-15 family. Proc Natl Acad Sci USA 2013;110(1):187–92.

[23] Tijsen AJ, van der Made I, van den Hoogenhof MM, Wijnen WJ, van Deel ED, de Groot NE, et al. The microRNA-15 family inhibits the TGFβ-pathway in the heart. Cardiovasc Res 2014;104(1):61–71.

[24] Wynendaele J, Böhnke A, Leucci E, Nielsen SJ, Lambertz I, Hammer S, et al. An illegitimate microRNA target site within the 3′ UTR of MDM4 affects ovarian cancer progression and chemosensitivity. Cancer Res 2010;70(23):9641–9.

[25] Montgomery RL, Yu G, Latimer PA, Stack C, Robinson K, Dalby CM, et al. MicroRNA mimicry blocks pulmonary fibrosis. EMBO Mol Med October 2014;6(10):1347–56.

[26] Kanasty R, Dorkin JR, Vegas A, Anderson D. Delivery materials for siRNA therapeutics. Nat Mater 2013;12(11):967–77.

[27] Bader AG. miR-34-a microRNA replacement therapy is headed to the clinic. Front Genet 2012;3:120.

[28] Asokan A, Samulski RJ. An emerging adeno-associated viral vector pipeline for cardiac gene therapy. Hum Gene Ther 2013;24(11):906–13.

[29] Jaski BE, Jessup ML, Mancini DM, Cappola TP, Pauly DF, Greenberg B, et al. Calcium upregulation by percutaneous administration of gene therapy in cardiac disease (CUPID Trial), a first-in-human phase 1/2 clinical trial. J Card Fail 2009;15(3):171–81.

[30] Jessup M, Greenberg B, Mancini D, Cappola T, Pauly DF, Jaski B, et al. Calcium Upregulation by Percutaneous Administration of Gene Therapy in Cardiac Disease (CUPID): a phase 2 trial of intracoronary gene therapy of sarcoplasmic reticulum Ca^{2+}-ATPase in patients with advanced heart failure. Circulation 2011;124(3):304–13.

[31] Ebert MS, Neilson JR, Sharp PA. MicroRNA sponges: competitive inhibitors of small RNAs in mammalian cells. Nat Methods 2007;4(9):721–6. [z&rendertype=abstract].

[32] Ebert MS, Sharp PA. MicroRNA sponges: progress and possibilities. RNA November 2010;16(11):2043–50.

[33] Unger E, Porter T, Lindner J, Grayburn P. Cardiovascular drug delivery with ultrasound and microbubbles. Adv Drug Deliv Rev 2014;72:110–26.

[34] Chen JF, Murchison EP, Tang R, Callis TE, Tatsuguchi M, Deng Z, et al. Targeted deletion of Dicer in the heart leads to dilated cardiomyopathy and heart failure. Proc Natl Acad Sci USA 2008;105(6):2111–6.

[35] Rao PK, Toyama Y, Chiang HR, Gupta S, Bauer M, Medvid R, et al. Loss of cardiac microRNA-mediated regulation leads to dilated cardiomyopathy and heart failure. Circ Res 2009;105(6):585–94.

[36] Da Costa Martins PA, Bourajjaj M, Gladka M, Kortland M, van Oort RJ, Pinto YM, et al. Conditional dicer gene deletion in the postnatal myocardium provokes spontaneous cardiac remodeling. Circulation 2008;118(15):1567–76.

[37] Van Rooij E, Sutherland LB, Liu N, Williams AH, McAnally J, Gerard RD, et al. A signature pattern of stress-responsive microRNAs that can evoke cardiac hypertrophy and heart failure. Proc Natl Acad Sci USA 2006;103(48):18255–60.

[38] Thum T, Galuppo P, Wolf C, Fiedler J, Kneitz S, van Laake LW, et al. MicroRNAs in the human heart: a clue to fetal gene reprogramming in heart failure. Circulation 2007;116(3):258–67.

[39] Ikeda S, Kong SW, Lu J, Bisping E, Zhang H, Allen PD, et al. Altered microRNA expression in human heart disease. Physiol Genomics 2007;31(3):367–73.

[40] Tatsuguchi M, Seok HY, Callis TE, Thomson JM, Chen JF, Newman M, et al. Expression of microRNAs is dynamically regulated during cardiomyocyte hypertrophy. J Mol Cell Cardiol 2007;42(6):1137–41.

[41] Cheng Y, Ji R, Yue J, Yang J, Liu X, Chen H, et al. MicroRNAs are aberrantly expressed in hypertrophic heart: do they play a role in cardiac hypertrophy? Am J Pathol 2007;170(6):1831–40.

[42] Elzenaar I, Pinto YM, van Oort RJ. MicroRNAs in heart failure: new targets in disease management. Clin Pharmacol Ther 2013;94(4):480–9.

[43] Da Costa Martins PA, De Windt LJ. MicroRNAs in control of cardiac hypertrophy. Cardiovasc Res 2012;93(4):563–72.

[44] Van Rooij E, Sutherland LB, Qi X, Richardson JA, Hill J, Olson EN. Control of stress-dependent cardiac growth and gene expression by a microRNA. Science 2007;316(5824):575–9.

[45] Van Rooij E, Quiat D, Johnson BA, Sutherland LB, Qi X, Richardson JA, et al. A family of microRNAs encoded by myosin genes governs myosin expression and muscle performance. Dev Cell 2009;17(5):662–73.

[46] Callis TE, Pandya K, Seok HY, Tang RH, Tatsuguchi M, Huang ZP, et al. MicroRNA-208a is a regulator of cardiac hypertrophy and conduction in mice. J Clin Invest 2009;119(9):2772–86.

[47] Montgomery RL, Hullinger TG, Semus HM, Dickinson BA, Seto AG, Lynch JM, et al. Therapeutic inhibition of miR-208a improves cardiac function and survival during heart failure. Circulation 2011;124(14):1537–47.

[48] Grueter CE, van Rooij E, Johnson BA, DeLeon SM, Sutherland LB, Qi X, et al. A cardiac microRNA governs systemic energy homeostasis by regulation of MED13. Cell 2012;149(3):671–83.

[49] Wahlquist C, Jeong D, Rojas-Muñoz A, Kho C, Lee A, Mitsuyama S, et al. Inhibition of miR-25 improves cardiac contractility in the failing heart. Nature 2014;508(7497):531–5.

[50] Dirkx E, Gladka MM, Philippen LE, Armand A-S, Kinet V, Leptidis S, et al. Nfat and miR-25 cooperate to reactivate the transcription factor Hand2 in heart failure. Nat Cell Biol November 2013;15(11):1282–93.

[51] Creemers EE, Pinto YM. Molecular mechanisms that control interstitial fibrosis in the pressure-overloaded heart. Cardiovasc Res 2011;89(2):265–72.

[52] Souders CA, Bowers SLK, Baudino TA. Cardiac fibroblast: the renaissance cell. Circ Res 2009;105(12):1164–76.

[53] van Rooij E, Sutherland LB, Thatcher JE, DiMaio JM, Naseem RH, Marshall WS, et al. Dysregulation of microRNAs after myocardial infarction reveals a role of miR-29 in cardiac fibrosis. Proc Natl Acad Sci USA 2008;105(35):13027–32.

[54] Abonnenc M, Nabeebaccus AA, Mayr U, Barallobre-Barreiro J, Dong X, Cuello F, et al. Extracellular matrix secretion by cardiac fibroblasts: role of microRNA-29b and microRNA-30c. Circ Res 2013;113(10):1138–47.

[55] Zhang Y, Huang XR, Wei LH, Chung AC, Yu CM, Lan HY. miR-29b as a therapeutic agent for angiotensin II-induced cardiac fibrosis by targeting TGF-β/Smad3 signaling. Mol Ther May 2014;22(5):974–85.

[56] Izumiya Y, Shiojima I, Sato K, Sawyer DB, Colucci WS, Walsh K. Vascular endothelial growth factor blockade promotes the transition from compensatory cardiac hypertrophy to failure in response to pressure overload. Hypertension 2006;47(5):887–93.

[57] Bonauer A, Carmona G, Iwasaki M, Mione M, Koyanagi M, Fischer A, et al. MicroRNA-92a controls angiogenesis and functional recovery of ischemic tissues in mice. Science 2009;324(5935):1710–3.

[58] Hinkel R, Penzkofer D, Zühlke S, Fischer A, Husada W, Xu QF, et al. Inhibition of microRNA-92a protects against ischemia/reperfusion injury in a large-animal model. Circulation 2013;128(10):1066–75.

[59] Bellera N, Barba I, Rodriguez-Sinovas A, Ferret E, Asín MA, Gonzalez-Alujas MT, et al. Single intracoronary injection of encapsulated antagomir-92a promotes angiogenesis and prevents adverse infarct remodeling. J Am Heart Assoc 2014;3(5):e000946.

[60] Devaux B, Scholz D, Hirche A, Klövekorn WP, Schaper J. Upregulation of cell adhesion molecules and the presence of low grade inflammation in human chronic heart failure. Eur Heart J 1997;18(3):470–9.

[61] Testa M, Yeh M, Lee P, Fanelli R, Loperfido F, Berman JW, et al. Circulating levels of cytokines and their endogenous modulators in patients with mild to severe congestive heart failure due to coronary artery disease or hypertension. J Am Coll Cardiol 1996;28(4):964–71.

[62] Heymans S, Corsten MF, Verhesen W, Carai P, van Leeuwen REW, Custers K, et al. Macrophage microRNA-155 promotes cardiac hypertrophy and failure. Circulation 2013;128(13):1420–32.

[63] Seok HY, Chen J, Kataoka M, Huang ZP, Ding J, Yan J, et al. Loss of MicroRNA-155 protects the heart from pathological cardiac hypertrophy. Circ Res 2014;114(10): 1585–95.

[64] Yue L, Xie J, Nattel S. Molecular determinants of cardiac fibroblast electrical function and therapeutic implications for atrial fibrillation. Cardiovasc Res 2011;89(4):744–53.

[65] Vrtovec B, Knezevic I, Poglajen G, Sebestjen M, Okrajsek R, Haddad F. Relation of B-type natriuretic peptide level in heart failure to sudden cardiac death in patients with and without QT interval prolongation. Am J Cardiol 2013;111(6):886–90.

[66] Davey PP, Barlow C, Hart G. Prolongation of the QT interval in heart failure occurs at low but not at high heart rates. Clin Sci (Lond) May 2000;98(5):603–10.

[67] Zhao Y, Ransom JF, Li A, Vedantham V, von Drehle M, Muth AN, et al. Dysregulation of cardiogenesis, cardiac conduction, and cell cycle in mice lacking miRNA-1-2. Cell 2007;129(2):303–17.

[68] Besser J, Malan D, Wystub K, Bachmann A, Wietelmann A, Sasse P, et al. MiRNA-1/133a clusters regulate adrenergic control of cardiac repolarization. PLoS ONE 2014;9(11):e113449.

[69] Terentyev D, Belevych AE, Terentyeva R, Martin MM, Malana GE, Kuhn DE, et al. miR-1 overexpression enhances Ca(2+) release and promotes cardiac arrhythmogenesis by targeting PP2A regulatory subunit B56alpha and causing CaMKII-dependent hyperphosphorylation of RyR2. Circ Res 2009;104(4):514–21.

[70] Karakikes I, Chaanine AH, Kang S, Mukete BN, Jeong D, Zhang S, et al. Therapeutic cardiac-targeted delivery of miR-1 reverses pressure overload-induced cardiac hypertrophy and attenuates pathological remodeling. J Am Heart Assoc 2013;2(2):e000078.

[71] Ai J, Zhang R, Gao X, Niu HF, Wang N, Xu Y, et al. Overexpression of microRNA-1 impairs cardiac contractile function by damaging sarcomere assembly. Cardiovasc Res 2012;95(3):385–93.

[72] Porrello ER, Johnson BA, Aurora AB, Simpson E, Nam Y-J, Matkovich SJ, et al. MiR-15 family regulates postnatal mitotic arrest of cardiomyocytes. Circ Res 2011;109(6):670–9.

[73] Van Kimmenade RRJ, Pinto YM, Januzzi JL. Importance and interpretation of intermediate (gray zone) amino-terminal pro–B-type natriuretic peptide concentrations. Am J Cardiol 2008;101(Suppl.):39A–42A.

[74] Mitchell PS, Parkin RK, Kroh EM, Fritz BR, Wyman SK, Pogosova-Agadjanyan EL, et al. Circulating microRNAs as stable blood-based markers for cancer detection. Proc Natl Acad Sci USA 2008;105(30):10513–8.

[75] Chen X, Ba Y, Ma L, Cai X, Yin Y, Wang K, et al. Characterization of microRNAs in serum: a novel class of biomarkers for diagnosis of cancer and other diseases. Cell Res October 2008;18(10):997–1006.

[76] Weber JA, Baxter DH, Zhang S, Huang DY, Huang KH, Lee MJ, et al. The microRNA spectrum in 12 body fluids. Clin Chem November 2010;56(11):1733–41.

[77] Creemers EE, Tijsen AJ, Pinto YM. Circulating microRNAs: novel biomarkers and extracellular communicators in cardiovascular disease? Circ Res March 3, 2012;110 (3):483–95.

[78] Rayner KJ, Hennessy EJ. Extracellular communication via microRNA: lipid particles have a new message. J Lipid Res 2013;54(5):1174–81.

[79] Fichtlscherer S, Zeiher AM, Dimmeler S. Circulating microRNAs: biomarkers or mediators of cardiovascular diseases? Arterioscler Thromb Vasc Biol 2011;31(11):2383–90.

[80] Tijsen AJ, Creemers EE, Moerland PD, de Windt LJ, van der Wal AC, Kok WE, et al. MiR423-5p as a circulating biomarker for heart failure. Circ Res 2010;106(6): 1035–9.

[81] Goren Y, Kushnir M, Zafrir B, Tabak S, Lewis BS, Amir O. Serum levels of microRNAs in patients with heart failure. Eur J Heart Fail 2012;14(2):147–54.

[82] Fan KL, Zhang HF, Shen J, Zhang Q, Li XL. Circulating microRNAs levels in Chinese heart failure patients caused by dilated cardiomyopathy. Indian Heart J 2013;65(1):12–6.

[83] Ellis KL, Cameron VA, Troughton RW, Frampton CM, Ellmers LJ, Richards AM. Circulating microRNAs as candidate markers to distinguish heart failure in breathless patients. Eur J Heart Fail 2013;15(10):1138–47.

[84] Tutarel O, Dangwal S, Bretthauer J, Westhoff-Bleck M, Roentgen P, Anker SD, et al. Circulating miR-423_5p fails as a biomarker for systemic ventricular function in adults after atrial repair for transposition of the great arteries. Int J Cardiol 2013;167(1):63–6.

[85] Bauters C, Kumarswamy R, Holzmann A, Bretthauer J, Anker SD, Pinet F, et al. Circulating miR-133a and miR-423-5p fail as biomarkers for left ventricular remodeling after myocardial infarction. Int J Cardiol 2013;168(3):1837–40.

[86] Dickinson BA, Semus HM, Montgomery RL, Stack C, Latimer PA, Lewton SM, et al. Plasma microRNAs serve as biomarkers of therapeutic efficacy and disease progression in hypertension-induced heart failure. Eur J Heart Fail June 2013;15(6):650–9.

[87] Goldraich LA, Martinelli NC, Matte U, Cohen C, Andrades M, Pimentel M, et al. Transcoronary gradient of plasma microRNA 423-5p in heart failure: evidence of altered myocardial expression. Biomarkers 2014;19(2):135–41.

[88] Watson CJ, Gupta SK, O'Connell E, Thum S, Glezeva N, Fendrich J, et al. MicroRNA signatures differentiate preserved from reduced ejection fraction heart failure. Eur J Heart Fail 2015;17(4):405–15.

[89] Wong LL, Armugam A, Sepramaniam S, Karolina DS, Lim KY, Lim JY, et al. Circulating microRNAs in heart failure with reduced and preserved left ventricular ejection fraction. Eur J Heart Fail 2015;17(4):393–404.

[90] Vogel B, Keller A, Frese KS, Leidinger P, Sedaghat-Hamedani F, Kayvanpour E, et al. Multivariate miRNA signatures as biomarkers for non-ischaemic systolic heart failure. Eur Heart J 2013;34(36):2812–22.

[91] Gupta MK, Halley C, Duan Z-H, Lappe J, Viterna J, Jana S, et al. miRNA-548c: a specific signature in circulating PBMCs from dilated cardiomyopathy patients. J Mol Cell Cardiol September 2013;62:131–41.

[92] Voellenkle C, van Rooij J, Cappuzzello C, Greco S, Arcelli D, Di Vito L, et al. MicroRNA signatures in peripheral blood mononuclear cells of chronic heart failure patients. Physiol Genomics 2010;42(3):420–6.

[93] Cakmak HA, Coskunpinar E, Ikitimur B, Barman HA, Karadag B, Tiryakioglu NO, et al. The prognostic value of circulating microRNAs in heart failure: preliminary results from a genome-wide expression study. J Cardiovasc Med 2015;16(6):431–7.

[94] Miyamoto SD, Karimpour-Fard A, Peterson V, Auerbach SR, Stenmark KR, Stauffer BL, et al. Circulating microRNA as a biomarker for recovery in pediatric dilated cardiomyopathy. J Heart Lung Transpl 2015;34(5):724–33.

[95] Derda AA, Thum S, Lorenzen JM, Bavendiek U, Heineke J, Keyser B, et al. Blood-based microRNA signatures differentiate various forms of cardiac hypertrophy. Int J Cardiol 2015;196:115–22.

[96] Taylor AC. Titration of heparinase for removal of the PCR-inhibitory effect of heparin in DNA samples. Mol Ecol 1997;6(4):383–5.

[97] Johnson ML, Navanukraw C, Grazul-Bilska AT, Reynolds LP, Redmer DA. Heparinase treatment of RNA before quantitative real-time RT-PCR. Biotechniques 2003;35(6):1140–2.

[98] Boeckel J-N, Thomé CE, Leistner D, Zeiher AM, Fichtlscherer S, Dimmeler S. Heparin selectively affects the quantification of microRNAs in human blood samples. Clin Chem 2013;59(7):1125–7.

[99] Devaux Y, Zangrando J, Schroen B, Creemers EE, Pedrazzini T, Chang CP, et al. Long noncoding RNAs in cardiac development and ageing. Nat Rev Cardiol 2015;12(7).

[100] Han P, Li W, Lin C-H, Yang J, Shang C, Nurnberg ST, et al. A long noncoding RNA protects the heart from pathological hypertrophy. Nature 2014;514(7520):102–6.

[101] Salmena L, Poliseno L, Tay Y, Kats L, Pandolfi PP. A ceRNA hypothesis: the Rosetta Stone of a hidden RNA language? Cell 2011;146(3):353–8.

[102] Hansen TB, Jensen TI, Clausen BH, Bramsen JB, Finsen B, Damgaard CK, et al. Natural RNA circles function as efficient microRNA sponges. Nature 2013;495(7441):384–8.

[103] Memczak S, Jens M, Elefsinioti A, Torti F, Krueger J, Rybak A, et al. Circular RNAs are a large class of animal RNAs with regulatory potency. Nature 2013;495(7441):333–8.

CHAPTER 12

The Role of Noncoding Rnas in Prostate Cancer

T. Hasegawa, H. Lewis, A. Esquela-Kerscher

Key Concepts

Prostate cancer is the most common cause of cancer in men and the second leading cause of male cancer-related death after lung cancer in the United States. No accurate biomarkers exist that can discriminate for aggressive and metastatic forms of prostate cancer and therapeutic options for this disease are limited.

MicroRNAs (miRNAs) are important negative regulators of gene expression that are often misexpressed in tissues and fluids from prostate cancer patients and closely correlate with disease progression. These small noncoding RNAs hold great promise as novel diagnostic and therapeutic tools for prostate cancer.

Accumulating evidence using human prostate cancer cell lines and mouse models has identified functional roles for specific miRNAs in the prostate that influence tumor progression, metastasis, and androgen signaling. A complete understanding of how miRNA networks regulate protein-coding tumor suppressor genes and oncogenic factors as well as other noncoding RNAs in the prostate will be needed to translate this work into the clinic.

Considerable effort has focused on the development of miRNAs as diagnostic tools that can identify patients earlier during the course of their disease who are at highest risk to develop aggressive and lethal forms of prostate cancer. Fluid-based tests that detect circulating miRNA biomarkers, including those packaged in membrane-bound exosomes, would allow physician to screen patient populations noninvasively to avoid repeat biopsies and recommend personalized treatment plans.

1. INTRODUCTION

Among men, prostate cancer is the most common cause of cancer and the second leading cause of cancer-related death after lung cancer in the United States [1]. An estimated 220,800 new cases of prostate cancer were diagnosed, and 27,540 patients died from this disease in the United State in 2015 (NCI: http://www.cancer.gov.cancertopics/types/prostate). Currently, no biomarkers exist that can differentiate between indolent and advanced forms of prostate cancer and detect men with the highest risk to

Translating MicroRNAs to the Clinic
ISBN 978-0-12-800553-8

seek aggressive treatment earlier during the course of their disease. Thera-peutic options for this disease are limited and metastatic prostate cancer remains incurable and lethal. miRNAs have recently emerged as promising diagnostic and therapeutic targets for prostate cancer. Evidence reviewed in this chapter discusses how these noncoding RNAs play multiple roles dur-ing cancer progression in the prostate. These findings could soon be trans-lated into the clinic to develop better diagnostic tools and therapeutic drugs for advanced prostate cancer.

miRNAs are endogenous single-stranded 19–25 nucleotide noncoding RNAs found throughout the animal kingdom that act to negatively regu-late gene expression posttranscriptionally. These small RNAs function by binding in a sequence-dependent manner to their target messenger RNAs (mRNAs) in association with the miRISC complex (miRNA-containing RNA-induced silencing complex), resulting in the repression of protein translation and/or mRNA degradation of the target. miRNAs in animal systems bind to their targets with incomplete complementarity. Bioinfor-matic studies suggest that a single miRNA can regulate upward of 100 targets, indicating that these small RNAs can modulate unrelated cellular pathways simultaneously. Multiple miRNAs can associate with a single mRNA target and therefore miRNA-based regulation of gene expression is extremely complex. To date, there are over 2500 miRNAs isolated in the human genome (www.mirbase.org, release 21), although the large major-ity of these small RNAs await functional characterization. The subset of miRNAs that have been well-studied are involved in many essential bio-logical processes in human cells during embryogenesis and in adulthood related to proliferation, differentiation, metabolism, apoptosis, and the immune response [2–4].

Since the initial discovery of miRNAs almost 20 years ago, research using in vitro cell culture systems and animal disease models indicates that aberrant miRNA expression can directly impact tumor initiation and disease progression, epithelia to mesenchymal (EMT) transitions, invasion, metastasis, and drug resistance by targeting a large range of cancer-related genes [4–6]. In this way, patterns of miRNA expression detected in diseased tissue and fluids from cancer patients compared to normal control popula-tions likely reflect abnormalities in mechanistic pathways related to cancer progression, patient survival, and response to clinical therapies. These miRNA expression signatures are being developed as diagnostic and prog-nostic biomarkers for detection of early stage cancer as well as discrimina-tory markers to identify patients at highest risk to advance to aggressive and lethal forms of cancer and who would benefit most from therapeutic and/ or surgical interventions. Functional characterization of individual miR-NAs in recent years has identified subclasses of these small RNAs that act as tumor suppressor genes to repress malignant growth by targeting factors such as MYC and RAS, or conversely act as oncogenes that accelerate tumor progression by blocking factors such as p53 and PTEN. Furthermore, miRNAs have also been identified to influence metastatic events in both negative and positive ways. This chapter will specifically focus on the role miRNAs play during prostate cancer and summarize how they are involved in tumor formation and disease progression in this tissue. This section will also discuss exciting developments in using miRNAs as novel diagnostic and therapeutic targets in the clinic for prostate cancer.

2. PROSTATE CANCER AND MICRORNAs
2.1 Prostate Cancer Is a Major Health Problem

Prostate cancer is the most common nonskin malignancy in men and the second leading cause of cancer-related deaths in US men. Prostate cancer is classified into four stages, ranging from small tumors confined in the prostate (Stage I) to metastatic disease (Stage IV) (Fig. 12.1). Factors that put men at a higher risk for prostate cancer include advanced age (men aged 55–75 years), family history of the disease, and being of African-American descent [7]. A biopsy is the only way a patient can be diagnosed with prostate cancer. Once diagnosed, a physician stages the patient's cancer based on prostate-specific antigen (PSA) serum levels (described later), digital rectal examination (DRE) results, and Gleason scoring of the biopsy to aid in prognosis and therapy selection. Patients diagnosed with prostate cancer can be divided into two groups: low-risk, indolent disease and high-risk, aggressive disease. Indolent (or insignificant) disease is defined as prostate tumors of low Gleason score (a grade of six or less) and/or small size that remain organ-confined and slow growing. Indolent disease, if left untreated, would have little impact on the patient's health or quality of life and the patient would likely die of other causes unrelated to the prostate cancer, ie, heart disease or old age. Approximately 20% of men diagnosed with prostate cancer will progress to more advanced disease, in which tumor cells in the prostate are fast growing. Men with advanced disease are at higher risk for their cancer to become invasive and spread outside of the prostate and metastasize to other regions of the body. The metastatic, aggressive form of the disease is currently incurable and lethal.

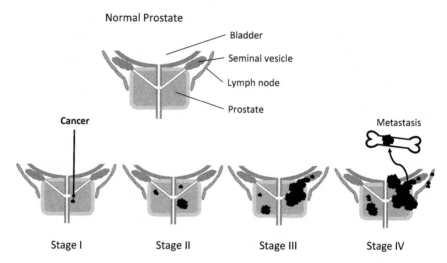

Figure 12.1 *Stages of prostate cancer.* Clinicians describe the extent of prostate cancer by staging where the prostate tumors are located and the extent of the cancer. Stage 1: Prostate tumors are confined to one lobe of the prostate. Stage II: Prostate tumors have increased in size/number and spread throughout the prostate. Stage III: Cancer has grown past the outer layer of the prostate gland and potentially spread to neighboring seminal vesicles. Stage IV: Cancer has metastasized to the lymph nodes and distant organs.

2.1.1 Treatment Options for Prostate Cancer Is Limited

Treatment for prostate cancer usually depends on the extent that malignant cells have spread throughout the body. For patients considered at lowest risk for prostate cancer, due to a low Gleason score and/or small tumor size, watchful waiting or active surveillance is encouraged. During watchful waiting and active surveillance, a patient undergoes frequent exams and biopsies, however, they do not receive treatment unless there is an indication that the tumor is becoming more aggressive. If the degree of cancer is more significant (Gleason grade of eight or more) but is still confined within the prostate, therapies for this form of the disease commonly include surgery (prostatectomy or pelvic lymphadenectomy), hormone therapy (androgen deprivation, since prostate cells rely on androgens for growth), ion or proton radiation (brachytherapy), and chemotherapy. For patients diagnosed with high-risk localized prostate cancer, multimodal treatment provides a higher likelihood of disease-free survival in comparison to monotherapy (eg, brachytherapy or surgery alone) and also decreases the incidence of biochemical recurrence [8]. However, if the cancer has spread outside of the prostate and has metastasized to distant sites such as bone, there is no cure for the disease [9]. In these instances, palliative care is performed.

Clinicians are faced with a lack of effective treatment options for patients who developed resistance to cancer therapies. As prostate cells depend on androgens and androgen receptor (AR) signaling for its growth and survival, androgen-derivation therapies (ie, castration and AR antagonists) are often the first line of treatment for most prostate cancer patients to slow tumor growth and are generally effective. However, within 2 years after initial treatment, a significant percentage of prostate cancer patients develop "castration-resistant" disease [10]. This form is often refractory to chemotherapy and results in disease progression and metastasis [11]. Bone metastases are common during late-stage prostate cancer, occurring in 70% of castration-resistant patients. Patients with castration-resistant prostate cancer have a median survival time of approximately 18 months [12]. Newer drugs such as Alpharadin and Enzalutamide increase survival of patients with castration-resistant prostate cancer by 3–5 months, not years, emphasizing the need for better treatment options.

The lack of sufficient prostate cancer therapies highlights the importance of investigating the molecular mechanisms that convert a slow-growing tumor cell phenotype to that of an aggressive, metastatic cell phenotype during the course of prostate disease. Furthermore, the identification of effective biomarkers that can discriminate for advanced disease will identify patients earlier requiring aggressive treatment options. The discovery that miRNAs closely correlate with prostate cancer progression and play functional roles in this tissue to modulate tumor growth and metastasis, indicates that this large class of noncoding RNAs are promising therapeutic and diagnostic targets for prostate cancer.

2.1.2 No Accurate Biomarkers Exist for Prostate Cancer

For many years, DRE and PSA screening represented the primary diagnostic tests for prostate cancer detection. The effectiveness of PSA as a biomarker to detect prostate cancer and decrease cancer mortality remains controversial. PSA is a protein product transcribed from the *KLKB1* gene [kallikrein B, plasma (fletcher factor) 1]; previously referred to as the Kallikrein-related peptidase 3 gene; and is secreted by the epithelial cells of the prostate. Short-comings of PSA as a prostate cancer biomarker is partly due to the fact that it is prostate organ specific, but not cancer specific [13–15]. PSA can be elevated in patient serum by various nonmalignant diseases, such as benign prostatic hyperplasia (BPH), prostatitis or pharmacological therapy. Serum PSA has been relied on as a diagnostic tool for prostate cancer and as a predictor of cancer recurrence following surgery. However, PSA is far from perfect. Population PSA screening has increased the frequency of prostate cancer detection worldwide. Many of these cases are negative when followed up by tissue biopsy, due to the low specificity of the PSA biomarker. PSA screening has also resulted in the cancer diagnosis of men with insignificant prostate disease, a condition that poses no life threat. It is estimated that only 30% of men with elevated PSA levels (4.1–9.9 ng/ml), and which are subjected to a prostate biopsy, will be diagnosed with prostate cancer [16]. Approximately 5% of this diagnosed group will die from their disease within 10 years of the initial diagnosis. A large proportion of diagnosed men will thus undergo radical prostatectomy and radiotherapy for a condition that posed no increased risk of death, procedures that negatively impact the patients' quality of life. PSA also lacks sensitivity at the lower end of its range as 15% of men with PSA values of 4.0 ng/ml or lower will have prostate cancer [17]. Thus, serum PSA has both a high false-positive and false-negative rate, and is not directly correlated with tumor aggressiveness or therapeutic sensitiveness. Indeed, after 17 years of routine population-wide PSA screening, the US Preventive Services Task Force recommended in 2012 that PSA testing be discontinued [18].

A novel factor that emerged in 1999 as a promising diagnostic marker for prostate cancer is the noncoding mRNA of the prostate cancer antigen 3 (PCA3) gene (formerly known as differential display clone 3). The expression of PCA3 is limited to prostatic tissue, which means no expression is seen in other normal or cancerous human tissues [19]. PCA3 is elevated in 95% of prostate tumor specimens and metastatic tissues from prostate cancer patients. Studies using patient urine enriched with prostatic fluids collected after prostatic massage indicate that PCA3 is superior to serum PSA [20,21]. A urine PCA3 score had an average sensitivity of 66% and specificity of 76% for the prediction of prostate cancer at biopsy, whereas serum PSA had a specificity of only 47% at the same sensitivity (65%). PCA3 score is also not influenced by age, prostatitis, or prostate volume such as PSA. It is unclear if PCA3 can accurately discriminate for advanced disease. Three independent studies found that there was no significant association between PCA3 score and prostate cancer prognostic factors (ie, clinical or pathological stage, Gleason score, tumor volume, and extracapsular prostatic extension) at the time of prostate biopsies. Therefore, the development of better tools, such as miRNA-based biomarkers, is needed that can reliably discriminate for advanced prostate cancer and aid clinicians in determining an effective treatment regimen.

2.2 MicroRNAs Are Strongly Correlated With Prostate Cancer

The underlying molecular causes of human prostate cancer have been studied extensively at the level of protein-coding genes. *SPOP*, *TP53*, and *PTEN* are frequently mutated in patients with this disease and more than 50% of all prostate cancers possess *TMPRSS2-ETS* family (eg, *TMPRSS2-ERG*) fusions [22]. Familial cases of hereditary prostate cancer are commonly associated with *BRCA2* mutations. Only recently has it been appreciated that the small noncoding RNA class of miRNA genes significantly contributes to prostate cancer progression. Global expression profiles of human miRNAs have been performed using microarrays, quantitative real-time PCR (qRT-PCR) and next-generation deep sequencing on prostate cancer cell lines differing in their androgen dependent and metastatic status and specimens isolated from prostate cancer patients compared to noncancer patients. These profiling studies clearly indicate that there is widespread deregulation of miRNA expression during prostate cancer progression and highlights the potential use of miRNAs as prognostic and diagnostic tools (discussed in detail later in this chapter). AR-mediated oncogenic signaling plays a central role in stimulating prostate growth, regulating the progression of prostate cancer, and the development of castration-resistant (androgen-independent) disease. Several studies show that miRNAs regulate AR signaling (ie, miR-21, miR-34, miR-205) or can themselves be regulated by androgen (ie, miR-21, miR-34, miR-101, miR-200b, miR-221, miR-221, *let-7c*, miR-99a, miR-125b) to modulate prostate cell growth and cancer progression.

Functional characterization of miRNAs that correlate with prostate cancer reveal that they can regulate and/or be regulated by oncogenes and tumor suppressor genes that control a wide range of processes; ie, cell cycle progression, differentiation, apoptosis, angiogenesis, hypoxia, chromatin remodeling; processes which directly effect prostate tumor formation and growth; as well as metastatic-related events such as EMTs, migration, invasion, and mesenchymal to epithelial transitions (METs; a process allowing colonization of metastasized cancer cells at distant sites) (Fig. 12.2). An individual miRNA can sometimes play opposing roles and function both as a tumor suppressor gene and an oncogene. miRNA activity is dependent on the cell or tissue type (and the specific mRNA target makeup of that particular cell) and on environmental cues. Context of the miRNA analysis is thus important when determining if an miRNA acts as a tumor suppressor gene or an oncogene in a particular tissue. Therefore, although a growing number of miRNAs have been identified to play roles in a wide range of human cancers, only a subset of these small RNAs have been extensively characterized specifically

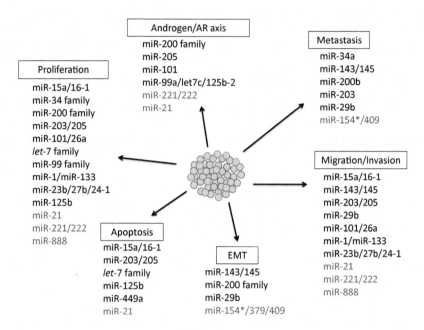

Figure 12.2 *MicroRNAs associated with prostate cancer progression.* MicroRNAs play multiple roles during cancer progression in both negative (*black*) and positive (*green* (gray in print versions)) ways to modulate processes such as tumor proliferation, apoptosis, migration/invasion, epithelial to mesenchymal transition, metastasis, and androgen response in the prostate.

in the prostate using both in vitro functional assays and animal models. Prostate-associated miRNAs will be the primary focus here.

miRNAs that contribute to the underlying etiology of prostate cancer disease can be divided into the abundant class of tumor suppressor miRNAs, for example, miR-15a/miR-16-1, miR-34, miR-143/145, miR-200 family, miR-203, miR-205, miR-29b, miR-101, miR-26, let-7 family, miR-99 family, and the oncogenic miRNAs, for example, miR-125, miR-21, miR-221/222, miR-888, miR-154*, miR-379, and miR-409. There is a growing understanding of miRNAs' influence on prostate cancer metastasis, a clinical condition currently incurable and lethal for patients. miRNAs such as miR-34a, miR-143/145, miR-200b, miR-203, and miR-29b are reported to suppress prostate metastasis, but there is less known how miRNAs promote metastasis in this tissue. Further research to determine the interaction of miRNAs with other protein-coding and noncoding RNAs will lead to a better understanding of the underlying mechanisms of prostate cancer and the development of innovative strategies for diagnosis and treatment of this disease.

In this section, specific examples of how tumor suppressor and onco-genic miRNAs regulate cancer progression and metastatic processes will be discussed in the context of the prostate.

2.3 Tumor Suppressive MicroRNAs in Prostate Cancer

2.3.1 miR-15 and miR-16

Nearly 15 years ago, *mir-15a* and *mir-16-1* (residing in a cluster on human chromosome 13q14.2) were the first miRNAs shown to play a role in human cancers and preferentially deleted or downregulated in patients with B-cell chronic lymphocytic leukemia [23]. Since this discovery, *mir-15a* and *mir-16-1* and *mir-16-2* (on chromosome 3q25.33) have been reported to function as tumor suppressor genes in many types of cancers including prostate cancer [24]. miR-15a and miR-16 are two of the best characterized miRNAs using in vivo mouse models for their role in suppressing prostate tumor growth and cell invasiveness. This cluster is consistently downregulated in prostate tumors compared to noncancer prostate tissue, particularly from patients with advanced disease [25]. Forced expression of miR-15a and miR-16 in LNCaP human prostate cancer cell lines and primary tumor cells result in decreased growth and apoptosis [25]. In mouse xenograft experiments, inactivation of miR-15a and miR-16 activity in prostate RWPE-2 cells implanted into NOD-SCID mice cause these cells to grow faster and exhibit enhanced prostate cell invasiveness [25]. Conversely, tumor size is reduced in mice upon exogenous overexpression of miR-15a/miR-16-1 in prostate xenografts [25]. AntimiRNA treatment to block miR-15a and miR-16 activity in the prostate of normal BALB/c mice result in marked prostate hyperplasia [25]. In the prostate, miR-15a/miR-16-1 likely exerts these effects in mice to suppress prostate tumor growth and invasion by targeting BCL2 (an antiapoptotic protein), Cyclin D1 (cell cycle progression protein), and WNT3A (development controlling protein) [25]. miR-16 also has a unique set of targets in the prostate from miR-15a that includes the cell cycle and proliferation regulators CK1 and cyclin-dependent kinase 2 (CDK2) [26]. Furthermore, an antelocollagen-based delivery method in mice also highlights the use of miR-16 treatment to inhibit the growth of prostate tumor cells located within the bone, suggesting a novel therapeutic approach for metastatic prostate cancer [26]. miR-15 and miR-16 might act as suppressor of cancer cell growth by altering the tumor microenvironment. Downregulation of these miRNAs in the stromal fibroblast cells of the prostate can induce tumor growth at the level of FGF-2 and FGFR1 [27]. Based on these findings, miR-15a/miR-16-1 drugs could be effective for treating organ-confined prostate disease as well as represent a personalized-therapy approach for advanced prostate cancer and bone metastasis.

2.3.2 miR-34 Family

The miR-34 family (consisting of *mir-34a* on chromosome 1p36.22 and *mir-34b* and *mir-34c* located on 11q23.1) are well-established tumor suppressors in multiple cell types, including the prostate. miR-34 family expression is activated in a p53-dependent manner by DNA damage and oncogenic stress and leads to cell cycle arrest and apoptosis by targeting cell cycle and apoptotic proteins [28,29]. Reduced expression of the miR-34 family in primary prostate tumors compared to noncancer prostate tissues indeed correlates with p53 loss [30,31]. However, recent data indicate that miR-34 expression can be regulated independently from p53, and p53 function does not depend on miR-34 [32]. In vivo experiments using mouse models indicate that miR-34 could be a powerful therapeutic target for prostate cancer. Overexpression of miR-34a in human prostate cell lines reduces tumor load and tumor number when these cells are implanted into immunocompromised mice. Furthermore, systemic delivery of miR-34a into tumor-bearing mice reduced lung metastasis and increased animal survival [31]. These antitumor effects in the prostate by miR-34a are mediated by suppression of targets related to cell cycle progression, ie, CDK6 and CCND1 [33]; apoptosis, ie, BCL2 and BIRC5 (baculoviral inhibitor of apoptosis repeat-containing 5) [25,34]; and metastatic factors, ie, Wnt signaling-related transcription factor TCF7 (transcription factor 7) that is associated with bone metastasis [35]. Mice generated that lack all three miR-34 family members do not spontaneously develop prostate cancer unless these mutations are combined with loss of p53 in the prostate epithelium [36]. These results indicate that miR-34 does not play a role in prostate cell initiation but is important during progression of prostate disease and metastasis. The miR-34 family is now recognized as an important regulator of prostate cancer stem cells (CSCs). CSCs are CD44-positive subpopulations of tumors that are particularly tumorigenic, aggressive, possess self-renewing characteristics, and are often resistant to chemotherapy. miR-34 is noted to be underexpressed in CD44-positive prostate CSC isolated from xenografts and primary tumors [31]. Overexpression of miR-34a decreases CD44-positive stemlike prostate cancer cells and inhibits tumor regeneration and metastasis. Conversely, inactivation of miR-34a activity increases tumor formation and invasion of CD44-negative prostate CSCs. Moreover,

CD44 has been identified to be a direct and functional target of this miR-34a [31]. Taken together, miR-34 is an important regulator of tumor proliferation, metastasis, and CSC maintenance and apparently acts cooperatively with p53 as a tumor suppressor. Development of miR-34 therapies could be an effective tool to treat p53 negative tumors as well as advanced, metastatic forms of prostate cancer and potentially effective for patients refractory to other drug treatments.

2.3.3 miR-143 and miR-145

mir-143 and *mir-145*, a cluster found on chromosome 5q32, are frequently underexpressed in prostate cancer and associated with advanced prostate disease [37]. Decreased levels of miR-143 and miR-145 correlate with higher Gleason score, elevated serum PSA levels, and bone metastasis [38,39]. Reduced miR-145 expression is also associated with high risk of biochemical recurrence and shorter disease-free survival [38]. In vitro studies show that miR-143 and miR-145 overexpression in PC-3 and LNCaP human prostate cell lines reduce cancer migration, invasion, and cell adhesion and conversely, inhibiting these miRNAs have opposite effects on these processes [39]. Using an intratibial injection mouse model, in which PC3 prostate cells treated with miR-143 and miR-145 are implanted into the bone, it was revealed that miR-143 and miR-145 act to reduce both tumor size and cancer invasion into bone [39,40]. miR-143 and miR-145 function to reduce the aggressive behavior of prostate cells potentially by inhibiting factors associated with EMT. As predicted, overexpression of these miRNAs in prostate cell can decrease vimentin and fibronectin and upregulate E-cadherin [39,41]. Validated targets for miR-143 in the prostate include the cell proliferation and differentiation factors ERK5 [40]; the cell motility regulator FNDC3B (fibronectin type III domain containing 3B) [42]; and the matrix metalloproteinase-13 (MMP13) [43]. miR-145 is also found to target the MYC oncogene and Fascin homolog 1 (FSCN1), which act to regulate cell proliferation, migration, and invasion [44,45]. Interestingly, miR-145 also targets the ERG oncogene, suggesting a potential therapeutic approach for this cluster in patients harboring TMPRSS2-ERG gene fusions [46]. Furthermore, miR-145 may function within the p53 network of tumor suppressors since miR-145 is transcriptionally induced by p53 [47]. Finally, these miRNAs play an inhibitory role in prostate stem cells, and can suppress tumor sphere formation as well as expression of CSC markers (ie, CD44) and pluripotency genes (ie, Oct4) [48]. Taken together, miR-143 and miR-145 show promise both as diagnostic and prognostic biomarkers for advanced prostate cancer and as a therapy for aggressive disease and bone metastasis.

2.3.4 miR-200 Family

The miR-200 family is composed of two distinct clusters; *mir-200a, mir-200b*, and *mir-429* located on human chromosome 1p36.33 and *mir-200c* and *mir-141* located on 12p13.31. These miRNAs are recognized as key suppressors of EMT in the prostate by targeting the prometastatic transcription factors ZEB1 and ZEB2 (zinc finger E-box binding homeobox 1 and 2) and SNAIL2 [49–51]. In vitro data verified that loss of miR-200 directly induces EMT in LNCaP and PC3 prostate cell lines. Reciprocally, restored expression of miR-200 induces MET in these prostate cancer cells [52]. Members of the miR-200 family (miR-200a, -200b, -200c) have also been shown to be repressed in PC3 cells overexpressing the EMT inducer, PDGF-D, and reintroduction of miR-200 expression results in blocked EMT marker expression (ZEB1, ZEB2, and SNAIL2) and reversal of the EMT phenotype [51]. The miR-200 family also functions outside the EMT pathway to regulate cell growth, and inactivation of miR-200b activity induces proliferation in androgen-dependent prostate cancer cells while the reverse is true in androgen-independent prostate cancer cells [52]. The tumor suppressor role of this family has been confirmed in vivo using both subcutaneous and orthotopic prostate xenograft mouse models. Overexpression of miR-200b in PC3 cells suppresses tumor load, prostate metastasis, and tumor angiogenesis in these animal studies [53]. Finally, the function of the miR-200 family in the AR axis is unclear. miR-200b is reported to be a downstream target of the AR and miR-141 can induce AR transcription indirectly by targeting the androgen corepressor SHP [54].

2.3.5 miR-203 and miR-205

mir-203 (located on human 14q32.33) and *mir-205* (located on 1q32.2) have emerged as important factors that suppress prostate cancer progression and metastasis. Both miR-203 and miR-205 are underexpressed in primary prostate tumors and metastases [55,56], and miR-203 expression is particularly decreased in bone metastatic tissues [57]. In vitro assays show that these miRNAs possess both tumor suppressor and antimetastatic activities such as blocking proliferation, EMT, invasion, and migration, as well as inducing apoptosis [55,56,58,59]. Forced expression of these miRNAs in prostate cells is consistent with promoting MET, and cells are shown to take on a more flattened, epithelial phenotype in cell culture [55,56]. In vivo experiments reveal that miR-203 treatment significantly decreases brain and bone metastasis and increases animal survival using an intracardial injection mouse model [57,58]. Although miR-205 has not been tested similarly for its direct in vivo role in metastasis, subcutaneous injection of DU145 prostate cells overexpressing miR-205 into the flanks of mice significantly lowers tumorigenicity and decreases tumor size [59].

Multiple targets related to prostate cancer progression for miR-203 and miR-205 have been identified. miR-203 negatively regulates prometastatic targets such as the EMT-inducing factors ZEB2 and SNAIL, the antiapoptotic factor survivin, and the bone metastasis-regulator RUNX2 [56,60,61]. miR-205 is similarly reported to target the EMT-factor ZEB2 as well as ZEB1 [62]. Therefore, overlapping targets/regulatory pathways for these two miRNAs might explain their similar phenotypes in MET and metastasis. Moreover, miR-203 is linked to prostate CSC maintenance by targeting the self-renewal-associated BMI1 polycomb ring ginger oncogene [60]. Both miR-203 and miR-205 (together with miR-103a) are found to regulate prostate growth via the AR, the MAPK pathway, and other important oncogenic signaling factors [63]. Recent reports for miR-205 show that it directly targets and blocks the AR and its signaling cascade, and functions to control the c-SRC oncogene and the antiapoptotic BCL2 protein [63–65]. Furthermore, miR-205 is noted to work together with an additional miRNA, miR-31, to promote apoptosis by blocking BCL2L2 and E2F6 [66]. These results indicated that combinatorial miRNA therapies for prostate cancer, such as those targeting miR-203 and miR-205, could be an effective strategy to treat prostate metastasis.

2.3.6 miR-29b

The miR-29b family (mir-29b-1 located on human chromosome 7q32.3 and mir-29b-2 on 1q32.2) are defined as suppressors of prostate cancer metastasis. miR-29b is downregulated in more aggressive castration-resistant PC3 compared to androgen-responsive LNCaP human prostate cancer cells and is also decreased in malignant compared to normal prostate tissue [67,68]. Functionally, miR-29b acts as a tumor suppressor by decreasing cell motility and invasiveness, and directly targets the prometastatic matrix metallopeptidase-2 (MMP-2) [69]. miR-29b is also noted to suppress metastasis by regulating EMT and decreasing the expression of the mesenchymal markers N-cadherin, TWIST, and SNAIL [67]. In mice, forced expression of miR-29b in PC3 cells blocks tumor metastasis in mice compared to PC3 cells transfected with a control miRNA that readily colonizes the lungs of the animals [67]. In addition to regulating EMT, miR-29b also inhibits angiogenesis and directly targets a key regulator of angiogenesis, vascular endothelial growth factor A [70]. Taken together, miR-29b has emerged as an important metastatic suppressor in the prostate.

2.3.7 miR-101 and miR-26

Aberrant gene regulation at the epigenetic level due to changes in DNA methylation and histone modifications is closely linked to prostate cancer and miRNA expression. Two examples are miR-101 (*mir-101-1* located on 1p31.3 and *mir-101-2* on 9p24) and miR-26a (*mir-26a-1* residing on 3p22.2 and *mir-26a-2* on 12q14.1). miR-101 and miR-26a are found to directly target histone methyltransferase and prostate oncogene enhancer of zeste homolog 2 (EZH2), which functions to suppress gene expression by methylating histone H3 lysine 27 residues on specific promoter sites. EZH2 acts in the prostate to promote proliferation, differentiation, and metastasis; and overexpression of EZH2 is associated with prostate cancer, particularly in patients with poor prognosis for survival [71]. Reciprocally, both miR-101 and miR-26a are downregulated in these prostate tissues compared to normal tissues, and downregulation is pronounced in patients with more progressed forms of prostate cancer [71–73]. In fact, data reveal that the miR-101/EZH2 expression ratio may be a useful marker for metastatic prostate cancer. In vitro analysis shows that overexpression of either miR101 or miR-26a results in suppression of cell proliferation and metastatic processes such as invasion. These effects can be attributed to miRNA-mediated inhibition of EZH2 as well as other targets such as the miR-101 target, COX2, an inflammation prostaglandin regulator that promotes prostate cell proliferation and growth [74]. miR-26a also targets the *let*-7 miRNA biogenesis suppressor LIN28, the metastatic factor ZCCHC11 (zinc finger CCHC domain containing 11) [75], and WNT5a, a secreted glycoprotein that promotes prostate cancer growth and invasiveness [73,76]. Furthermore, miR-101 treatment leads to decreased tumor size in mouse xenografts using DU145 human prostate cancer cells. miR-101 is induced in prostate cells upon androgen treatment as well as in the presence of HIF1-alpha, a key factor that works together with HIF1-beta to induce prostate cancer angiogenesis and progression genes during hypoxia [72]. miRNAs such as miR-101 and miR-26a could thus be innovative targets for aggressive prostate cancer that could suppress oncogenic factors in this tissue both via epigenetic, inflammatory, and androgen signaling pathways.

2.3.8 let-7 Family

The *let-7* family, comprised of 12 closely related members in humans, acts globally to suppress tumor growth in many tissues types (ie, lung, colon) and downregulation of this family is closely associated with a wide array of human cancers. This miRNA family has also emerged as an important cellular regulator in the prostate. Profiling studies using microarray and quantitative RT-PCR first revealed that *let-7a* and *let-7c* are downregulated in prostate cancer cell lines and tissues from tumors relative to normal prostate epithelial cells and noncancerous adjacent tissue, respectively [77,78]. This family is also associated with advanced prostate disease. *let-7c* is preferentially downregulated in castration-resistant prostate cancer cells and *let-7a*, −7b, and −7c are significantly more reduced in tumors with Gleason scores 7–9 compared to tumors with lower Gleason scores [79,80]. Lewis et al. also demonstrated that *let-7c* is preferentially downregulated in urine enriched with prostatic fluids (collected following prostatic massage) in patients with high-grade prostate cancer [81]. A tumor suppressor role in the prostate has been validated for the *let-7* family using in vitro and in vivo assays. [78,79] Overexpression of *let-7a* decreases proliferation and induces cell cycle arrest in prostate cancer cell lines and can reduce tumor growth in mice. These affects are attributed to the ability of *let-7a* to target and suppress E2F2 and CCND2, both of which are cell cycle progression factors [78]. The *let-7c* miRNA is also shown to decrease proliferation and colony formation and increase apoptosis in vitro as well as decrease prostate tumor growth in mice [79]. A predicted target for *let-7c* is the MYC oncogene, which interestingly also regulates the transcription of the AR [79]. Finally, *let-7b* expression is significantly decreased in prostate CSCs and its reduction could serve as a prognostic marker for biochemical recurrence and metastasis [82,83]. *let-7b* is found to target the oncogene HMGA1 (high-motility group AT-hook gene 1), a nonhistone nuclear binding protein that is associated with high-grade and advanced prostate cancer [84]. Therefore, this family has the potential to be both effective biomarkers and therapeutics for reducing tumor load and treatment of aggressive disease.

2.3.9 miR-99 Family

The miR-99 family (miR-99a, miR-99b, and miR-100) is hypothesized to act as tumor suppressors in the prostate and family members are noted to be downregulated in advanced prostate cancer cell lines (C4-2 and WPE1-NB26) relative to less-aggressive cell lines (LNCaP and RWPE-1) and decreased in human prostate adenocarcinomas compared to nonmalignant tissue [85]. Indeed, when C4-2 prostate cells are treated with miR-99a, -99b, and/or miR-100, cell growth is inhibited and the growth inhibitory effects are also noted in the absence of androgens [85]. This indicates that the miR-99 family could be an effective treatment for prostate cancers that are androgen independent. Targets for miR-99a in the prostate include the growth regulator mTOR as well as the chromatin remodeling genes, SMARCA5 and SMARCD1. Additionally, the miR-99 family posttranscriptionally regulates PSA and overexpression of miR-99/100 decreases PSA mRNA levels [85]. Evidence supports the notion that the miR-99 family functions as prostate tumor suppressors and may be an effective treatment option for patients with castration-resistant prostate cancer.

2.3.10 Additional Tumor Suppressor MicroRNAs in the Prostate

An accumulating number of miRNAs, particularly those residing in clusters, continue to be identified that are downregulated in prostate cancer patients and posses functional roles to curb prostate tumor growth and progression. For example, the miR-1/miR-133 clusters (*mir-1-1* and *mir-133a-2* on human chromosome 20q13.33 and *mir-1-2* and *mir-133a-1* on 18q11.2) are downregulated in primary human prostate tumors and metastatic tissues [86,87]. Although mouse studies have not yet been performed, these miRNAs in vitro suppress proliferation, migration, and invasion activities in PC3 and DU145 human prostate cancer cell lines likely by sharing overlapping oncogenic targets, such as purine nucleoside phosphorylase, and functioning in common metastatic pathways, ie, EGFR [87]. In fact, miR-133 is shown to target EGFR in prostate cells to exert these effects [88] and EGFR is noted to repress miR-1 expression [89], therefore existing in a complex feedback loop. The closely related miR-133 member, *mir-133b* (clustered with *mir-206* on 6p12.2) also has tumor suppressor roles in the prostate and enhances tumor necrosis factor-α-induced cell death and directly targets the antiapoptotic protein FAS [90].

Another group of miRNAs, belonging to the *mir-23b/27b/24-1* cluster on human chromosome 9q2.32, are reduced in prostate cancer patients and miR-27b particularly could be an effective prognostic marker for patient progression to castration-resistant disease [91]. Members of this cluster act to reduce cell growth, migration and invasion of PC3 and DU145 prostate cell lines, that likely involve the ability of miR-24 to target the proapoptotic Fas-associated factor 1 [92] and miR-27b to suppress the prostate cancer-associated protein, Golgi membrane protein 1 (GOLM1) [91].

It is not clear why there have been more miRNAs functionally characterized to date that possess tumor suppressor roles in the prostate compared to those discussed in the next section, which act to promote cancer pathways in this tissue. The challenge going forward is to understand how all of these factors contribute to prostate disease and tissue homeostasis.

2.4 Oncogenic MicroRNAs in Prostate Cancer

2.4.1 miR-125

The members of the miR-125 family (miR-125a and miR-125b) reside in two distinct human miRNA clusters with the *let-7* and miR-99 families and these miRNAs are thus likely cotranscribed; namely the *mir-99a/let-7c/mir-125b-2* cluster on human chromosome 21q21.1; miR-99b/let-7e/miR-125a on 19q13.4; and *mir-125b-1* on chromosome 11q24.2. The regulatory mechanisms directed by the tumor suppressor *let-7* and miR-99 families (discussed in the section earlier) is likely very complex, especially due to their association with the miR-125 family. Recent work by Sun et al. indicates that *mir-99a/let-7c/mir-125b-2* cluster expression is repressed by androgen. Therefore, prostate cell growth stimulated by the AR axis might be mediated indirectly by turning off the *mir-99a/let-7c/mir-125b-2* cluster and thus allowing *mir-99a/let-7c/mir-125b-2* cluster targets that promote prostate cancer growth, such as IGF1R, to be expressed in the prostate in the presence of androgen [93]. This work supports the notion that the *let-7*, miR-99, and miR-125 families work together to suppress tumor growth in the prostate. Additional evidence in other organ systems, such as breast, supports a tumor suppressive role for all three families. Indeed, significant downregulation of miR-125b in prostate cancer tumor samples versus normal prostate has been reported [77,94].

Contrary to the findings above, growing functional characterization of the miR-125 family (predominantly miR-125b) indicates that miR-125 acts as an oncogenic factor in the prostate. miR-125b is noted to be upregulated in prostate tumors with high Gleason grade [77,95]. In vitro studies show that miR-125b treatment of androgen-dependent LNCaP prostate cell and androgen-independent PC3 cells enhances their growth and blocks apoptosis; and conversely inactivation of this miRNA results in prostate cancer cell repression [95,96]. In addition, these oncogenic effects on tumor growth can be mediated androgen independently [95]. Tumor-promoting effects for miR-125 were also confirmed using mouse models. When PC-346C prostate cancer cell lines stably expressing miR-125b are injected subcutaneously into immunocompromised mice, they show accelerated tumor formation (7–10 days vs 21 days for untreated cells) and these tumors grow larger in size [97]. Furthermore, tumors overexpressing miR-125b also exhibit accelerated growth in a castration-resistant mouse model [97]. Proposed targets for miR125 to exert these oncogenic effects included the tumor suppressor p53, and the apoptosis-regulators PUMA and BAK1 [95]. Recent work reveals that anti-miR-125b treatment plus cisplatin increases apoptosis of androgen-responsive prostate cancer cells [97]. These results indicate that unlike the cluster tumor suppressor *let-7* and miR-99 families, miR-125 acts in the prostate to stimulate cancer cell growth. More work needs to be done before this miRNA can be used as a therapeutic agent due to the conflicting data showing that miR-125 can act as a tumor suppressor and an oncogene in prostate cells.

2.4.2 miR-21

mir-21 (located on human chromosome 17q23.2) is considered a global oncogenic factor in a wide array of tissue types and is the most overexpressed miRNA noted in human cancers [98,99]. Li et al. reported that upregulated expression of miR-21 in prostate tissues from radical prostatectomy is associated with pathological stage, lymph node status, extracapsular extension, and biochemical recurrence [100]. However, there are several contradictory reports evaluating if miR-21 upregulation tightly correlates with prostate cancer [99,101–103], and so it is unclear how effective this miRNA would be as a prostate biomarker. Functional evidence clearly supports an oncogenic role for miR-21 in the prostate to promote tumor growth through proliferation, antiapoptotic and invasion pathways [104]. AR, which regulates prostate cancer growth and progression, is shown to directly bind to the miR-21 promoter to activate its expression in prostate cells [102]. However, although miR-21 is androgen responsive, it is not dependent on the presence of androgen for its expression. LNCaP cells treated with antiandrogens (Casodex) can still express miR-21 and promote cell growth [102]. This oncogenic miRNA is also implicated in promoting castration-resistant disease in vivo. Immunocompromised mice implanted subcutaneously with androgen-responsive LNCaP cells overexpressing miR-21, result in 100% of the animals developing palpable tumors 17 days earlier than the controls. Furthermore, castration of mice harboring LNCaP tumors treated with miR-21 continued to grow compare to untreated prostate tumors [100,102,105]. Factors targeted for downregulation by miR-21 include genes important in suppressing apoptosis, cell migration, and invasion; ie, PDCD4 (programmed cell death protein 4), PTEN (phosphatase and tension homolog), TPM1 (tropomyosin 1), SPRY2 (sprout homolog), and the metalloprotease inhibitors TIMP3 and RECK (reversion-inducing-cysteine-rich protein with kazal motifs) [106–110]. Continued understanding of how miR-21 can promote cancer progression and castration resistance may lead to future targeted therapies for late-stage disease.

2.4.3 miR-221 and miR-222

Like miR-21, the miRNAs *mir-221* and *mir-222* clustered on human chromosome Xp11.3 are oncogenic factors associated with many different types of cancer, ie, breast and pancreatic cancers and leukemia. This cluster has also been correlated with prostate cancer and initially classified as a tumor suppressing factor in this tissue. However, conflicting reports have also suggested an oncogenic role in the prostate. The miR-221/222 cluster is elevated in prostate tissues and cell lines representing advanced castration-resistant disease [111,112]. Conversely, other groups have noted decreased miR-221 and miR-222 levels in specimens from prostate cancer patients and miR-221 downregulation is often observed in patients with prostate tumors carrying *TMPRSS2:ERG* fusions [113–115].

Accumulating functional data indicate that miR-221 and miR-222 promote cancer progression by regulating cell growth. Overexpression of miR-221/222 in LNCaP prostate cancer cells increases proliferation by promoting the G1-S phase cell cycle transition and increases colony formation in soft agar, an in vitro assay for tumorigenesis [116]. These activities are attributed to miR-221/222's ability to target and block the expression the CDK inhibitors p27 (KIP1) and p57 (KIP2) [111,117]. miR-221 and miR-222 can also influence tumor growth in mice. miR-221 and miR-222 overexpression in LNCaP prostate cell lines increases tumor growth when injected subcutaneously into mice. Reciprocally, antimiRNA treatment to block the activity of the miR-221/222 cluster in aggressive PC3 cancer cells reduces tumor growth in SCID mice by releasing the repression of the p27 target [116]. miR-221 and miR-222 have also been shown to promote cell cycle progression, invasion, migration, and block apoptosis in PC3 cells at the level of the NAD-dependent protein deacetylase sirtuin-1 (SIRT1), however, neither miRNA directly targets SIRT1 in a 3′ UTR-dependent manner in prostate cells. This cluster also likely intersects with the AR axis to promote castration resistance. miR-221 overexpression in LNCaP prostate cells can induce their growth even in the absence of androgen, presumably by targeting HECTD2 (HECT domain containing E3 ubiquitin protein ligase 2) and RAB1A (RAS-related protein Rab-1A) [112].

Contrary to the studies above, a recent study by Kneitz et al. used similar human prostate cancer cell lines (eg, LNCaP and PC3 cells) and found that miR-221 inhibited proliferation, invasion, and apoptosis by targeting the oncogenes SOCS3 (suppressor of cytokine signaling 3) and IRF2 (interferon regulatory factor 2) that function to negatively regulate JAK/STAT signaling [118]. This suggests that miR-221 functions as a tumor suppressor in the prostate. Therefore, the precise role of the miR-221/222 cluster in human prostate cancer is still not clear and the activities of miR-221 and miR-222 are evidently extremely complex in the prostate. More studies will need to be done before this cluster could be considered as diagnostic or therapeutic tools for human prostate cancer.

2.4.4 miR-888

mir-888 is a novel oncogenic miRNA in the prostate identified by our laboratory and maps to human chromosome Xq27.3. miR-888 was initially identified as a prostate cancer-associated miRNA in a profiling screen for miRNAs that were highly enriched in aggressive, metastatic PC3-ML cells compared to noninvasive, nonmetastatic PC3-N prostate cancer cell lines [81]. This miRNA holds promise as a discriminating biomarker for high-grade prostate cancer. miR-888 is upregulated in castration-resistant prostate cancer cells relative to androgen-dependent cell lines (ie, LNCaP), in prostate tumors (particularly those taken from metastatic patients exhibiting seminal vesicle invasion), as well as in EPS urine (expressed prostatic secretions in urine enriched with prostatic fluids) obtained from patients with high-grade disease [81]. This miRNA also exhibits a functional role in the prostate and acts to promote proliferation, migration, and colony formation in vitro using PC3 and LNCaP cells, but does not appear to modulate apoptotic processes [81]. Our work indicates that miR-888 regulates proliferation and migration at the level of the cell cycle inhibitor RBL1 (retinoblastoma-like 1) and SMAD4 (SMAD family member 4), a TGF-beta signaling molecule that is associated with metastatic disease in mice. Interestingly, *mir-888* maps to *HPCX1*, a region on human chromosome X (Xq27-28) that is linked to 16% of hereditary forms of prostate cancer. Going forward it will be important to determine whether miR-888 can alter tumor formation and metastasis in animal models.

2.4.5 miR-154*, miR-379, and miR-409

Finally, *mir-154**, *mir-379*, and *mir-409* reside within the imprinted locus deltalike 1 homologue-deiodinase, iodothyronine 3 (*DLK1-DIO3*) on human 14q32, a cluster previously associated with pluripotency and stemliness. These miRNAs are elevated in serum and tissue specimens from patients with high-Gleason score and metastatic prostate cancer [119–121]. miR-154*, miR-379, and miR-409 are all noted in vitro to promote EMT and metastasis in human prostate cancer cells [120,121]. In mouse models, antimiR-154* treatment of ARCaPM prostate cells led to decreased bone metastasis and increased survival in mice and miR-409-5p inhibitors have similar properties when tested in mice [120,121]. Both miR-154* and miR-409-5p have independently been found to directly target the tumor suppressor STAG2 (stromal antigen 2), a cohesion complex protein associated with aneuploidy, tumor initiation, and cancer progression [120,122]. This could indicate a role for both of these miRNAs in the tumor microenvironment. Indeed, miR-409 is expressed both in the stromal fibroblasts and prostate tumor cells and its overexpression in prostate fibroblasts influences tumor growth in an athymic mouse model [120]. In addition, miR-154* targets SMAD7, which functions to interrupt the TGF-beta signaling pathway involved in EMT and miR-409 targets include the Ras suppressor protein 1 (RSU1) [120,123]. Taken together, the *DLK1-DIO3* cluster could be developed as a novel antimiRNA therapy that blocks EMT and cancer progression that targets both prostate cells and the microenvironment.

3. MICRORNAs AS PROSTATE CANCER BIOMARKERS

All of the miRNAs implicated to function as tumor suppressor genes and oncogenes in the prostate (discussed earlier) are also misexpressed in human prostate tissues and/or body fluids. This supports the notion that miRNA expression reflects the molecular etiology of prostate cancer and disease progression [28,94,124]. miRNAs are currently being developed as innovative diagnostic and prognostic tools for prostate cancer that possess better selectivity and sensitivity than PSA or PCA3 [10]. Many groups in recent years have performed global miRNA expression profiles using human prostate cancer cell lines differing in their androgen dependent and metastatic status and tissue samples from patients with prostate carcinomas compared to those exhibiting BPH and noncancer patients. These studies indicate that there is widespread deregulation of miRNA expression in prostate disease and individual miRNAs (or a select panel of miRNAs) can accurately discriminate between prostate cancer versus noncancer, and predict disease severity and outcome when tested in patient cohorts. Furthermore, circulating miRNA found in body fluids, such as plasma, serum, and urine, as well as those secreted in membrane-bound exosomes, also show great promise as discriminating diagnostic and prognostic biomarkers for prostate cancer (Table 12.1) [125].

3.1 MicroRNAs as Tissue Biomarkers

A biomarker is generally defined as a feature that is measured and evaluated as an indicator of normal physiological function, pathogenic progression, or response to therapy. Effective miRNA-based biomarkers for prostate cancer should differentiate malignant versus noncancer and responsive versus refractory to anticancer therapies. An ideal biomarker for prostate cancer should also discriminate indolent tumors from aggressive ones to reduce the risk of overdiagnosis and avoid unnecessary surgical/radiation therapies for patients with insignificant disease [10].

The first attempt to segregate patients through a miRNA profile screen came from primary prostate tumor analyses obtained from prostatectomy and compared to normal prostate tissues. These studies indicate that specific miRNAs are misexpressed in prostate tumors and could correlate with clinical characteristics such as pathogenesis, disease aggressiveness, and risk of recurrence. For example, in 2006 Volinia et al. analyzed the expression profile of 228 miRNAs in 56 prostate tumor tissues and 7 normal controls [126]. This group showed that 39 miRNAs were upregulated and 6 miRNAs were specifically downregulated in prostate tumors. Ambs et al. demonstrated that miR-32, -26a, -181a, -93, -196a, -25, -92, and let-7i were upregulated in a cohort of 60 tumor specimens and 16 controls [103]. Porkka et al. demonstrated a significant downregulation of miRNA expression correlating with prostate cancer progression and observed 37 downregulated miRNAs and 14 upregulated miRNAs in prostate tumors [101]. Specifically, levels of miR-205, miR-100, and the miR-30 family were reduced in castration-resistant prostate specimens, which overlapped with a similar study performed by Ozen et al. [77] These early studies of global miRNA expression profiles in prostate cancer were often contradictory to one another; for instance Volinia et al. found widespread upregulation of miRNA expression, whereas Porkka and Ozen et al. reported that the majority of miRNA genes analyzed were downregulated in tumor samples relative to noncancerous prostate tissue [77,101,126]. The disparities across these data sets could be due in part to the use of small patient cohorts and poorly defined clinical subgroups for prostate cancer and comparing expression profiles derived from different miRNA expression platforms.

Specific panels of miRNA genes are currently being validated in patient cohorts as biomarkers that can discriminate for disease progression and

Table 12.1 MicroRNA expression in prostate cancer

Reference	Study design	Sample size	↑ (elevated)	↓ (decreased)
Tissue				
Porkka et al. (2007)	Hormone-naïve PCa vs HRPC vs BPH tissues	13		Pca:miR–16, miR–99, *let-7* family, HRPC: miR–205, miR–100, miR–30
Ambs et al. (2008)	PCa vs normal tissues	76	miR–32, miR–26a, miR–181a, miR–93, miR–196a, miR–25, miR–92, *let-7i*	
Ozen et al. (2008)	PCa vs normal tissues	26		*let-7*, miR–30, miR–16
Tong et al. (2009)	Without BCR vs with BCR tissues	40	miR–135, miR–194	miR–145, miR–221, miR–222
Schaefer et al. (2010)	PCa vs adjacent normal tissues	76	miR–96	
Peng et al. (2011)	Primary PCa vs bone metastatic tissues	13		miR–508-5p, miR–143, miR–145, miR–33a, miR–100
Saini et al. (2011)	PCa vs metastatic vs normal tissues	52		miR–203
Leite et al. (2011)	Frozen BCR vs Frozen disease–free tissues	49	miR–100	
Martens–Uzunova et al. (2012)	PCa vs adjacent normal tissues	102	miR–19a, miR–130a, miR–20a, miR–106, miR–93	miR–27, miR–143, miR–221, miR–222
Jalava et al. (2012)	Primary PCa vs CRPC vs BPH tissues	54	miR–32, miR–148a, miR–590-5p, miR–21	miR–99a, miR–99b, miR–221
Hulf et al. (2013)	PCa vs matched normal tissues	179		miR–205

Continued

Table 12.1 MicroRNA expression in prostate cancer

Reference	Study design	Sample size	↑ (elevated)	↓ (decreased)
Strivastava et al. (2013)	PCa vs normal adjacent tissues	80		miR–205, miR–214
Larne et al. (2013)	PCa vs normal tissues	74	miR–96-5p, miR–183-5p	miR–145-5, miR–221-5p
Karatas et al. (2014)	BCR vs disease-free tissues	82		miR–1, miR 133b
Mortensen et al. (2014)	PCa: miR–449b vs no miR–449b tissues	163	miR–449b	
Goto et al. (2015)	49 PCa tissue vs 41 noncancerous tissues	90		miR–27b
Bian et al. (2015)	PCa vs normal prostate glands	185	miR–200a, miR–370, miR–31 (metastatic PCa)	miR–200a, miR–31 PCa
Song et al. (2015)	High-grade PCa vs low-grade PCa vs BPH tissues	36	miR–125b-5p, miR–126-5p, miR–151a-5p, miR–221-3p, miR–222-3p	miR–486-5p
Plasma				
Heneghan et al.(2010)	PCa vs healthy controls	83	*let-7a*	miR–145, miR–155
Yaman Agaoglu et al. (2011)	Metastatic PCa vs localized/local advanced PCa vs healthy controls	71	miR–21, miR–141, miR–221	
Bryant et al. (2012)	PCa vs healthy controls	106	miR–107, miR–130b, miR–141, miR–301a, miR–326, miR–331-3p, miR–432, miR–574-3p, miR–625★, miR–2110	miR–181a-2★
Zheng et al. (2012)	PCa vs healthy controls	43	miR–221	

Study	Comparison	n		
Chen et al.(2012)	PCa vs BPH and healthy controls	178	miR–622, miR–1285	*let-7c, let-7e*, miR–30c
Shen et al. (2012)	PCa: High risk vs intermediate risk vs low risk	82	miR–20a, 21, miR–145, miR–221	
Kachakova et al. (2014)	PCa vs BPH vs young asymptomatic men	86		Let-7c, miR–30c, miR–141, miR–375
Huang et al. (2014)	CRPC patients	123	miR–375, miR–1290	
Serum				
Mitchell et al. (2008)	Metastatic PCa vs healthy controls	50	miR–141	
Lodes et al. (2009)	Stage III and IV PCa vs healthy controls	13	miR–16, miR–34b, miR–92a, miR–92b, miR–103, miR–107, miR–197, miR–328,miR–485–3p, miR–486–5p, miR–574–3p,miR–636, miR–640, miR–766, miR–885–5p	
Mahm et al. (2011)	Localized PCa vs BPH and healthy control	83	*let-7i*, miR–26a, miR–195	
Zhang et al. (2011)	Hormone-refractory PCa vs androgen–dependent PCa vs localized PCa vs BPH	56	miR–21	
Moltzahn et al. (2011)	PCa vs healthy controls	48	miR–93, miR–106a, miR–874, miR–1207–5p, miR–1274a	miR–24, miR–26b, miR–30c, miR–223

Continued

Table 12.1 MicroRNA expression in prostate cancer

Reference	Study design	Sample size	↑ (elevated)	↓ (decreased)
Agalogu et al. (2011)	PCa vs healthy controls	71	miR–21, miR–221	
Gonzales et al. (2011)	Metastatic PCa	21	miR–141	
Brase et al. (2011)	Localized PCa vs metastatic PCa	28	miR–141, miR–375	
Bryant et al. (2012)	Metastatic PCa vs localized PCa	119	miR–141, miR–375	
Brase et al. (2012)	Metastatic PCa vs localized PCa	116	miR–141, miR–375	
Selth et al. (2012)	Metastatic PCa vs healthy controls	50	miR–141, miR–298, miR–346, miR–375	
Nguyen et al. (2013)	Localized PCa vs metastatic CRPC	84	miR–141, miR–375, MiR–378*	miR–409–3p
Urine				
Bryant et al. (2012)	PCa vs healthy controls	135	miR–107, miR–574–3p	
Strivastava et al. (2013)	PCa vs healthy controls	130		miR–205, miR–214
Haj-Ahmad et al. (2014)	PCa vs BPH vs healthy controls	30	miR–1825(only PCa)	miR–484(in PCa and BPH)
Sapre et al. (2014)	High-risk PCa vs low-risk PCa	36	miR–16, miR–21, miR–222	
Korzeniewski et al. (2015)	PCa vs healthy controls	89	miR–483–5p	
Prostate secretion				
Guzel et al. (2015)	PCa vs BPH	48	miR–203	miR–361-3p, miR–133b, miR–221

BPH, Benign prostatic hyperplasia; *PCa*, prostate cancer; *CRPC*, castration–resistant prostate cancer; *BCR*, early biochemical recurrence; *HRPC*, hormone refractory prostate cancer.

patient outcome. For instance, Tong et al. analyzed the miRNA expression profiles of 40 FFPE (formalin-fixed, paraffin-embedded) tumor tissues divided into early biochemical relapse and nonrelapse patient groups [127]. Sixteen miRNAs were able to distinguish between relapsed patients (75%) and nonrelapse patients (85%). In addition, single-assay qRT-PCR showed reproducible upregulation of miR-16, miR-135b, miR-194, and miR-218 and downregulation of miR-140 in these groups. Thus, this particular miRNA panel could be refined for identifying disease relapse. In an independent study, Schaefer et al. identified an miRNA signature consisting of 10 downregulated miRNAs (miR-16, -31, -125b, -145, -149, -181b, -184, -205, -221, and -222) and 5 upregulated (miR-96, -182, -182★, -183, and -375) genes in prostate tumor tissues versus normal adjacent tissues [128]. Further validation revealed that just five of these miRNAs, miR-31, -125b, -205, -222, and -96, significantly correlated with Gleason score and tumor stage. In addition, miR-96 expression alone represented a promising prognostic biomarker for biochemical recurrence.

Despite the progress made to identify miRNAs as tissue-based biomarkers for prostate cancer, major limitations remain. First, an ideal prostate cancer biomarker should be readily obtained in the clinic noninvasively and without need for specialized surgical procedures such as a tissue biopsy, which is also not amendable to repeat collection. Second, there is huge intratumoral heterogeneity for prostate cancer and a conventional tissue biopsy will likely result in the incomplete characterization of the tumor phenotype and of the extent of the patient's disease. Therefore, miRNA biomarkers obtained from patient fluids are more ideal for the clinical setting.

3.2 Circulating MicroRNA Biomarkers

Circulating miRNAs (ie, isolated from patient plasma, serum, or urine) are excellent candidates for diagnostic and prognostic purposes, since they are easy to obtain, amenable to repeat collection, and their expression is noted to reflect the profile of tumor origin and disease status [129,130]. It has also been demonstrated that circulating miRNAs are surprisingly stable at various conditions (temperature, high/low pH level, and freezing-thaw cycles) and they are well preserved in long-term banked biorepositories [10,131].

Cell-free miRNAs were first isolated in human plasma from a healthy donor by Mitchell et al, and they found that these secreted miRNAs were remarkably stable in this fluid source [131]. This group made the pivotal observation that human prostate tumor cells secrete tumor-associated miRNAs (namely the human-specific miRNAs, miR-629★ and miR-660) into the blood of xenograft mice [131]. They also reported that miR-141 levels were elevated in the serum of 25 metastatic patients compared to normal age-matched controls using TaqMan qRT-PCR.

After this report, many researchers have identified new diagnostic circulating miRNAs in body fluids for prostate cancer. Lodes et al. used microarray profiling and identified a panel of upregulated miRNAs (miR-16, -92a, -103, -107, -197, -34b, -28, -485-3p, -486-5p, -92b, -574-3p -636, -640, -766, -885-5p) in the sera of prostate cancer patients versus healthy controls [132]. Furthermore, Moltzahn et al. analyzed 36 prostate cancer patients compared to 12 healthy controls and identified 10 differentially expressed miRNAs (specifically miR-223, -26b,-30c, and -24 that were downregulated and miR-874, -1274a, -1207-5p, -93, and -106a that were upregulated in the cancer group) using qRT-PCR [133]. In addition, Yaman Agaoglu et al. investigated three miRNAs in serum in 51 patients in prostate cancer and 20 healthy controls. They demonstrated that miR-21 and miR-221 were upregulated in the prostate cancer group compared to the controls [134].

Much work has been done to identify circulating miRNAs as markers for advanced forms of prostate cancer. Gonzales et al. identified elevated miR-141 as an effective biomarker for metastatic prostate cancer [135]. Zhang et al. screened serum for increased levels of miR-21 in 20 patients with localized prostate cancer, 20 with hormone-sensitive metastatic prostate cancer, 10 with hormone-resistant disease, and 6 controls with BPH. They noted that miR-21 was elevated in castration-resistant and androgen-sensitive metastatic prostate cancer [136]. In a study of 667 miRNAs in sera of patients with metastatic and locally advanced cancer, Brase et al. noted that miR-375 and miR-141 levels could most significantly discriminate for high-risk prostate cancers and might be effective predictive biomarkers for lymph node positive prostate cancer [137]. Finally, Selth et al. identified miR-141, miR-298, miR-346, and miR-375 using a mouse model of prostate cancer and demonstrated that they were also upregulated in the sera of men with castration-resistant prostate cancer compared to control noncancer patients [138]. They also found that miR-141, as well as miR-146-3p and miR-194, could predict biochemical recurrence in patients with intermediate-risk prostate cancer.

In addition to plasma and serum, other fluid-based biomarkers for miRNA detection, such as urine, are also being tested for prostate cancer. Bryant et al. was the first to report using miRNAs in urine and showed that miR-107 and miR-573-3p levels were significantly higher in urine of patients with prostate cancer, compared with controls [139]. Sapre et al. also demonstrated that the combination of miR-16, miR-21, and miR-222 in the urine was a good predictor of high-risk prostate disease [140].

Because of the anatomical localization of the prostate in relation to the urethra, a DRE will result in the release of prostate cells and prostatic fluids into the urethra that is subsequently excreted into the urine. These fluids are also referred to as expressed prostatic secretions in urine (EPS urine) [81,141]. Guzel et al. compared miRNA expression in these prostatic fluids from prostate cancer compared to BPH patients and noted that miR-361-3p, miR-133b, and miR-221 were downregulated and miR-203 was upregulated in prostate cancer patients [142]. Furthermore, Lewis et al. tested 51 miRNAs is EPS urine from patients with high-risk prostate cancer versus lower risk disease and compared to noncancer patients. They found that 10 miRNAs could discriminate for high-risk disease that included elevated miR-888 expression and decreased *let-7c* and miR-200b expression in these patients.

Many studies have now demonstrated that circulating miRNA isolated from body fluids (plasma, serum, urine, prostate secretion) have great potential as prostate cancer biomarkers. Since biomarkers detectable in body fluids can be obtained in a noninvasive manner, they are excellent alternatives to tissue-based biomarkers that require a biopsy [10]. The detection and validation of miRNAs as blood and urine-derived biomarkers are still in the early stages of research development. The majority of studies to date used only small patient cohorts (<200 individuals), and thus the specificity and selectivity of many newly identified miRNA markers for prostate cancer still need to be validated in well-controlled and standardized large-scale prospective clinical studies [143].

3.3 Exosomal MicroRNAs as Biomarkers

Small membrane-bound vesicles (30–100 nm), called exosomes (sometimes referred to as microvesicles), are secreted by various cell types including tumor cells. Exosomes are released into the extracellular environment via the endosome pathway or shed from the plasma membrane and then can enter into the general circulation of the body. Exosomes have been detected in body fluids, such as blood, serum and plasma, bronchoalveolar lavage, urine, bile, ascites, breast milk, and cerebrospinal fluid [144]. Although exosomes were previously considered to be a cellular waste product, recent studies reveal that exosomes play important roles in cell-to-cell communication, tumor progression, and metastasis [125,145]. The protein and RNA content of exosomes are noted to represent their tissue origin and cellular condition [146]. An exciting breakthrough in exosome biology is the discovery that in addition to coding RNAs and protein, miRNAs are also present in exosomes [147]. Hunter et al. first identified miRNAs expressed in circulating plasma microvesicles from normal subjects [148]. Subsequently, Kogure et al. noted that the miRNA profiles of exosomes secreted from cancer cell lines were unique from exosomes isolated from noncancerous cell lines of the same tissue origin. Indeed, just as the miRNA signature of cancer patients are unique from the miRNA signatures of noncancer patients, the exosomal miRNA signature more closely resembles the miRNA makeup of the tumor cells [149]. Therefore, these studies indicate that miRNAs selectively packaged in exosomes and secreted from tumor cells may be developed as fluid-based biomarkers that reflect a patient's cancerous state and disease prognosis.

Bryant et al. sought to specifically evaluate whether circulating miRNAs within exosomes could be used as diagnostic and prognostic biomarkers for prostate cancer [139]. They found elevated miR-375 and miR-141 levels in the sera-derived exosomes from patients with metastatic disease following radical prostatectomy compared to patients with no disease recurrence after surgery. Furthermore, miR-107 and miR-574-3p were elevated in the urine-derived exosomes from prostate cancer patients versus noncancer controls. A recent publication by Huang et al. indicated the prognostic value of miR-1290 and miR-375 isolated from sera-derived exosomes. These miRNAs were significantly elevated in patients with castration-resistant prostate cancer and closely associated with poor overall survival [150]. As this research quickly moves forward, exosomal miRNAs represent an important class of noninvasively obtained biomarkers for prostate disease that also have immense therapeutic potential.

4. CONCLUSIONS AND FUTURE DIRECTIONS FOR MICRORNAs IN THE CLINIC

Since the discovery of miRNAs in 1993, it is clear that miRNAs play important roles in the prostate as tumor suppressor genes and oncogenes to regulate cancer processes associated with proliferation, differentiation, apoptosis, migration, invasion, metastasis, angiogenesis, chromatin remodeling, hypoxia, metabolism, immune response, CSC maintenance, and cross talk between the tumor and the microenvironment. MiRNAs hold great translational potential in the clinic as novel diagnostic and prognostic biomarkers for prostate cancer, a widespread disease in males responsible for a large fraction of cancer-related deaths worldwide. Large-scale miRNA expression profiling screens, including those touched upon in this chapter, indicate that miRNAs are specifically elevated or repressed in the disease state and represent the molecular etiology of prostate cancer. Thus, miRNAs, which are highly stable in tissues and body fluids, could be developed as better discriminating biomarkers and prognostic tools than the currently used but flawed biomarker, PSA. The advent of miRNAs as cancer-prevention tools will likely have an impact on deciding better treatment options in the era of personalized medicine. Several ongoing clinical studies are validating the benefits of tissue and circulating miRNAs for prostate cancer diagnosis and disease monitoring. These current studies include evaluating how well specific miRNA profiles can determine cancer diagnosis, tumor progression, therapeutic response, and disease prognosis (Table 12.2).

Therapeutic drugs for advanced forms of prostate cancer have been ineffective in curing the disease or improving long-term prognosis for metastatic patients. For hormone-sensitive prostate cancer, androgen-deprivation therapy is the first line of treatment. However, the majority of these patients will progress to castration-resistant prostate cancer and become refractory to other treatment options, ie, chemotherapy and radiation therapy. As miRNAs are shown to regulate androgen signaling and modulate castration-resistant disease, miRNA treatment might be an effective strategy against aggressive prostate cancer. Currently, there are no human clinical trials using miRNAs as drugs to treat prostate cancer. As mentioned in this chapter when describing the functional characterization of miRNAs as tumor suppressor genes or oncogenic factors in the prostate, preclinical animals studies indicate that miRNA-based therapeutic for prostate cancer is a realistic possibility [151]. For example, miRNA replacement therapies in mice for the tumor-suppressor miRNAs miR-16, miR-34a, *let-7*, and miR-199a-3p successfully reduced tumor growth and disease progression in vivo [26,31,79,152]. Similar strategies using liposome-encapsulated miRNA mimics for miR-34 (trade name MRX34) are currently being studied in a multicenter phase I clinical trial in patients with liver cancer (**NCT01829971;** http://clinicaltrials.gov/). If successful, this approach could be adopted for the treatment of human prostate cancer. Conversely, miRNAs identified as oncogenic factors or prometastatic agents in the prostate could also be targeted for inactivation using synthetic antisense oligonucleotides, referred to as antagomirs or antimiRNA oligonucleotides. Antagomirs are specifically designed to bind with perfect or near-perfect complementarity with its target miRNA, and are less toxic than other therapies such as chemotherapy [4].

Table 12.2 Clinical trials with microRNAs in prostate cancer

Title	Purpose	Study type	Sample	Trial number
MicroRNAs to predict response to androgen deprivation therapy	Identify exosomal miRNAs that predict responses to androgen–derivation therapy	Observational, prospective	Blood, serum–derived exosomal RNAs	NCT02366494
Focal brachytherapy in patients with selective "low–risk" prostate cancer—a phase II trial	Feasibility and toxicity of focal brachytherapy in patients with low–risk prostate cancer	Interventional, phase II clinical trial	Focal brachytherapy	NCT02391051
Abiraterone acetate in treating patients with metastatic hormone–resistant prostate cancer	miRNA expression in tumor metastases as biomarker for sensitivity and resistance; correlation of miRNA with response and progression	Interventional, phase II clinical trial	Blood (biopsy)	NCT01503229
Blood and tissue samples from patients with locally recurrent or metastatic prostate cancer	Sequencing of genomic tumor DNA, genotyping; gene expression, and miRNA/noncoding RNA	Observational, prospective	Blood, soft tissue or bone metastases	NCT01050504
Bicalutamide and goserelin or leuprolide acetate with/without cixutumumab in patients with newly diagnosed metastatic prostate cancer	Correlation of miRNA expression and circulating tumor cells (CTC) count	Interventional, phase II clinical trial	Serum	NCT01120236

For instance, antimiRNA treatment against the oncogenic miR-21 significantly reduced tumor growth in mice [116].

Exosomes are also promising clinical tools for prostate cancer because of their ability to transport miRNAs between cells and modulate tumor progression in vivo. Akao et al. recently indicated that miR-143BPs, a chemically modified form of this miRNA, was secreted in exosomes from THP-1 macrophages after miR-143BP transfection. Furthermore, miR-143BP levels increased in the serum, tumor, and kidney when the mice were intravenously injected with the THP-1 macrophages transfected with miR-143BP [153]. These data suggest that manipulating exosomal miRNAs ex vivo may be an effective therapeutic tool to deliver tumor suppressor miRNAs to the prostate. The future is bright for translational research using miRNAs to treat prostate cancer.

LIST OF ACRONYMS AND ABBREVIATIONS

ADT Androgen-derivation therapy
AMOs AntimiRNA oligonucleotides
AR Androgen receptor
BAK Bcl-2 homologous antagonist/killer
BCL2 B-cell CLL:lymphoma2
BCL2L2 Bcl-2-like protein 2
BIRC5 Baculoviral inhibitor of apoptosis repeat-containing 5
BMI1 Self-renewal-associated BMI1 polycomb ring ginger oncogene
BPH Benign prostatic hyperplasia
CCND Cyclin D
CDK Cyclin-dependent kinase
COX2 Cyclooxygenase-2
CSC Cancer stem cells
DD3 Differential display clone 3
DLK1-DIO3 Delta-like 1 homologue-deiodinase, iodothyronine 3
DRE Digital rectal examination
E2F2 E2F transcription factor 2
E2F6 E2F transcription factor 6
EGFR Epidermal growth factor receptor
EMT Epithelial to mesenchymal transition
ERK5 Extracellular-signal-regulated kinase 5
EZH2 Enhancer of zeste homolog 2
FAF1 Fas-associated factor 1
FFPE Formalin-fixed, paraffin-embedded
FGF2 Fibroblast growth factor 2 (Basic)
FGFR1 Fibroblast growth factor receptor 1
FNDC3B Fibronectin type III domain containing 3B
FSCN1 Fascin homolog 1

GOLM1 Golgi membrane protein 1
HECTD2 HECT domain containing E3 ubiquitin protein ligase 2
HEF1 Human enhancer of filamentation 1
HIF1 Hypoxia-inducible factor 1
HMGA1 High-mobility group AT-hook gene 1
IRF2 Interferon regulatory factor 2
KIP Kinesin-like protein
KLK3 Kallikrein-related peptidase 3
KLKB1 Kallikrein B, plasma (fletcher factor) 1
KRAS V-Ki-ras2 Kirsein rat sarcoma viral oncogene homolog
MAPK Mitogen-activated protein kinases
MET Mesenchymal to epithelial transition
miRISC complex miRNA-containing RNA-induced silencing complex
miRNA MicroRNA
MMP-2 Matric metallopeptidase-2
mRNA Messenger RNA
MYC Myelocytomatosis viral oncogene homolog
NO0B2 Nuclear receptor subfamily 0, group B
PCa Prostate cancer
PCA3 Prostate cancer antigen 3
PDCD4 Programmed cell death
PNP Purine nucleoside phosphorylase
PSA Prostate-specific antigen
PTEN Phosphatase and tension homolog
PUMA p53 upregulated modulator of apoptosis
qRT-PCR Quantitative real-time PCR
RAB1A RAS-related protein Rab-1A
RBCC1 RB1-inducible coiled-coli 1
RBL1 Retinoblastoma-like 1
RECK Reversion-inducing-cysteine-rich protein with kazal motifs
RUNX2 Runt-related transcription factor 2
SIRT1 Sirtuin-1
SMAD4 SMAD family member 4
SMAD7 SMAD family member 7
SMARCA5 SWI/SNF-related matrix-associated actin-dependent regulator of chromatin subfamily A member 5
SMARCD1 SWI/SNF-related matrix-associated actin-dependent regulator of chromatin subfamily D member 1
SNAIL2 Snail family zinc finger 2 gene (slug)
SOCS3 Suppressor of cytokine signaling 3
SPRY2 Sprout homolog 2
STAG2 Stromal antigen 2
TCF7 Transcription factor 7
TIMP Tissue inhibitor of metalloproteinases
TMPRSS2-ERG fusion Transmembrane protease, serine 2 promoter fused to the erythroblast transformation-specific (ETS)-related gene
TNF Tumor necrosis factor
TPM1 Tropomyosin 1

VEGF-A Vascular endothelial growth factor A
WNT3A Wingless-type MMTV integration site family-member 3A
ZCCHC11 Zinc finger CCHC domain containing 11
ZEB Zinc finger E-box binding homeobox

CONFLICT OF INTEREST

There are no financial or other interests with regard to the submitted manuscript that might be construed as a conflict of interest.

REFERENCES

[1] Siegel R, et al. Cancer statistics, 2014. CA Cancer J Clin 2014;64(1):9–29.
[2] Stefani G, Slack FJ. Small non-coding RNAs in animal development. Nat Rev Mol Cell Biol 2008;9(3):219–30.
[3] Friedman RC, et al. Most mammalian mRNAs are conserved targets of microRNAs. Genome Res 2009;19(1):92–105.
[4] Esquela-Kerscher A, Slack FJ. Oncomirs - microRNAs with a role in cancer. Nat Rev Cancer 2006;6(4):259–69.
[5] Garzon R, Calin GA, Croce CM. MicroRNAs in cancer. Annu Rev Med 2009;60:167–79.
[6] Ma J, Dong C, Ji C. MicroRNA and drug resistance. Cancer Gene Ther 2010;17(8):523–31.
[7] Reed A, et al. Current age and race adjusted prostate specific antigen threshold values delay diagnosis of high grade prostate cancer. J Urol 2007;178(5):1929–32. Discussion 1932.
[8] Lowrance WT, et al. Locally advanced prostate cancer: a population-based study of treatment patterns. BJU Int 2012;109(9):1309–14.
[9] Brumovsky P, Watanabe M, Hokfelt T. Expression of the vesicular glutamate transporters-1 and -2 in adult mouse dorsal root ganglia and spinal cord and their regulation by nerve injury. Neuroscience 2007;147(2):469–90.
[10] Cannistraci A, et al. MicroRNA as new tools for prostate cancer risk assessment and therapeutic intervention: results from clinical data set and patients' samples. Biomed Res Int 2014;2014:146170.
[11] Yu EY, et al. Duration of first off-treatment interval is prognostic for time to castration resistance and death in men with biochemical relapse of prostate cancer treated on a prospective trial of intermittent androgen deprivation. J Clin Oncol 2010;28(16):2668–73.
[12] Petrylak DP. Future directions in the treatment of androgen-independent prostate cancer. Urology 2005;65(6 Suppl.):8–12.
[13] Schroder FH. Review of diagnostic markers for prostate cancer. Recent Results Cancer Res 2009;181:173–82.
[14] Lin K, et al. Benefits and harms of prostate-specific antigen screening for prostate cancer: an evidence update for the U.S. Preventive Services Task Force. Ann Intern Med 2008;149(3):192–9.
[15] Lilja H, Ulmert D, Vickers AJ. Prostate-specific antigen and prostate cancer: prediction, detection and monitoring. Nat Rev Cancer 2008;8(4):268–78.
[16] Smith DS, Humphrey PA, Catalona WJ. The early detection of prostate carcinoma with prostate specific antigen: the Washington University experience. Cancer 1997;80(9):1852–6.

[17] Thompson IM, et al. Prevalence of prostate cancer among men with a prostate-specific antigen level < or = 4.0 ng per milliliter. N Engl J Med 2004;350(22):2239–46.

[18] Moyer VA. Screening for prostate cancer: U.S. Preventive Services Task Force Recommendation Statement. Ann Intern Med 2012;157(2):120–34.

[19] Bussemakers MJ, et al. DD3: a new prostate-specific gene, highly overexpressed in prostate cancer. Cancer Res 1999;59(23):5975–9.

[20] Tosoian J, Loeb S. PSA and beyond: the past, present, and future of investigative biomarkers for prostate cancer. Scientific World J 2010;10:1919–31.

[21] Tosoian JJ, et al. Accuracy of PCA3 measurement in predicting short-term biopsy progression in an active surveillance program. J Urol 2010;183(2):534–8.

[22] Attard G, et al. Prostate cancer. Lancet 2015;387(10013):70–82.

[23] Calin GA, et al. Frequent deletions and down-regulation of micro- RNA genes miR15 and miR16 at 13q14 in chronic lymphocytic leukemia. Proc Natl Acad Sci USA 2002;99(24):15524–9.

[24] Aqeilan RI, Calin GA, Croce CM. miR-15a and miR-16-1 in cancer: discovery, function and future perspectives. Cell Death Differ 2010;17(2):215–20.

[25] Bonci D, et al. The miR-15a-miR-16-1 cluster controls prostate cancer by targeting multiple oncogenic activities. Nat Med 2008;14(11):1271–7.

[26] Takeshita F, et al. Systemic delivery of synthetic microRNA-16 inhibits the growth of metastatic prostate tumors via downregulation of multiple cell-cycle genes. Mol Ther 2010;18(1):181–7.

[27] Musumeci M, et al. Control of tumor and microenvironment cross-talk by miR-15a and miR-16 in prostate cancer. Oncogene 2011;30(41):4231–42.

[28] Coppola V, De Maria R, Bonci D. MicroRNAs and prostate cancer. Endocr Relat Cancer 2010;17(1):F1–17.

[29] Fujita Y, et al. Effects of miR-34a on cell growth and chemoresistance in prostate cancer PC3 cells. Biochem Biophys Res Commun 2008;377(1):114–9.

[30] Yamamura S, et al. MicroRNA-34a modulates c-Myc transcriptional complexes to suppress malignancy in human prostate cancer cells. PLoS ONE 2012;7(1):e29722.

[31] Liu C, et al. The microRNA miR-34a inhibits prostate cancer stem cells and metastasis by directly repressing CD44. Nat Med 2011;17(2):211–5.

[32] Concepcion CP, et al. Intact p53-dependent responses in miR-34-deficient mice. PLoS Genet 2012;8(7):e1002797.

[33] Sun F, et al. Downregulation of CCND1 and CDK6 by miR-34a induces cell cycle arrest. FEBS Lett 2008;582(10):1564–8.

[34] Kojima K, et al. MiR-34a attenuates paclitaxel-resistance of hormone-refractory prostate cancer PC3 cells through direct and indirect mechanisms. Prostate 2010;70(14):1501–12.

[35] Chen WY, et al. MicroRNA-34a regulates WNT/TCF7 signaling and inhibits bone metastasis in Ras-activated prostate cancer. Oncotarget 2015;6(1):441–57.

[36] Cheng CY, et al. miR-34 cooperates with p53 in suppression of prostate cancer by joint regulation of stem cell compartment. Cell Rep 2014;6(6):1000–7.

[37] Kojima S, et al. The tumor-suppressive microRNA-143/145 cluster inhibits cell migration and invasion by targeting GOLM1 in prostate cancer. J Hum Genet 2014;59(2):78–87.

[38] Avgeris M, et al. The loss of the tumour-suppressor miR-145 results in the shorter disease-free survival of prostate cancer patients. Br J Cancer 2013;108(12):2573–81.

[39] Peng X, et al. Identification of miRs-143 and -145 that is associated with bone metastasis of prostate cancer and involved in the regulation of EMT. PLoS ONE 2011;6(5):e20341.

[40] Clape C, et al. miR-143 interferes with ERK5 signaling, and abrogates prostate cancer progression in mice. PLoS ONE 2009;4(10):e7542.

[41] Guo H, et al. The regulation of toll-like receptor 2 by miR-143 suppresses the invasion and migration of a subset of human colorectal carcinoma cells. Mol Cancer 2013;12:77.

[42] Fan X, et al. Up-regulated microRNA-143 in cancer stem cells differentiation promotes prostate cancer cells metastasis by modulating FNDC3B expression. BMC Cancer 2013;13:61.

[43] Wu D, et al. MicroRNA-143 inhibits cell migration and invasion by targeting matrix metalloproteinase 13 in prostate cancer. Mol Med Rep 2013;8(2):626–30.

[44] Sachdeva M, et al. p53 represses c-Myc through induction of the tumor suppressor miR-145. Proc Natl Acad Sci USA 2009;106(9):3207–12.

[45] Fuse M, et al. Restoration of miR-145 expression suppresses cell proliferation, migration and invasion in prostate cancer by targeting FSCN1. Int J Oncol 2011;38(4): 1093–101.

[46] Hart M, et al. The proto-oncogene ERG is a target of microRNA miR-145 in prostate cancer. FEBS J 2013;280(9):2105–16.

[47] Suzuki HI, et al. Modulation of microRNA processing by p53. Nature 2009;460(7254):529–33.

[48] Huang S, et al. miR-143 and miR-145 inhibit stem cell characteristics of PC-3 prostate cancer cells. Oncol Rep 2012;28(5):1831–7.

[49] Bullock MD, et al. MicroRNAs: critical regulators of epithelial to mesenchymal (EMT) and mesenchymal to epithelial transition (MET) in cancer progression. Biol Cell 2012;104(1):3–12.

[50] Mongroo PS, Rustgi AK. The role of the miR-200 family in epithelial-mesenchymal transition. Cancer Biol Ther 2010;10(3):219–22.

[51] Kong D, et al. miR-200 regulates PDGF-D-mediated epithelial-mesenchymal transition, adhesion, and invasion of prostate cancer cells. Stem Cells 2009;27(8):1712–21.

[52] He M, et al. Down-regulation of miR-200b-3p by low p73 contributes to the androgen-independence of prostate cancer cells. Prostate 2013;73(10):1048–56.

[53] Williams LV, et al. miR-200b inhibits prostate cancer EMT, growth and metastasis. PLoS ONE 2013;8(12):e83991.

[54] Xiao J, et al. miR-141 modulates androgen receptor transcriptional activity in human prostate cancer cells through targeting the small heterodimer partner protein. Prostate 2012;72(14):1514–22.

[55] Gandellini P, et al. miR-205 exerts tumor-suppressive functions in human prostate through down-regulation of protein kinase cepsilon. Cancer Res 2009;69(6): 2287–95.

[56] Viticchie G, et al. MiR-203 controls proliferation, migration and invasive potential of prostate cancer cell lines. Cell Cycle 2011;10(7):1121–31.

[57] Saini S, et al. Regulatory role of mir-203 in prostate cancer progression and metastasis. Clin Cancer Res 2011;17(16):5287–98.

[58] Siu MK, et al. Loss of EGFR signaling regulated miR-203 promotes prostate cancer bone metastasis and tyrosine kinase inhibitors resistance. Oncotarget 2014;5(11): 3770–84.

[59] Wang C, et al. miR-203 inhibits cell proliferation and migration of lung cancer cells by targeting PKCalpha. PLoS ONE 2013;8(9):e73985.

[60] Yu J, et al. miR-200b suppresses cell proliferation, migration and enhances chemosensitivity in prostate cancer by regulating Bmi-1. Oncol Rep 2014;31(2):910–8.

[61] Qu Y, et al. MiR-182 and miR-203 induce mesenchymal to epithelial transition and self-sufficiency of growth signals via repressing SNAI2 in prostate cells. Int J Cancer 2013;133(3):544–55.

[62] Gregory PA, et al. The miR-200 family and miR-205 regulate epithelial to mesenchymal transition by targeting ZEB1 and SIP1. Nat Cell Biol 2008;10(5):593–601.

[63] Boll K, et al. MiR-130a, miR-203 and miR-205 jointly repress key oncogenic pathways and are downregulated in prostate carcinoma. Oncogene 2013;32(3):277–85.

[64] Verdoodt B, et al. MicroRNA-205, a novel regulator of the anti-apoptotic protein Bcl2, is downregulated in prostate cancer. Int J Oncol 2013;43(1):307–14.

[65] Wang N, et al. miR-205 is frequently downregulated in prostate cancer and acts as a tumor suppressor by inhibiting tumor growth. Asian J Androl 2013;15(6):735–41.

[66] Bhatnagar N, et al. Downregulation of miR-205 and miR-31 confers resistance to chemotherapy-induced apoptosis in prostate cancer cells. Cell Death Dis 2010;1:e105.

[67] Ru P, et al. miRNA-29b suppresses prostate cancer metastasis by regulating epithelial-mesenchymal transition signaling. Mol Cancer Ther 2012;11(5):1166–73.

[68] Walter BA, et al. Comprehensive microRNA profiling of prostate cancer. J Cancer 2013;4(5):350–7.

[69] Steele R, Mott JL, Ray RB. MBP-1 upregulates miR-29b that represses Mcl-1, collagens, and matrix-metalloproteinase-2 in prostate cancer cells. Genes Cancer 2010;1(4):381–7.

[70] Szczyrba J, et al. Identification of ZNF217, hnRNP-K, VEGF-A and IPO7 as targets for microRNAs that are downregulated in prostate carcinoma. Int J Cancer 2013;132(4):775–84.

[71] Varambally S, et al. Genomic loss of microRNA-101 leads to overexpression of histone methyltransferase EZH2 in cancer. Science 2008;322(5908):1695–9.

[72] Cao P, et al. MicroRNA-101 negatively regulates Ezh2 and its expression is modulated by androgen receptor and HIF-1alpha/HIF-1beta. Mol Cancer 2010;9:108.

[73] Zhang K, et al. MicroRNA-101 inhibits the metastasis of osteosarcoma cells by downregulation of EZH2 expression. Oncol Rep 2014;32(5):2143–9.

[74] Hao Y, et al. Enforced expression of miR-101 inhibits prostate cancer cell growth by modulating the COX-2 pathway in vivo. Cancer Prev Res (Phila) 2011;4(7):1073–83.

[75] Fu X, et al. miR-26a enhances miRNA biogenesis by targeting Lin28B and Zcchc11 to suppress tumor growth and metastasis. Oncogene 2014;33(34):4296–306.

[76] Shojima K, et al. Wnt5a promotes cancer cell invasion and proliferation by receptor-mediated endocytosis-dependent and -independent mechanisms, respectively. Sci Rep 2015;5:8042.

[77] Ozen M, et al. Widespread deregulation of microRNA expression in human prostate cancer. Oncogene 2008;27(12):1788–93.

[78] Dong Q, et al. MicroRNA let-7a inhibits proliferation of human prostate cancer cells in vitro and in vivo by targeting E2F2 and CCND2. PLoS ONE 2010;5(4):e10147.

[79] Nadiminty N, et al. MicroRNA let-7c is downregulated in prostate cancer and suppresses prostate cancer growth. PLoS ONE 2012;7(3):e32832.

[80] Kong D, et al. Loss of let-7 up-regulates EZH2 in prostate cancer consistent with the acquisition of cancer stem cell signatures that are attenuated by BR-DIM. PLoS ONE 2012;7(3):e33729.

[81] Lewis H, et al. miR-888 is an expressed prostatic secretions-derived microRNA that promotes prostate cell growth and migration. Cell Cycle 2014;13(2):227–39.

[82] Liu C, et al. Distinct microRNA expression profiles in prostate cancer stem/progenitor cells and tumor-suppressive functions of let-7. Cancer Res 2012;72(13):3393–404.

[83] Schubert M, et al. Distinct microRNA expression profile in prostate cancer patients with early clinical failure and the impact of let-7 as prognostic marker in high-risk prostate cancer. PLoS ONE 2013;8(6):e65064.

[84] Wei JJ, et al. Regulation of HMGA1 expression by microRNA-296 affects prostate cancer growth and invasion. Clin Cancer Res 2011;17(6):1297–305.

[85] Sun D, et al. miR-99 family of microRNAs suppresses the expression of prostate specific antigen and prostate cancer cell proliferation. Cancer Res 2011;71(4):1313–24.

[86] Hudson RS, et al. MicroRNA-1 is a candidate tumor suppressor and prognostic marker in human prostate cancer. Nucleic Acids Res 2012;40(8):3689–703.

[87] Kojima S, et al. Tumour suppressors miR-1 and miR-133a target the oncogenic function of purine nucleoside phosphorylase (PNP) in prostate cancer. Br J Cancer 2012;106(2):405–13.

[88] Tao J, et al. microRNA-133 inhibits cell proliferation, migration and invasion in prostate cancer cells by targeting the epidermal growth factor receptor. Oncol Rep 2012;27(6):1967–75.

[89] Chang YS, et al. Egf receptor promotes prostate cancer bone metastasis by downregulating miR-1 and activating TWIST1. Cancer Res 2015;75(15):3077–86.

[90] Patron JP, et al. MiR-133b targets antiapoptotic genes and enhances death receptor-induced apoptosis. PLoS ONE 2012;7(4):e35345.

[91] Goto Y, et al. The microRNA-23b/27b/24-1 cluster is a disease progression marker and tumor suppressor in prostate cancer. Oncotarget 2014;5(17):7748–59.

[92] Qin W, et al. miR-24 regulates apoptosis by targeting the open reading frame (ORF) region of FAF1 in cancer cells. PLoS ONE 2010;5(2):e9429.

[93] Sun D, et al. Regulation of several androgen-induced genes through the repression of the miR-99a/let-7c/miR-125b-2 miRNA cluster in prostate cancer cells. Oncogene 2014;33(11):1448–57.

[94] Lu J, et al. MicroRNA expression profiles classify human cancers. Nature 2005;435(7043):834–8.

[95] Shi XB, et al. An androgen-regulated miRNA suppresses Bak1 expression and induces androgen-independent growth of prostate cancer cells. Proc Natl Acad Sci USA 2007;104(50):19983–8.

[96] Lee YS, et al. Depletion of human micro-RNA miR-125b reveals that it is critical for the proliferation of differentiated cells but not for the down-regulation of putative targets during differentiation. J Biol Chem 2005;280(17):16635–41.

[97] Shi XB, et al. miR-125b promotes growth of prostate cancer xenograft tumor through targeting pro-apoptotic genes. Prostate 2011;71(5):538–49.

[98] Selcuklu SD, Donoghue MT, Spillane C. miR-21 as a key regulator of oncogenic processes. Biochem Soc Trans 2009;37(Pt 4):918–25.

[99] Folini M, et al. miR-21: an oncomir on strike in prostate cancer. Mol Cancer 2010;9:12.

[100] Li T, et al. miR-21 as an independent biochemical recurrence predictor and potential therapeutic target for prostate cancer. J Urol 2012;187(4):1466–72.

[101] Porkka KP, et al. MicroRNA expression profiling in prostate cancer. Cancer Res 2007;67(13):6130–5.

[102] Ribas J, et al. miR-21: an androgen receptor-regulated microRNA that promotes hormone-dependent and hormone-independent prostate cancer growth. Cancer Res 2009;69(18):7165–9.

[103] Ambs S, et al. Genomic profiling of microRNA and messenger RNA reveals deregulated microRNA expression in prostate cancer. Cancer Res 2008;68(15):6162–70.

[104] Hassan O, et al. Recent updates on the role of microRNAs in prostate cancer. J Hematol Oncol 2012;5:9.

[105] Mishra S, et al. Androgen receptor and microRNA-21 axis downregulates transforming growth factor beta receptor II (TGFBR2) expression in prostate cancer. Oncogene 2014;33(31):4097–106.

[106] Asangani IA, et al. MicroRNA-21 (miR-21) post-transcriptionally downregulates tumor suppressor Pdcd4 and stimulates invasion, intravasation and metastasis in colorectal cancer. Oncogene 2008;27(15):2128–36.

[107] Meng F, et al. MicroRNA-21 regulates expression of the PTEN tumor suppressor gene in human hepatocellular cancer. Gastroenterology 2007;133(2):647–58.

[108] Zhu S, et al. MicroRNA-21 targets the tumor suppressor gene tropomyosin 1 (TPM1). J Biol Chem 2007;282(19):14328–36.

[109] Sayed D, et al. MicroRNA-21 targets Sprouty2 and promotes cellular outgrowths. Mol Biol Cell 2008;19(8):3272–82.

[110] Gabriely G, et al. MicroRNA 21 promotes glioma invasion by targeting matrix metalloproteinase regulators. Mol Cell Biol 2008;28(17):5369–80.

[111] Galardi S, et al. miR-221 and miR-222 expression affects the proliferation potential of human prostate carcinoma cell lines by targeting p27Kip1. J Biol Chem 2007;282(32):23716–24.

[112] Sun T, et al. The role of microRNA-221 and microRNA-222 in androgen-independent prostate cancer cell lines. Cancer Res 2009;69(8):3356–63.

[113] Zheng Q, et al. Investigation of miR-21, miR-141, and miR-221 expression levels in prostate adenocarcinoma for associated risk of recurrence after radical prostatectomy. Prostate 2014;74(16):1655–62.

[114] Fuse M, et al. Tumor suppressive microRNAs (miR-222 and miR-31) regulate molecular pathways based on microRNA expression signature in prostate cancer. J Hum Genet 2012;57(11):691–9.

[115] Gordanpour A, et al. miR-221 Is down-regulated in TMPRSS2:ERG fusion-positive prostate cancer. Anticancer Res 2011;31(2):403–10.

[116] Mercatelli N, et al. The inhibition of the highly expressed miR-221 and miR-222 impairs the growth of prostate carcinoma xenografts in mice. PLoS ONE 2008;3(12):e4029.

[117] Fornari F, et al. MiR-221 controls CDKN1C/p57 and CDKN1B/p27 expression in human hepatocellular carcinoma. Oncogene 2008;27(43):5651–61.

[118] Kneitz B, et al. Survival in patients with high-risk prostate cancer is predicted by miR-221, which regulates proliferation, apoptosis, and invasion of prostate cancer cells by inhibiting IRF2 and SOCS3. Cancer Res 2014;74(9):2591–603.

[119] Nguyen HC, et al. Expression differences of circulating microRNAs in metastatic castration resistant prostate cancer and low-risk, localized prostate cancer. Prostate 2013;73(4):346–54.

[120] Josson S, et al. miR-409-3p/-5p promotes tumorigenesis, epithelial-to-mesenchymal transition, and bone metastasis of human prostate cancer. Clin Cancer Res 2014;20(17):4636–46.

[121] Gururajan M, et al. miR-154* and miR-379 in the DLK1-DIO3 microRNA megacluster regulate epithelial to mesenchymal transition and bone metastasis of prostate cancer. Clin Cancer Res 2014;20(24):6559–69.

[122] Kim MS, et al. Mutational and expressional analyses of STAG2 gene in solid cancers. Neoplasma 2012;59(5):524–9.

[123] Lenferink AE, et al. Transcriptome profiling of a TGF-beta-induced epithelial-to-mesenchymal transition reveals extracellular clusterin as a target for therapeutic antibodies. Oncogene 2010;29(6):831–44.

[124] Fang YX, Gao WQ. Roles of microRNAs during prostatic tumorigenesis and tumor progression. Oncogene 2014;33(2):135–47.

[125] Goto Y, et al. Functional significance of aberrantly expressed microRNAs in prostate cancer. Int J Urol 2015;22(3):242–52.

[126] Volinia S, et al. A microRNA expression signature of human solid tumors defines cancer gene targets. Proc Natl Acad Sci USA 2006;103(7):2257–61.

[127] Tong AW, et al. MicroRNA profile analysis of human prostate cancers. Cancer Gene Ther 2009;16(3):206–16.

[128] Schaefer A, et al. Diagnostic and prognostic implications of microRNA profiling in prostate carcinoma. Int J Cancer 2010;126(5):1166–76.

[129] Sita-Lumsden A, et al. Circulating microRNAs as potential new biomarkers for prostate cancer. Br J Cancer 2013;108(10):1925–30.

[130] Fabbri M. miRNAs as molecular biomarkers of cancer. Expert Rev Mol Diagn 2010;10(4):435–44.

[131] Mitchell PS, et al. Circulating microRNAs as stable blood-based markers for cancer detection. Proc Natl Acad Sci USA 2008;105(30):10513–8.

[132] Lodes MJ, et al. Detection of cancer with serum miRNAs on an oligonucleotide microarray. PLoS ONE 2009;4(7):e6229.

[133] Moltzahn F, et al. Microfluidic-based multiplex qRT-PCR identifies diagnostic and prognostic microRNA signatures in the sera of prostate cancer patients. Cancer Res 2011;71(2):550–60.

[134] Yaman Agaoglu F, et al. Investigation of miR-21, miR-141, and miR-221 in blood circulation of patients with prostate cancer. Tumour Biol 2011;32(3):583–8.

[135] Gonzales JC, et al. Comparison of circulating MicroRNA 141 to circulating tumor cells, lactate dehydrogenase, and prostate-specific antigen for determining treatment response in patients with metastatic prostate cancer. Clin Genitourin Cancer 2011;9(1):39–45.

[136] Zhang HL, et al. Serum miRNA-21: elevated levels in patients with metastatic hormone-refractory prostate cancer and potential predictive factor for the efficacy of docetaxel-based chemotherapy. Prostate 2011;71(3):326–31.

[137] Brase JC, et al. Circulating miRNAs are correlated with tumor progression in prostate cancer. Int J Cancer 2011;128(3):608–16.

[138] Selth LA, et al. Discovery of circulating microRNAs associated with human prostate cancer using a mouse model of disease. Int J Cancer 2012;131(3):652–61.

[139] Bryant RJ, et al. Changes in circulating microRNA levels associated with prostate cancer. Br J Cancer 2012;106(4):768–74.

[140] Sapre N, et al. Curated microRNAs in urine and blood fail to validate as predictive biomarkers for high-risk prostate cancer. PLoS ONE 2014;9(4):e91729.

[141] Hessels D, et al. DD3(PCA3)-based molecular urine analysis for the diagnosis of prostate cancer. Eur Urol 2003;44(1):8–15. Discussion 15-6.

[142] Guzel E, et al. Identification of microRNAs differentially expressed in prostatic secretions of patients with prostate cancer. Int J Cancer 2015;136(4):875–9.

[143] Kuner R, et al. microRNA biomarkers in body fluids of prostate cancer patients. Methods 2013;59(1):132–7.

[144] Rupp AK, et al. Loss of EpCAM expression in breast cancer derived serum exosomes: role of proteolytic cleavage. Gynecol Oncol 2011;122(2):437–46.

[145] Pegtel DM, et al. Functional delivery of viral miRNAs via exosomes. Proc Natl Acad Sci USA 2010;107(14):6328–33.

[146] Gonzalez-Begne M, et al. Proteomic analysis of human parotid gland exosomes by multidimensional protein identification technology (MudPIT). J Proteome Res 2009;8(3):1304–14.

[147] Mittelbrunn M, et al. Unidirectional transfer of microRNA-loaded exosomes from T cells to antigen-presenting cells. Nat Commun 2011;2:282.

[148] Hunter MP, et al. Detection of microRNA expression in human peripheral blood microvesicles. PLoS ONE 2008;3(11):e3694.

[149] Kogure T, et al. Intercellular nanovesicle-mediated microRNA transfer: a mechanism of environmental modulation of hepatocellular cancer cell growth. Hepatology 2011;54(4):1237–48.

[150] Huang X, et al. Exosomal miR-1290 and miR-375 as prognostic markers in castration-resistant prostate cancer. Eur Urol 2015;67(1):33–41.

[151] Pickl JM, et al. Novel RNA markers in prostate cancer: functional considerations and clinical translation. Biomed Res Int 2014;2014:765207.

[152] Qu Y, et al. miR-199a-3p inhibits aurora kinase A and attenuates prostate cancer growth: new avenue for prostate cancer treatment. Am J Pathol 2014;184(5):1541–9.

[153] Akao Y, et al. Microvesicle-mediated RNA molecule delivery system using monocytes/macrophages. Mol Ther 2011;19(2):395–9.

INDEX

'Note: Page numbers followed by "f" indicate figures and "t" indicate tables.'